THE CRYSTALLINE STATE

AN INTRODUCTION

PETER GAY M.A., PH.D.

Fellow of Downing College, Cambridge
University Lecturer in Mineralogy and Petrology

THE
CRYSTALLINE
STATE

AN INTRODUCTION

OLIVER & BOYD

EDINBURGH

OLIVER & BOYD
Tweeddale Court
14 High Street
Edinburgh EH1 1YL
A Division of Longman Group Limited
ISBN 0 05 002433 7
First published 1972
© 1972 Peter Gay

Printed in Great Britain by
William Clowes & Sons, Limited
London, Beccles and Colchester

PREFACE

Studies on the nature and properties of crystalline solids are increasing across a widening spectrum of scientific disciplines and technical applications. A common interest in the crystalline state has emerged to provide a unifying link between scientists, technologists and many others who would have found little in common in earlier generations. Professional crystallographers can still be identified as such, but are usually to be found applying their skills and knowledge to problems in disparate and apparently unrelated fields such as molecular biology, metallurgy, mineralogy, pharmacology, petrology and so on. This evolution has meant that an elementary knowledge of the fundamental language of the crystalline state is important to many scientists with different specialisations, and this, in turn, has stimulated the teaching of the elements of crystallography and the properties of crystalline matter in undergraduate courses.

The present book aims to provide a modern introduction to the crystalline state for students in a broad range of scientific disciplines early in their university and college courses; more specialised interests can subsequently be built on the foundations provided in crystallography, crystal physics and crystalline imperfections. The presentation and emphasis are developed from the essential underlying atomic regularity of crystalline matter; the strict rigour of more advanced texts is relaxed in order to allow greater continuity in subject development, and mathematical techniques are deliberately held back to an elementary level to ensure that the approach is comprehensible to readers with a limited mathematical background. In general, the book is based on the author's experience of teaching such an introductory course to first-year undergraduates at Cambridge for over twenty years; such experience owes much to colleagues and students, and to all of them I am greatly indebted for their major contribution to the preparation of this book.

After an introductory chapter, the earlier sections develop the concepts of geometrical crystallography through the analysis of two- and three-dimensional patterns. Later chapters give an introductory account of the interaction of X-rays with crystalline matter; the treatment is unsophisticated (e.g. the reciprocal lattice, although mentioned, is not used as an interpretative tool), and is directed towards an appreciation of applications and potentialities of X-ray diffraction methods rather than a description of the detailed techniques used by a crystallographer. The influence of symmetry relationships on physical properties is discussed in an elementary manner (i.e. without the use of tensor notation) in a later section, whilst the final chapter outlines the nature of the commoner imperfections which occur in real crystals. Progressive exercises and problems are provided for most chapters, which also contain suggestions for further more advanced reading and reference.

Much of the text of the book was completed during a period of leave of absence from duties in Cambridge, and I must express my gratitude to the General Board of the University and the Governing Body of Downing College for this opportunity. Among my colleagues, I am particularly indebted to Dr. M. G. Bown for reading the final manuscript and providing many detailed suggestions for its improvement. Finally I must record my thanks to my wife for her invaluable assistance at every stage in the preparation of this book.

P. GAY

Cambridge,
1970

CONTENTS

1

INTRODUCTION

1.1. The character of crystalline matter

Our common experience of matter comes through our senses, which are very selective in what they allow us to perceive about the real nature of our surroundings. They distinguish readily enough those materials which we call solids from those called liquids and gases, and they tell us much that is familiar and useful about such objects; we recognise size, shape, colour, opacity, texture and so on, but these are not necessarily the most suitable qualities for separating one type of solid matter from another, for any subdivisions are really related to changes on an atomic scale. By observation alone we cannot always distinguish the imitation cut-glass gemstone from the crystalline mineral; of course, we can avoid being misled by such a substitution by examining certain less obvious properties which are quite different for the two kinds of solids represented by a glass and a crystal. Indeed distinctive properties of various classes of solids were recognised long before their origins in divergent atomic arrangements were appreciated; although the crystalline state can be characterised by the behaviour of the solids in which it occurs, we shall start by establishing those features of the internal atomic structure which allow it to be differentiated from other forms of solid matter.

As its name implies, the crystalline state was originally associated with crystals, i.e. what we normally understand by the descriptive term confined to homogeneous solids bounded by naturally formed planar faces, often quite regularly arranged. The beauty of many crystalline minerals attracted man's attention from earliest times, and eventually crystallographic science began to emerge from studies of their shapes. In seeking explanation of the visible features of their specimens, the earliest crystallographers became convinced that naturally developed external faces were related to underlying regularities of internal arrangements. Almost two centuries ago, Haüy showed that the many different shapes exhibited by crystals of the same substance could all be constructed by regular repetition of a small fundamental unit of a characteristic size and shape, though at that time he had little conception of the nature of such units. Crystallography, as the study of the external shapes of natural and synthetic crystals, continued to advance during the nineteenth century with the development of detailed relationships between the observable geometry of faces and the sizes and shapes of the internal units. Speculations as to the constitution of units continued during this period until they were rationalised by the foundations of modern atomic theory at the beginning of this century. Crystalline solids were then recognised as regular and repeating arrangements of the constituent atoms to fill the whole volume of a crystal; the fundamental internal unit could then be identified with that distinctive small volume, containing groups of atoms in a particular configuration, which can be repeated to build up the solid. Subsequent developments in the present century have made it possible to determine the size and shape of these units and the relative dispositions of the various atoms within them; in doing this we are said to define the *crystal structure* which distinguishes one crystalline solid from another.

An ordered arrangement of the constituent atoms into a repetitive three-dimensional pattern is, then, the prime characteristic of the ideal crystalline state. With this conception, advancing experimental methods freed the study of crystalline materials from the restrictions imposed by the necessity of a regular external shape; crystals, in this older sense, may be developed by some materials, but other solids which have never formed natural external plane faces are just as representative of the crystalline state of matter. Moreover, it soon became clear that the strict atomic regularity of the perfect crystalline state is really only one extreme limit of the structural arrangements that are to be found in solid matter; it can merge by degrees to an ideal amorphous state in which the whole volume is occupied by a random jumble of the constituent atoms with no repetitive regularity. Apart from the many materials in which the atomic arrays approximate closely to the ideal crystalline state, some solids retain a degree of crystallinity which, although considerably less than perfect, is still sufficient to allow their properties to be investigated and described in crystallographic terms. Crystallographic studies now overflow into organic, inorganic, metallic, physical and biological fields in ways that were inconceivable in the circumscribed horizons less than a century ago, and to such an extent that the modern crystallographer tends to be a specialist within one of the more readily identifiable disciplines.

1.2. The degree of crystallinity in solids

The orderly arrangement of the crystalline state is a consequence of attractive binding forces which form permanent attachments between the constituent atoms; the development of perfect regularity corresponds to a minimisation of the energy of the solid at the change of state in which it is formed. In crystallisation from a liquid, for example, the constant re-shuffling of atoms is changed in the solid to fixed atomic positions, with only small vibrations due to thermal motion; permanent attachments (or bonds) are formed between adjacent atoms to give the solid its rigidity, and regular and symmetrical configurations of neighbouring atoms develop, where possible, as a reflection of an attempt to achieve the lowest potential energy. Wave-mechanical treatments of the interactions of atoms in close proximity can predict quantitatively the minimum energy configurations for some systems of simple atoms, and although the calculations become too difficult to be handled quantitatively in more complicated systems, they can be extended in a rough qualitative manner. In crystalline matter the configuration of first nearest neighbours of any atom (or its *co-ordination*) is developed by the primary attractive forces and corresponds to a local minimum energy arrangement; the periodic regularity of the structural continuum grows as permanently bonded neighbours influence more and more atoms into the most energetically favourable positions.

Sometimes these primary attractive forces are of such a kind that they are responsible for the rigidity of the atomic array throughout the whole volume of the solid, with all the constituent atoms directly linked into a closely packed framework. Sometimes they form only sub-units (of complexes of atoms) within the structural pattern; these must then be cross-linked to other similar sub-units by secondary bonds, often of a weaker and different kind, to build up the stability of a crystalline solid. These sub-units may be recognisable chemical molecules (as, for example, in many organic crystals) or separate groups of some of the constituent atoms (as in the complex ion groupings of carbonates, nitrates, etc.) or even infinite repetitive chains, sheets or frameworks of some atoms (as, for example, in many silicates). The essential integrity of such sub-units is demonstrated by observations which show that they often persist after the long-range regularity of the crystalline solid has been destroyed by melting or solution; their formation and the peculiarities of their structure can play an important part in determining the degree of crystallinity achieved in the process of solidification.

In solidification many materials achieve a condition resembling the ideal crystalline state with a repetitive regularity of structural pattern over volumes containing at least millions of atoms, but there are a number of other materials which do not develop this long-range order; these include many solids so familiar to us in everyday life that we might be tempted to think of their low degree of crystallinity as more common than it really is. Nearly all of these materials contain some degree of a local short-range atomic order; some never develop this regularity over more than nearest neighbours or perhaps a few atomic distances, and approximate to an amorphous state, which may be compared to the 'freezing' of a liquid structure; others show a greater tendency to develop a crystalline form of lower energy, and their particular degree of crystallinity depends on the conditions of their solidification

and subsequent treatment. Among such 'non-crystalline solids'* are pla-
stics and resins formed from long chain polymers, natural and synthetic
rubbers, glasses, wood, etc., groups of materials whose structures and properties are often very different. Some of them contain ordered molecules,
sub-units built by primary attractive forces into complexes of a highly irregular shape, sometimes with little mechanical stability; in the process of
solidification these sub-units become hopelessly intertwined, and they have
insufficient mobility to re-arrange themselves with any long-range regularity
under the influence of weak secondary forces, though small ordered domains
are sometimes realised. Others, mainly the glasses, have linked three-dimensional frameworks in which sub-units of first nearest neighbours are identical
but the network into which they are linked is irregular with no long-range repetition; they tend to occur in materials for which any crystalline configuration of the constituent atoms is relatively open with each atom having a small
number of nearest neighbours and bonds between sub-units which are somewhat flexible; glass structures are inherently unstable and will sometimes
revert to a crystalline (or devitrified) form spontaneously unless inhibited by
chemical additives during manufacture. Other solids have structures which
are, in a sense, intermediate between the rigidity of the frameworks of simple
sub-units in glasses and the more flexible tangled complexes of the polymers;
the particular form of their non-crystallinity depends on the stable permanent
attachments which arise between the constituent atoms on solidification. All
of these materials are important in the technology of solids, but we shall
not be concerned further with them.

1.3. Crystal chemistry and the description of structures

All our subsequent discussions will be confined to the very large number of
solids which are crystalline and which have a high degree of long-range order
in their atomic arrangements; each substance has its own crystal structure
which determines many of its characteristic properties. The description of
crystal structures and the rationalisation of the infinitely variable arrangements adopted by different combinations of atoms is the province of crystal
chemistry, which is blended from a judicious mixture of theoretical expecta-
tions tempered by the experimental observations of structure determination
to provide some reasonable principles for the construction of the atomic
architecture of crystalline matter. Unfortunately at present these can be little
more than sophisticated guide-lines, for we have no means of calculating *ab*

---

* We have not included the gels in which very small particles of a pseudo-crystalline
character (as in a colloidal solution) join together to entrap small volumes of liquid
within the framework; their important mechanical properties depend on the materials
involved, the linking of the particles and the size and distribution of the pores, and range
from elastic (as in gelatine) to rigid (as in the setting of cement).

initio the minimum energy configurations of anything other than small groups of atoms of relatively simple extranuclear structure; crystal structure determination is still essentially an experimental exercise. The wave-mechanical treatments of the simple systems, to which we have already referred, suggest that the nearest neighbour attractive forces can, with some justification, be described as ionic or covalent or metallic for particular atomic linkages, but such a division cannot always be entirely realistic, and in many combinations the primary cohesive forces must be assumed to be mixtures, in some way, of binding forces with these idealised limiting characteristics.

The only rational systematic approach to crystal chemistry must start from the influences of such idealised primary bonds in determining the crystal structure adopted by a particular collection of atoms. Most important among these is the directed characters of covalent forces, which require permanently linked neighbouring atoms to be restricted numerically and to be disposed in only certain kinds of spatial array; this may be contrasted with the undirected nature of ionic and metallic bonds where the packing of neighbours is essentially a geometrical operation of minimising the occupied volume (whilst maintaining electrical neutrality in ionic structures). The development of sub-units consisting of tightly bound atomic complexes is one of the consequences of binding forces which are predominantly covalent in character. The resulting crystal structures with clearly distinguishable sub-units linked together by secondary attractive forces* are often said to be *heterodesmic*; most structures are of this kind, and the shape and form of the atomic groupings in the sub-units are important as they determine the nature of repeatable motifs in the pattern theory that we shall discuss later. When atoms in combination are exclusively bound by predominantly ionic or metallic forces, crystal structures tend to be *homodesmic*, i.e. they have no recognisable sub-units; the rigidity of the structural framework depends on identical nearest-neighbour interactions for all atoms and we cannot identify any separate molecular unit formed by groups of atoms. Such structures are relatively rare and are confined to elements and simple compounds containing only a few chemically different atoms (or ions); occasionally they are formed in covalent solids (diamond is a notable example, in which the whole framework can be regarded as an infinite molecular complex), and in inert gases simple homodesmic structures are crystallised by the weak influence of residual attractive forces.

Crystal chemistry seeks to explain the many different co-ordinations which

---

* These secondary attractive forces may be relatively weak as when they are van der Waal's (or residual) bonds or perhaps hydrogen bonds; they can also be forces with different primary character, as when complex anion groups are linked by ionic attraction through cations. If they are undirected the resultant structure is determined by the attempt to pack the sub-units (often of an irregular shape) and any other atoms in an economic space-filling manner; hydrogen bonds, however, can demand specific orientations related to the formation of suitable bridges between molecular complexes.

can arise from the interactions between neighbouring atoms, and how such co-ordinations can be assembled to form the repetitive unit of a particular structure. From the outlines of the relationships between co-ordination and bonding forces that have been sketched above, it develops into a fascinating but involved field of study, and can be pursued by an interested reader through the references in the bibliography at the end of this chapter. Our immediate

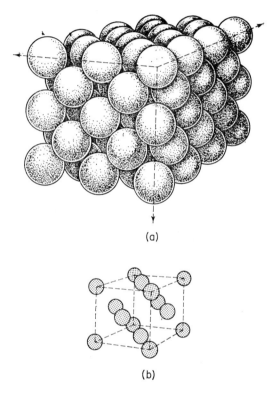

(a)

(b)

Fig. 1.1. The crystal structure of a metallic element (Au). (a) The packing of atoms within a small volume. (b) Representation of the atomic arrangement in a repetitive volume; this volume is $\frac{1}{8}$ of that shown in (a), and its edges are parallel to the dashed lines in (a).

concern, however, is not with such principles of atomic architecture but rather with the basic descriptions of three-dimensional atomic patterns as they occur. Fig. 1.1(a) is a sketch of a small volume of the structure of a crystalline metallic element; the close proximity of neighbouring atoms is obvious, but details of co-ordination and repetitive packing are obscured by the overlapping of atoms in this representation. Fig. 1.1(b) shows a smaller repetitive volume of the same structure in which the spheres denoting atoms have been reduced in size; the arrangement of the atoms is much clearer, but even so this kind of representation would be very confusing in more complex structures. One obvious way of showing atomic arrangements is to build scale models of structures; such models are often available from commercial organisations and some examples will be found in most crystallographic laboratories. But it is also important that we are able to describe any crystal structure by a systematic representation on the two dimensions of a sheet of paper; this requires a descriptive method which will produce some kind of standardised projection.

All descriptions start with the recognition of a repetitive volume in the atomic pattern. In any structure there are any number of different unit volumes whose repetition will build up the crystalline solid; the three-dimensional periodicity of the atomic arrangement means that any parallelipiped whose edges are parallel to three non-co-planar lines joining identically situated atoms will suffice. Possible unit volumes will have a wide variety of shapes and sizes, but, as we shall see in later chapters, there are great advantages when one of minimum volume is chosen in conformity with certain aspects of the symmetry displayed by the atomic arrangement; the criteria by which this is done need not concern us for the present. The conventional unit volume (or cell) for the metallic element of Fig. 1.1 is the cube shown in (b) whose edges are parallel to the three perpendicular lines indicated in (a). With the choice of cell, three intersecting edges can be selected as reference axes; its contents are then listed by the co-ordinates of the individual atoms within it, usually expressed as fractions of the lengths of the cell edges. For example, the atom in Fig. 1.2(a) within a cell with edges a, b, c is at a distance from the origin whose components parallel to these edges are x', y' and z' respectively; it is described as having *fractional cell co-ordinates* x, y, z (where $x = x'/a$, $y = y'/b$ and $z = z'/c$) with values between zero and unity. From a listing of the fractional co-ordinates of all atoms in the chosen unit volume, the structure is described and standardised projections can be constructed. These projections are usually made on to one of the cell faces, as shown in Fig. 1.2(b) for the metallic element. The fractional heights of atoms above the plane of projection are written adjacent to them, though values of 0 and 1 corresponding to positions on the lower and upper faces of the cell are often omitted. All the essential geometrical features of the crystal structure of our metallic element are described by stating that the unit repetitive volume is a cube of a certain size, and that gold atoms are to be found with co-ordinates $0, 0, 0; \frac{1}{2}, \frac{1}{2}, 0; \frac{1}{2}, 0, \frac{1}{2}; 0, \frac{1}{2}, \frac{1}{2}$. Notice that co-ordinates of atoms which would be introduced by the presence of identical adjacent unit volumes are not listed; the co-ordinates $0, 0, 0$ specify an atom at the origin on one corner of the cube; those at the seven other corners of this cube are automatically inserted by the presence of this atom in the adjacent cells which must meet at these corners in the crystalline solid. In fact the number of listed co-ordinates always corresponds to the number of atoms within the chosen unit volume; if we displace the origin of the cell, we see that there are four metallic atoms contained within the unit cube.

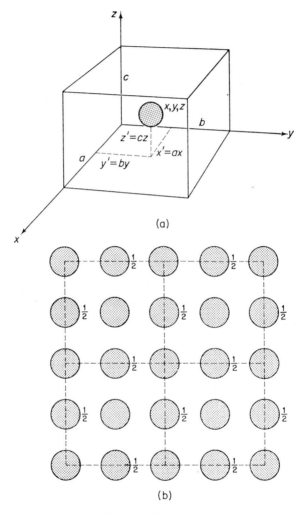

Fig. 1.2. Atomic co-ordinates and projections of crystal structures. (a) Fractional cell co-ordinates x, y, z. (b) A projection of four cells of the metallic element of Fig. 1.1 on to a face of the unit cube.

Atomic co-ordinates (and cell projections) are an essential part of the language of the structural crystallographer; they allow the general nature of the atomic pattern to be visualised and provide a record of the interatomic distances and spatial arrangements within co-ordination polyhedra for each type of atom, all of which is vital to any understanding of the crystal chemistry of a particular material. Whilst we shall not be much concerned with the detailed configurations of individual structures, it is nevertheless important that in our studies of crystalline matter we should be familiar with this method of representing its atomic patterns. It is for this purpose that we quote two further examples of structural descriptions; they are both compounds and both have conventional unit volumes which are right prisms with a square base; the reference axes are orthogonal with x and y as the edges of the square and the z-axis as another edge of the parallelipiped normal to this plane. Tin dioxide, SnO_2 (the mineral cassiterite) has a unit cell of this shape within which the atomic co-ordinates are:

$$Sn: 0, 0, 0; \quad \tfrac{1}{2}, \tfrac{1}{2}, \tfrac{1}{2}$$
$$O: \pm(u, u, \tfrac{1}{2}); \quad \pm(\tfrac{1}{2} + u, \tfrac{1}{2} - u, 0)*$$

where $u = 0.2$. Fig. 1.3 shows the contents of four cells projected down the y-axis on to rectangular faces indicated by the broken lines; the network of

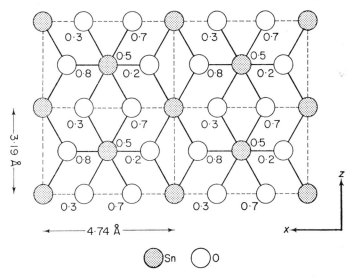

Fig. 1.3. A projection of four cells of the cassiterite (SnO_2) structure down the y-axis. 1 Å $\equiv 10^{-8}$ cm.

* If a fractional atomic co-ordinate x_1 has a value between $\tfrac{1}{2}$ and 1, this is often listed in terms of the co-ordinate below the upper face of the cell as $-x_2$ (written $\bar{x}_2$); clearly $x_1 + x_2 = 1$. When this applies to all three co-ordinates of an atom, they are usually written $-(x, y, z)$; $\pm(u, u, \tfrac{1}{2})$ in the O co-ordinates here indicates two oxygen atoms, one with an entirely positive set and one with an entirely negative set.

primary bonds (in full lines) gives rigidity to an almost homodesmic frame-work of Sn and O atoms. From the figure we see that each Sn atom has 6 O neighbours at the vertices of an octahedron and that each O atom has 3 Sn neighbours in planar co-ordination at the corners of a triangle. Calculations from the dimensions of the cell show that the octahedron is slightly irregular and that the triangle is not equilateral; both deviations from regularity must reflect a departure from uniformity in the primary attractive forces which develop between tin and oxygen atoms in this structure.

In our second example, $CO(NH_2)_2$, urea, has a similarly shaped conven-tional cell though it has different dimensions; atomic co-ordinates are quoted as:

$$C: 0, \tfrac{1}{2}, u; \quad \tfrac{1}{2}, 0, \bar{u}, \quad \text{with } u = 0.32$$
$$O: 0, \tfrac{1}{2}, v; \quad \tfrac{1}{2}, 0, \bar{v}, \quad \text{with } v = 0.57$$
$$N: w, \tfrac{1}{2} - w, t; \quad \tfrac{1}{2} - w, \bar{w}, \bar{t}; \quad \bar{w}, \tfrac{1}{2} + w, t; \quad \tfrac{1}{2} + w, \bar{w}, \bar{t},$$
$$\text{with } w = 0.14 \text{ and } t = 0.17.$$

In Fig. 1.4, atoms in four cells have been projected down the z-axis on to square faces outlined by dashed lines. Primary and secondary linkages be-tween atoms can now be recognised; primary bonds (shown as full lines) are responsible for sub-units consisting of individual urea molecules in a hetero-desmic structure. These molecules are planar and triangular in outline with vertices at oxygen atoms pointing alternately up and down; interatomic dis-tances within a molecule are much shorter than those between molecules, indicating the greater strength of the primary attractive forces. Sub-units are assembled into the regular ordered three-dimensional pattern of a urea crystal by the weaker secondary bonds between molecules (indicated by dotted lines); these are bridges which join the sub-units in mutually perpendicular orienta-tions with the N atoms of one molecule linked to oxygen atoms in neigh-bouring molecules through H atoms located somewhere between them.

Apart from their value in illustrating the kinds of atomic patterns found in crystalline matter, these examples (and all other structural descriptions) de-monstrate one fundamental physical relationship that is common to all crys-tals, viz. the unit cell weight for a particular material must be constant. We can therefore write that

volume of the chosen cell × density = sum of the atomic weights of the atoms within the cell × mass of the hydrogen atom

= Z × atomic weight of a for-mula unit × mass of the hydrogen atom where Z is the number of formula units per cell for any compound with an accepted chemical formula. In SnO_2, for example, the atomic weights of Sn and O are 118·7 and 16·0 respectively, the mass of the hydrogen atom is $1·66 \times 10^{-24}$ gm, and the density is 6·99 gm/cc so that

$$Z = \frac{(4·74)^2 \times 3·19 \times 10^{-24} \times 6·99}{(118·7 + 32) \times 1·66 \times 10^{-24}} = 2.$$

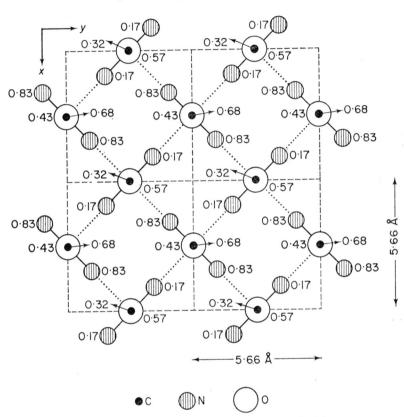

Fig. 1.4. A projection of four cells of the urea $(CO(NH_2)_2)$ structure down the z-axis. The planes of the urea molecules are parallel to the z-axis which has a repeat length of 4·71 Å. 1 Å $\equiv 10^{-8}$ cm.

There are two formula units of SnO_2 within the conventional cell, which corresponds, as we would expect, to the two Sn and four O co-ordinates uniquely specified in the description earlier. In a similar way, the cell of the metallic element contains four atoms, whilst that of urea contains two molecules of $CO(NH_2)_2$; in other circumstances, this equality can provide us with a calculated density which should match the observed value if the chemical constitution of an unknown material has been correctly determined.

1.4. Scope of the present book

The catholic nature of modern crystallographic work diffuses through many scientific disciplines, and in an introductory text it is not practicable to pursue the methods and purposes of investigation for all the many disparate materials of varying crystallinity that can be encountered; we can only hope to provide some account of the fundamental common principles on which all these investigations are based. Even so, there are problems in deciding limits for our discussions, for the study of perfect crystalline matter quickly develops from simple foundations into expanding and interlocking branches each demanding individual treatment. These foundations are formed by a knowledge of the general nature of crystal structures, which, as we have already suggested in this introduction, represent the long-range order of crystalline solids by describing the geometry and contents of small repetitive volumes. Many of the more elementary aspects of geometrical crystallography and crystalline properties are common to groupings of structures containing materials which are apparently unconnected on most physical or chemical criteria; among our examples the chemically dissimilar urea and tin dioxide at least share a common shape of unit cell. In fact, all the arrays of atoms which are developed by the various binding forces are subject to certain geometrical rules irrespective of the variety of constituents in any particular crystalline solid. Our interest will be in the different pattern types which can be constructed according to these rules; each of the sub-groups of pattern theory defines a permissible repetition of a unit motif to fill space, and since we can identify this motif with an atom (or the atomic complex of a structural sub-unit), all the atomic arrangements of crystalline matter must conform to one or other of these pattern types. In this book we shall be concerned with the general properties of these pattern types, either separately or in groups; we shall mention only rarely the details of individual structures that are important in crystal chemistry or in the fullest investigation of some particular substance.

Our objective in this text, both in content and presentation, is to provide a basic knowledge of those concepts which are essential to an understanding of crystalline matter in whatever context it is found. To do this we must analyse the principles of the construction of repetitive patterns, first for the more familiar planar patterns, and subsequently, with the aid of projections, for the 230 space patterns basic to the internal atomic architecture of all crystalline solids. Inevitably this involves the introduction of the symbolic language of the subject and its use to describe those essential geometrical features which are our concern. These are the abstract foundations of all subsequent developments, but it is unwise, even in an elementary treatment, to divorce them from the practical problems of determining the pattern sub-group to which a particular material belongs. This requires some discussion of the external shapes of crystals and an elementary account of the principles of diffraction by crystalline matter. It is in areas such as these that the most arbitrary limitations on subject matter have to be imposed; these have been set, so far as possible, at unsophisticated levels of treatment which, nevertheless, allow the determination of the cell shapes and symmetries which characterise individual space patterns. We do not, however, give much attention to the methods of structure determination; these, like other branches of many solid state studies, are firmly rooted in the elements of pattern theory and their practical application to crystalline solids that are presented here.

SELECTED BIBLIOGRAPHY

General background reading

BOWEN, H. J. M. 1967. *Properties of solids and their atomic structures.* McGraw-Hill.
BUNN, C. W. 1964. *Crystals.* Academic Press.
CHALMERS, B., HOLLAND, J. G., JACKSON, K. A. and WILLIAMSON, R. B. 1965. *Introduction to crystallography; a programmed course.* Appleton-Century-Crofts, New York.
CRACKNELL, A. P. 1969. *Crystals and their structures.* Pergamon.
Materials, a collection of articles reprinted from the *Scientific American.* 1967. W. H. Freeman.
WELLS, A. F. 1956. *The third dimension in chemistry.* Clarendon Press.

Crystal chemistry

ADDISON, W. E. 1961. *Structural principles in inorganic compounds.* Longmans.
EVANS, R. C. 1964. *Crystal chemistry.* Cambridge University Press.
HUME-ROTHERY, W. and RAYNOR, G. V. 1962. *The structure of metals.* Institute of Metals, London.
WELLS, A. F. 1962. *Structural inorganic chemistry.* Clarendon Press.

Source books for crystal structures

KITAIGORODSKII, A. I. 1961. *Organic chemical crystallography.* Translated by Consultants Bureau, New York.
WYCKOFF, R. W. G. 1963–6. *Crystal structures* (five volumes). Interscience.

2

TWO-DIMENSIONAL PATTERNS

2.1. The nature of repetitive operators

In the introductory chapter we saw that crystalline matter is formed by the continuous repetition in three dimensions of a particular fundamental atomic pattern. Many aspects of the crystallographic properties of a substance depend on the nature of this repetition and the operators which describe it, and so it is important to develop the ideas of three-dimensional pattern theory. At first the student often finds such concepts difficult, usually because he is much more familiar with the two dimensions provided by a piece of paper or a blackboard. This chapter will set out the basic elements of two-dimensional pattern theory, so avoiding, for the moment, the more difficult mental step into three dimensions. Many of the ideas that are developed here may then be transferred to the discussions of the geometrical features of crystalline matter in later chapters, so providing a more certain foundation for the greater complexity of three-dimensional patterns.

Two-dimensional repetitive patterns are a familiar part of our everyday experience, as in wallpapers, fabrics, etc.; at first sight they appear to have infinite variation but most are manufactured by repetitive operations which severely restrict the number of distinct pattern types. The variety in design that we see is due to human ingenuity in changing both the form of the basic motif (or unit of pattern) and the spacing of its periodic repetition; such ingenuity is of little importance in pattern theory, which is concerned rather with the ways in which a motif (of whatever form) can be repeated, i.e. the nature of repetitive operators and their combinations. We can begin our study of the operators that develop patterns using the very simple example in Fig. 2.1(a); only a limited part of an infinite pattern is shown, but this is enough for any analysis. The motif (or unit) consists of two perpendicular lines of unequal length; in a more sophisticated version of the same basic pattern it could be a complex geometrical design or a floral arrangement. In an analysis of the nature of this pattern, all such possibilities may be represented quite generally by a simple unit whose only requirement in the first instance is that it is *asymmetric* (or without any kind of symmetry). The pattern is derived from one single motif by two repetitive processes; firstly, repeated translation by a vector operator *b* would build up an infinite horizontal row, which is then transformed into the two-dimensional pattern by repeated translation by a vector operator *a*. This pattern may therefore be described as arising from a single motif by the movements due to two vector operators *a*

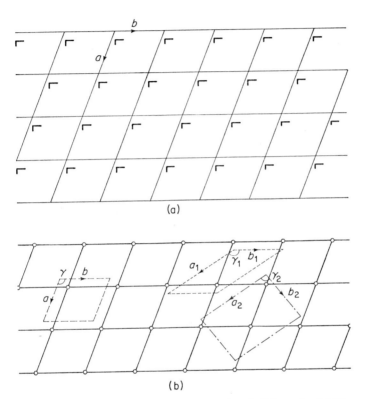

Fig. 2.1. A simple pattern with translational periodicity only. (a) The pattern of motifs with lattice translations superposed. (b) The equivalent array of lattice points; unit cells have been displaced to show the number of lattice points within each cell.

and *b* (inclined at angle γ). The essential features of this pattern type are specified by these two translational operators which mark out a grid of identical points in two dimensions; in Fig. 2.1(a) this grid is shown superimposed upon the pattern. Since it is of infinite extent its origin may be taken at any position, and for generality in the figure the origin (and all other points at the intersections of the grid) are displaced from the motif. In all two-dimensional patterns (however complex) we shall have to describe translational periodicity in terms of two inclined vector operators which lead to a grid of identical points; this regular repetitive pattern of points with identical environment about every point is known as a two-dimensional (or planar) lattice. The nature of the *lattice type* is a fundamental feature of any pattern.

The description of the lattice is, then, an important step in characterising a pattern type, and we have seen above how the array of identical points can be built up by the repetitive operations of two non-parallel vectors. However, the pair of operators a and b that were used to generate the lattice points of the example in Fig. 2.1(a) is not unique. The same pattern of points could be produced by any other pair of translational vector operators which join lattice points; for example we can regard it as derived from the translational vectors a_1 and b_1 (inclined at γ_1) or a_2 and b_2 (inclined at γ_2) as shown in Fig. 2.1(b).

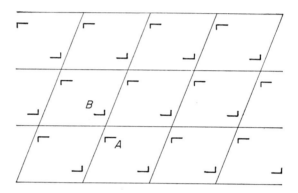

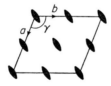

Fig. 2.2. A simple pattern with translational periodicity and rotational symmetry. The rotation diad points associated with each cell are shown below.

Such pairs of vectors outline a repeated area of the pattern in each case, and are said to define a *unit cell* (or unit mesh) for the lattice. The lattice of Fig. 2.1 could be described in terms of each of the alternative cells we have outlined, though there are certain criteria which, as we shall see later, enable us to select the most suitable unit cell. Comparing the three cells of Fig. 2.1(b), that defined by a_2 and b_2 is different from the others in that it is twice as large; the other two cells are of minimum area. As a reflection of this we can see from the diagram that the larger cell contains two points of the lattice, whereas the other two have only one. Cells with more than one lattice point are said to be *multiply primitive* (doubly primitive in this case); the cells defined by a and b, or a_1 and b_1 are said to be singly primitive (or, more usually, just *primitive*). Wherever possible, lattices are described in terms of primitive unit cells of minimum area, though, as the discussion on possible lattice types in the next section will make clear, there are some occasions where the use of non-primitive cells cannot be avoided without considerable inconvenience. In this example, however, the cell outlined by a_2 and b_2 would not normally be utilised to describe the pattern of lattice points. Both the other two cells are primitive and of minimum area, and any preference could only be decided by an arbitrary choice. The basis of any conventions need not concern us here, and we will proceed to further developments using the cell enclosed by a and b, remembering that an essential feature of any two-dimensional pattern is the lattice described by two non-parallel translation vectors; the distinctive types of planar lattices are examined in greater detail in Chapter 2.2.

However, in addition to lattice operators, patterns can contain further repetitive operators of a different kind. In Fig. 2.2 the same motif has been repeated by the same lattice translations, but there are two motifs in each unit cell. It is important to recognise that the lattice remains primitive, for there is still only one lattice point associated with each unit cell; there are now two motifs associated with each lattice point, and this has arisen because of an additional operator. In this example, the operator has repeated a motif by rotation through 180° about a direction normal to the paper (e.g. A and B are related in this way), so doubling the number of motifs in the cell. The kind of symmetry operator that repeats a motif by rotation about a direction is known as a *rotation point* (of symmetry). For self-consistency, rotation points must ensure coincidence with the original motif after a complete revolution, so that $n\phi = 2\pi$, where ϕ (in radians) is the angular interval between each repetition, and n is an integer defining the degree of the rotation. In Fig. 2.2, $\phi = \pi$ so that $n = 2$, and the pattern is showing a two-fold rotation operation normal to the paper; it is possible to have other rotation points with $n = 3, 4, \ldots$ to give three-, four- $\ldots$ fold repetition of the motif in a complete revolution, although we shall see shortly that the combination of rotation and translation operators restricts the values of n.

There are still further repetitional operators which can occur in two-dimensional patterns; one of these is seen in the pattern of Fig. 2.3. The lattice reproducing the same motifs now has a primitive rectangular cell ($\gamma = 90°$), and there are again two motifs associated with each cell, i.e. with each lattice point. This pattern differs fundamentally from those in the previous figures in that it is impossible to relate the motifs by operations involving either the translations or the rotations that have been described (e.g. consider the motifs A and B in Fig. 2.3). This is due to the asymmetric nature of the original motif, which can exist in two different but related forms. A pair of human hands shows the most familiar example of two such forms, and motifs related in this way (sometimes called enantiomorphous pairs) are said to be left- and right-handed; in the pattern, A and B are a left- and right-handed enantiomorphous pair. Since in general the left-handed motif (A) cannot be brought into congruence with the right-handed motif (B) by any combination of translation and pure rotation operators whilst remaining at all times in the plane of the pattern, there must be some further operator which can cause such an inversion. In two dimensions the change of hand in the inversion can only be produced by reflection across a line, and the *reflection line* is another kind of repetitive operator which can occur in planar patterns. The pattern of Fig. 2.3 is produced by the translational repetitions of the lattice together with inversion to the opposite hand by a reflection line parallel to the *b*-lattice translation.

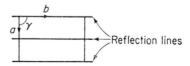

Fig. 2.3. A simple pattern with translational periodicity and reflection symmetry. The reflection lines associated with each cell are shown below.

So far, we have established the nature of the basic operations (translation, rotation and reflection) which control the repetition of a motif to form the two-dimensional pattern. Often the pattern is the result of the combination of these operators, and there is an interdependence which limits the number of permissible combinations. We shall investigate this interdependence in detail later, but for the present the discussion will be confined to those general features which enable us to see whether any new types of operators are encountered and what restrictions are placed on rotation points. In particular, since lattice translations must always be present it is important to examine the effect of compounding these with the reflection and rotation operators. Firstly, the effect of such a combination for translation and reflection operations is shown in Fig. 2.4(a); the left-handed motif (A) is reflected across the symmetry line to change its hand, and subsequently translated to give the right-handed motif (B). This is then reflected again to give the left-handed motif (A′) and so on. In effect this provides a new kind of operator, the *glide*

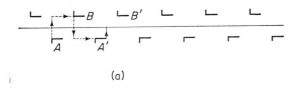

(a)

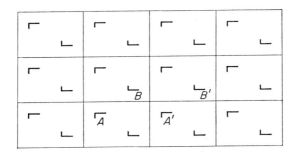

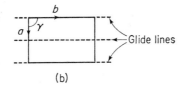

(b)

Fig. 2.4. A simple pattern with translational periodicity and glide symmetry. (a) Combination of a lattice translation and a reflection line to give a glide line; the periodicity of the lattice translation is *AA′*. (b) The simple pattern; the glide lines associated with each cell are shown below.

line, in which right- and left-handed forms of the motif are successively repeated on either side of the glide line with a periodicity of half the lattice translation in the direction of glide. Fig. 2.4(b) shows part of a pattern produced using the same lattice translation as Fig. 2.3, but with the reflection line replaced by a glide line.

Next, the combination of an *n*-fold rotation operation with a lattice translation is considered in Fig. 2.5. In (a), the symmetry operator, normal to the paper, passing through 0_1 is repeated at 0_2, 0_3, etc. by the lattice translation. The operation through 0_1 in a clockwise sense will demand repetition of 0_2, 0_3, etc. after an angular interval ϕ to give the row $0_10'_2$, etc.; simultaneously we must consider the effect of the other operators, where, for example, after

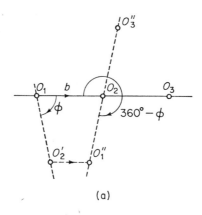

(a)

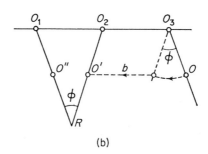

(b)

Fig. 2.5. Combination of a lattice translation and a rotation. (a) The effect upon the original lattice row. (b) The effect upon a lattice point off the original row.

rotation through $360° - \phi$ in the same sense, the rotation point at 0_2 will produce the row $0''_10_20''_3$, etc. The result of these operations will be to produce a planar mesh of symmetry related identical points; the mesh will have point-rows parallel to the original row $0_10_20_3$, etc. developed by the lattice translation. From the simple geometry the closest spacing of points on these rows is $0'_20''_1$ $(=b(1-2\cos\phi))$, but for mutual consistency of the mesh this spacing must be an integral multiple of the original translation operator, b, in the direction of the rows. Thus

$$1 - 2\cos\phi = \text{an integer},$$

and since the limits of cos ϕ are ± 1, the only values of this integer are 3, 2, 1, 0 and -1. In turn, this implies that the only permissible values of ϕ are 180°, 120°, 90°, 60° and 0° respectively, showing the restrictions on the degrees of rotational symmetry imposed by their combination with translational operators. In our repetitive patterns, therefore, we can only have two-, three-, four- and six-fold rotation points of symmetry; the value of 0° (or 360°), of course, indicates repetition only after a complete revolution, i.e. the absence of what would conventionally be recognised as a symmetry point. Operators for which $n = 2$, 3, 4, and 6 are known as *diad*, *triad*, *tetrad* and *hexad* points respectively (repetition only after a complete revolution is sometimes spoken of as a *monad*). A rotation diad point repeats a motif after 180° rotation, a

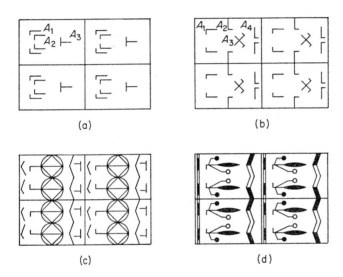

(a)   (b)

(c)   (d)

Fig. 2.6. Patterns showing the same translational periodicity and reflection symmetry. (a) and (b) use combinations of the simple motif in Fig. 2.3; there are three and four motifs per lattice point respectively. (c) and (d) show more complex patterns derived using a variety of motifs associated with each lattice point.

rotation triad after 120° and so on. In addition to demonstrating these limitations, simple considerations also show that no new symmetry operator (analogous to the glide line) is formed so long as the translation is confined to a plane. In three dimensions, as we see later, a translation could occur in the same direction as the analogous operator, a rotation axis, and this can produce a new kind of symmetry operator; in two dimensions, with possible lattice translations restricted to a plane, the same rotational symmetry is repeated in new positions in the resultant mesh. This repetition is illustrated in a general way in Fig. 2.5(b), which shows the effect of such a combination upon a lattice point 0 displaced from the original row $0_1 0_2 0_3$, etc. The symmetry axis through 0_3 together with the lattice translation b repeats 0 at $0'$; on the same row (parallel to $0_1 0_2 0_3$, etc.) there must also be $0''$ obtained from 0 by lattice translations alone. Thus $0''$ and $0'$ (and the lines on which they lie) are related by a clockwise rotation through an angle ϕ about the point R away from the original row; since the operations which produced them must be generally valid for the whole of the resultant mesh, this net will show the development of the original symmetry point at all positions like R.

To summarise, the discussion so far has been concerned with the nature of the repetitive operators to be found in two-dimensional patterns. In planar repeating patterns, the recognition of translations defining the unit cell of the lattice always begins any analysis; to these may be added other repetitive operators, usually called symmetry elements, which can be rotation points (diad, triad, tetrad, hexad) and symmetry lines (reflection or glide). We have seen already some general indications of the interdependence of lattice translations and symmetry elements, and we must now examine the manner of their relationship in more detail. However, it is important to emphasise again at this point that limitations on pattern type in no way restrict the varieties of design by the use of different motifs. This is shown in a simple way in the four patterns of Fig. 2.6; these all belong to the same basic pattern type, in which the lattice translations give a primitive rectangular cell and a reflection line parallel to a cell edge is also present.

2.2. Two-dimensional lattice types

In the preceding section the concept of a lattice as a mesh of points formed by the repetitive operations of two non-parallel translational vectors has been described; each lattice point in the mesh can be associated with a simple or complex element of the pattern, and the points are identical in the sense that an observer viewing his surroundings from any point could not decide the exact location of his particular point within the mesh. An infinite lattice does not need to have a particular origin in the pattern, although for convenience it is often chosen to coincide with a particular feature, usually an associated symmetry element (e.g. the diad point in Fig. 2.2). The lattice is described by a unit cell, whose shape and size are determined by the magnitudes of the lattice translations a and b and the interaxial angle γ; the chosen unit cell is usually primitive with one lattice point per cell, though occasionally a multiply primitive cell with more than one lattice point is selected. We must now enquire into the number of distinctive lattice types in two dimensions, and the combinations of symmetry elements with which they are likely to be found. Before answering these questions, the meaning and significance of a lattice type must be clarified. It may be argued that any two non-parallel rows of lattice points could be used to define lattice translations and a unit cell. For example, in Figs. 2.3 and 2.4, a rectangular cell has been chosen; it is possible to describe these patterns in terms of lattice translations parallel to the diagonals of the rectangular cell so employing an oblique rhomboidal cell. However, this is not a desirable choice for not only would the oblique cell be larger (two lattice points per cell), but its shape would be out of harmony with the symmetry shown by the pattern. Although it would be possible to proceed in this arbitrary way to describe any arrangement of lattice points in terms of a general oblique cell, there are serious disadvantages in attempting systematic developments on this basis. Unit cells are therefore chosen to be of minimum area for a particular pattern of lattice points and in accord with any symmetry displayed by the pattern. Different lattice types are distinguished by the general shapes of the cells and whether they are simply or multiply primitive; thus two primitive rectangular cells of different lattice translations are of the same general lattice type, but we recognise a distinction between lattices with square and rectangular primitive cells.

In two dimensions there are only four distinctive shapes (Fig. 2.7); (a) shows an *oblique cell* with no particular relationship between a and b and a general value of γ, (b) shows a *rectangular cell*, again with no relationship between a and b, but with $\gamma = 90°$, (c) shows the *square cell* with $a = b$, $\gamma = 90°$, while (d) shows a *hexagonal cell*, rhombus-shaped with $a = b$, $\gamma = 120°$. Each of these cells could be used to mark out a different two-dimensional primitive (p) lattice type with lattice points at each of the cell corners. But since they are to be used in association with particular symmetry elements, one must ask what symmetry elements are consistent with the arrangement of lattice points that they define; in the lower diagrams of Fig. 2.7, the *maximum symmetry* consistent with each distribution of lattice points is shown. It is important to realise that in any pattern all of these symmetry elements need not necessarily be present; the patterns of Fig. 2.6 have the rectangular cell of (b) with one reflection line only. Nevertheless, once a particular symmetry element is displayed in the pattern, the most suitable cell shape is determined; thus the existence of the reflection lines in the patterns of Fig. 2.6 allows the choice of a rectangular cell. The association of cell shape and *essential symmetry elements* is simply expressed. The oblique p-lattice is suitable for the introduction of any rotational symmetry up to a diad; a rectangular p-lattice is demanded once a reflection line is present, a square p-lattice once a tetrad is present and the hexagonal p-lattice once triad or hexad points are introduced. For each lattice type there may be several different permissible combinations of symmetry operators each containing the essential symmetry element; the possible groupings of symmetry elements for each lattice type will be discussed later. At this time we can recognise that it is the presence of the essential symmetry elements in the pattern which requires a particular unit cell shape, not the converse; thus a pattern without symmetry elements could have any specialised cell shape, though the pattern type on which it is based would be represented formally using the general shape of the oblique cell.

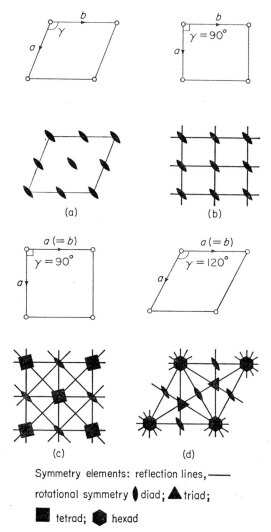

Symmetry elements: reflection lines, ——

rotational symmetry ❨ diad; ▲ triad;

■ tetrad; ⬢ hexad

Fig. 2.7. Cell shapes in two dimensions and the associated maximum symmetry elements. The upper diagrams show the cells for primitive lattice types; the lower diagrams show the maximum number of symmetry elements and their mutual orientations consistent with lattices based on each shape.

So far, the existence of four distinctive primitive two-dimensional lattice types is established; but it is also necessary to ask if any of these cell shapes can occur in a multiply primitive form. This may be answered by first seeing if additional lattice points can be added without destroying any essential symmetry or the identity of all lattice points, and then examining whether a smaller primitive cell may be chosen with the same general shape; if a new primitive cell cannot be found, one must recognise a new multiply primitive lattice type. As an illustration we will consider the possibilities of multiply primitive oblique lattices. The symmetry elements of an oblique cell are shown in Fig. 2.7(a); in addition to the diad points at the corners of the cell, there are other diads at both the centres of the cell edges and of the cells. If extra lattice points are inserted separately or in any combinations of these positions, multiply primitive cells of the same shape and symmetry will be produced; some of the possibilities are shown in Fig. 2.8. In (a) the extra lattice point is at the centre of an edge at A to give a doubly primitive cell; but

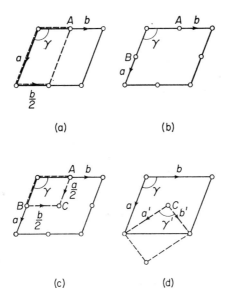

(a)   (b)

(c)   (d)

Fig. 2.8. The addition of extra lattice points to a primitive oblique cell. Alternative primitive cells are shown with broken lines.

this is not a new lattice type, for it merely leads to the halving of a lattice translation to give a primitive oblique cell defined by a, $b/2$ and γ. A similar argument would apply if the extra lattice point had been added at the mid-point of the other cell edge, but in (b) two extra lattice points are added at A and B, the centres of both cell edges, to give a triply primitive cell. However, this does not constitute a lattice, for the environments of A and B are now different, and this attempt to construct a triply primitive oblique cell fails on this criterion. In (c), with all possible positions occupied by extra lattice points at A, B and C, the lattice points are once again identical, giving a quadruply primitive oblique cell; but this is not a new lattice type, for both lattice translations can be halved to give a primitive oblique cell defined by $a/2$, $b/2$ and γ. In the last example, (d), a doubly primitive cell is formed by the addition of an extra lattice point at the centre of the cell at C. At first it might be thought that this represents a new non-primitive lattice type, but we must remember that we are concerned only with a general cell shape and that the magnitudes of a, b and γ are unimportant in the oblique cell. By changing the lattice translations a smaller primitive oblique cell (with new dimensions a', b' and γ') can be found to describe this arrangement of lattice points, and so again no multiply primitive lattice type is formed. By such arguments it is clear that there is only one distinctive oblique lattice type, for all multiply primitive oblique cells may be reduced to a simple primitive form by a suitable choice of lattice translations.

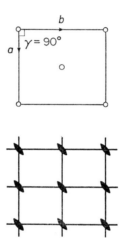

Fig. 2.9. The centred rectangular lattice cell. The upper diagram shows the distribution of lattice points and the lower diagram the arrangement of the maximum number of symmetry elements consistent with lattices based on such a cell.

The implications for the other cell shapes could be systematically explored in this way, when it will be found there is only one two-dimensional lattice type with a non-primitive cell. This is the *centred* (c) *rectangular cell* (Fig. 2.9); it is doubly primitive, consistent with symmetry and lattice requirements, and it cannot be described by a primitive cell in which the orthogonality ($\gamma = 90°$) of the lattice translations is preserved. This may be added to the list of planar lattice types to bring the total to five, the oblique p-lattice, the rectangular p- and c-lattices, the square p-lattice and the hexagonal p-lattice; the translational operators in any two-dimensional pattern are described in terms of one of these lattice types.

2.3. Two-dimensional point groups

In the last section the translational aspects of pattern theory have been described, but the repetitive nature of certain patterns may also be due to the presence of symmetry elements; only when both facets are combined is it possible to complete the analysis of a given pattern. It is convenient to take this last step in stages, and in this section we consider only those combinations of symmetry elements which could occur if the repetition demanded by translational vectors is neglected. Rotation points and reflection lines are the only symmetry elements that need consideration, for glide lines involve a translation. Strictly speaking, with translational repetition suspended, a restriction on the degrees of rotational symmetry need no longer be imposed; but, since we shall wish to relate this discussion to the symmetry elements found in

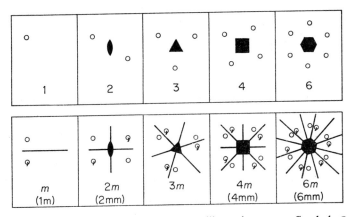

Fig. 2.10. The ten two-dimensional crystalline point groups. Symbols ○ and ☉ are used to represent left- and right-handed objects.

repeating two-dimensional patterns, we must only consider those combinations with rotation operations for which $n = 1, 2, 3, 4$ or 6. The combinations of symmetry elements without translational repetition repeat an object about a point, and so each distinct combination is said to form a *point group*. Point group theory can be developed without reference to planar pattern theory, and it can be argued that a restricted discussion which refers only to a limited number of point groups implies that some other descriptive term should be used. If this is necessary, it is difficult to propose a suitable term except by analogy with the discussion of the three-dimensional point groups in Chapter 4; in this, crystalline matter can be divided into a finite number of classes, in each of which there is the same point group symmetry. We can therefore refer to the restricted number of point groups associated with crystals as the crystalline point groups, so that the present discussion concerns only *two-dimensional crystalline point groups*. To avoid the repetition of this cumbersome expression, it will be understood that the point groups referred to in subsequent sections are the limited number that are compatible with translational repetition as it exists in planar patterns and crystalline matter.

The number of such planar point groups is very simply derived. Each symmetry element can exist independently. Thus there will be point groups for each kind of rotation point, and there must also be a group with a reflection line alone; symbolically these groups are represented as 1, 2, 3, 4, 6 and m respectively. All other point groups are combinations of rotation points and reflection lines; in each of these groups the n-fold rotational symmetry repeats the reflection line at the appropriate angular interval. This provides four more point groups, symbolically $2m$, $3m$, $4m$ and $6m$, to bring the total of two-dimensional crystalline point groups to ten, illustrated in Fig. 2.10. Groups common to a particular cell shape are classed together into *two-dimensional crystal systems* (again by analogy with the discussion of crystalline materials in Chapter 4.2); thus 4 and $4m$, which both require a square cell, are said to belong to the same system.

2.4. Two-dimensional space groups

To complete the analysis of planar patterns, repetitions demanded by point group symmetry must be combined with those of the lattice type to produce one of the basic two-dimensional pattern types. These are called two-dimensional space groups, for they show the repetition of units of pattern to fill an infinite plane, and in this section their formation is discussed. In considering the nature of these space groups, however, it must be remembered that the combination of the point group operators with lattice translations does allow the development of new symmetry elements; in two dimensions, as we have seen, this leads only to the glide line (represented symbolically g) which can replace or supplement reflection lines. Moreover, in associating point group symmetry with a particular lattice cell, the relationships between symmetry and cell shape discussed in 2.2 must be maintained, so that, for example, the oblique p-lattice is only found with the point groups 1 and 2; higher symmetry demands more regularly shaped unit cells. A space group is expressed symbolically by stating both the lattice type and the point group symbol (modified if necessary to include glide elements); thus the simplest planar space groups are $p1$ and $p2$, indicating a primitive oblique cell with no symmetry and a diad point respectively.

Before describing the systematic derivation of all two-dimensional space groups, it is valuable to examine one group in some detail to clarify their significance. For this we shall use $p2mm$, in which the symmetry elements demand that the primitive lattice cell is rectangular in shape. Space groups are conventionally represented on two diagrams as in Fig. 2.11. Both show a unit cell of the lattice; in the right-hand cell the symmetry elements are displayed, whilst the left-hand cell shows the distribution of *equivalent points* produced within the cell by all the operators. To construct these diagrams we start with the single lattice point of the primitive cell in some general position, A, on the left; the symmetry elements given by the point group symbol (diads at a, reflection lines at b and c) are marked on the right-hand diagram. The symmetry elements are then operated in turn on the lattice point; since the point group 2*mm* produces four related points (Fig. 2.10) and we start with one lattice point per cell, the final number of equivalent points within the cell must be $4 \times 1 = 4$. Rotation of A about the diads at a gives A_1 (points in adjacent cells must be repeated in the original cell by the operation of the lattice translations). A will also be reflected across the symmetry line at b to A_2; this involves an inversion, conventionally shown by placing a comma in the circle of the lattice point, indicating that if any motif associated with A is left-handed, the repetition at A_2 will be right-handed. Finally, reflection of A across the symmetry line c to give A_3 (again with a change of hand) completes

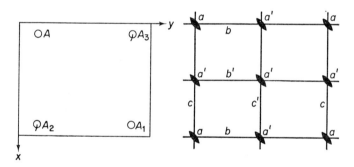

Fig. 2.11. Space group diagrams for *p2mm* (*pmm*). On the left is the diagram showing the general equivalent positions; on the right the diagram showing the distribution of symmetry elements.

the four equivalent points in the cell of this space group. The distribution of points in the cell on the left is consistent with the development of extra symmetry elements; there are further diads at a', and reflection lines at b' and c' to be inserted on the diagram of symmetry elements on the right. The significance of the *general equivalent positions* is that, if some object (an abstract motif, a rose, etc.) is placed at A it must be repeated at A_1, A_2 and A_3 in the appropriate hand to produce this kind of pattern. For each space group such positions are listed in terms of fractional co-ordinates; if A is at (x, y), the four general equivalent positions (G.E.P.'s) are listed as $\pm(x, y)$; $\pm(x, \bar{y})$. The number of equivalent positions per cell may often be reduced, by siting A in a special position with respect to the symmetry elements, e.g. if A is placed on the symmetry line b it will coalesce with A_2, whilst A_1 and A_3 will also coincide to reduce the number of equivalent positions per cell to two. These locations are known as *special equivalent positions* (S.E.P.'s) and are also listed for each space group according to the manner in which they occur; in the example above, they would be shown as *m* (indicating the symmetry of the chosen special position), $\pm (0, y)$. A pattern based on this space group using symmetric and asymmetric motifs in general and special positions is shown in Fig. 2.12.

To derive all the possible two-dimensional space groups, one must consider systematically all combinations of lattice type with permissible groupings of symmetry elements. With an oblique cell only *p1* and *p2* can occur;

Fig. 2.12. A pattern based on the space group *p2mm* (*pmm*). The reader can analyse this pattern to see how many of each kind of motif there are in each cell and whether they are in general or special positions.

but for a rectangular cell there are many more combinations, for it can be both primitive and centred and have symmetry associated with the point groups *m* and *2mm*. Thus we might expect space groups *pm, pg, cm, cg* to be derived from point group *m*, and *pmm, pmg* ($\equiv$*pgm*), *pgg, cmm, cmg* ($\equiv$*cgm*), *cgg* from point group *2mm* (the diad symbol 2 is often omitted from this point group and its associated space groups, for its presence is always implied by the other symmetry elements). Looking at the list of two-dimensional space groups in Table 2.1, we see that only 7 out of these 10 possibilities are listed;

Table 2.1. *The relationships between two-dimensional lattice types, point groups and space groups*

Cell shape (crystal system)	Oblique		Rectangular		Square		Hexagonal			
Lattice type	*p*		*p* and *c*		*p*		*p*			
Point group (crystal class)	1	2	*m*	*mm*	4	4*m*	3	3*m*	6	6*m*	
Space groups	*p*1	*p*2	*pm* *pg* *cm*	*pmm* *pmg* *pgg* *cmm*	*p*4	*p*4*m* *p*4*g*	*p*3	*p*3*m*1[a] *p*31*m*[a]	*p*6	*p*6*m*	

[a] The existence of two distinct groups is related to the two alternative positions for the three reflection lines associated with the triad rotation; in *p*3*m*1, the initial position of the reflection line bisects γ, in *p*31*m*, it is taken along an edge of the cell.

the reason for this is that not all combinations form new and different space groups. For example, Fig. 2.13 demonstrates the equivalence between *cm* and *cg*; the symmetry elements are the same in both cases, interleaving *m* and *g* lines, and the distributions of G.E.P.'s can be made identical by a displacement of the origin of the cell. Similarly *cmm* ≡ *cmg* ≡ *cgg*, and only one

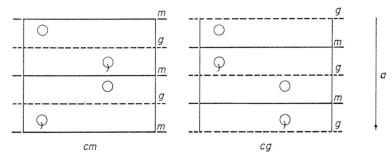

Fig. 2.13. Diagrams to show the equivalence of *cm* and *cg*. For easier comparison in both cases the distributions of general equivalent positions and symmetry elements are shown on a single diagram; identity is achieved if the origin of the cell is displaced through *a*/4 in either case.

distinctive space group symbol need be listed. By this kind of argument it can be deduced that there are only the 17 different two-dimensional space groups set out in Table 2.1, where their relation to symmetry and lattice operators is shown. All repetitive planar patterns must be based upon one or other of these space groups.

2.5. Extension to three-dimensional patterns

The elements of two-dimensional pattern theory given in this chapter are relevant to any discussion of the three-dimensional regularity shown by crystalline matter; naturally the analysis with an extra dimension is more complex. Three translational vector operators will be needed to mark out the pattern of identical points which form the lattice. Some of the other repetitional operations encountered in planar patterns are comparable with those which exist in crystalline matter; rotational symmetry around a point is replaced by rotation about a line to give axes of symmetry, whilst inversion across a line becomes reflection across a symmetry plane. But, as will be seen in later chapters, there are possibilities of new symmetry elements in both point groups and space groups that do not arise for planar platterns. Nevertheless, the classification of the pattern types to be found in crystals is generally similar to that which we have established in this chapter for two-dimensional patterns. The ultimate sub-division is the space group of which there are 230 (compared with 17 in two dimensions). Each space group is associated with a particular point group (or crystal class) of which there are 32 (compared with 10 in two dimensions). Particular groups of crystal classes can be associated with particular cell shapes (or crystal systems); depending on certain conventions, there are said to be 6 or 7 systems (compared with 4 in two dimensions). Each cell shape may define a primitive or multiply primitive lattice type, giving a total of 14 (compared with 5 in two dimensions).

Despite its complications, an analysis of three-dimensional patterns can follow the outlines laid down by the simpler two-dimensional analogy. In succeeding chapters, we shall discuss the nature of the lattice and symmetry operations leading to the description of crystallographic lattice types, systems, point groups and space groups, though in an introductory text it is not possible to pursue detailed studies of each of the 230 space groups. Our objective must be to provide sufficient understanding of the essential principles in order that the reader may continue further developments quite readily; for convenience the order of presentation may differ slightly from that in this chapter.

2.6. Exercises and problems

1. (i) Examine wall and floor coverings, fabrics, etc. to identify motifs and their translational repetitions so as to recognise the lattice type.

   (ii) Analyse the symmetries shown by the arrangement of motifs in such patterns so as to recognise their point groups. [Further practice in recognising point group symmetry can be obtained from individual symbols, e.g. the letters of the alphabet in Roman capitals, a swastika, the star of David and so on, though it should be remembered that there will be no limitation on the degree of rotational symmetry.]

(iii) In repetitive patterns combine your observations under (i) and (ii) to determine the two-dimensional space groups to which they belong.

(iv) Reverse the process by designing your own patterns with various motifs to illustrate individual two-dimensional space groups.

2. Demonstrate that any multiply primitive lattices with (a) triad symmetry and cell shapes in which $a = b, \gamma = 120°$ and (b) tetrad symmetry and cell shapes in which $a = b, \gamma = 90°$ can be reduced to a suitable primitive form.

3. Demonstrate that the two-dimensional space groups $p3g1$, $p31g$, $p6g$ are already represented by alternative symbols in Table 2.1.

4. In a repetitive pattern in which the unit cell has $a \neq b, \gamma = 90°$ a characteristic motif was identified at general equivalent positions whose fractional co-ordinates are of the form $\pm(x, y)$; $\pm(\frac{1}{2} - x, y)$. Identify the space group to which this pattern belongs. Other motifs occur in the pattern in special equivalent positions; list their possible fractional co-ordinates.

SELECTED BIBLIOGRAPHY

Planar pattern analysis
BUERGER, M. J. 1956. *Elementary crystallography*. Wiley.
INTERNATIONAL UNION OF CRYSTALLOGRAPHY. 1965. *International tables for X-ray crystallography*, vol. I. Kynoch Press, Birmingham.

Illustration of planar patterns
ESCHER, M. C. 1960. *Grafiek en tekeningen*. Zwolle, J. J. Tijl.
MACGILLAVRY, C. H. 1965. *Symmetry aspects of M. C. Escher's periodic drawing*. International Union of Crystallography.

3

STEREOGRAPHIC PROJECTION

3.1. The nature of the problem

The three-dimensional character of crystalline materials often makes it difficult to visualise linear and angular relationships between various features of an atomic pattern, and it is essential to supplement any mental picture by a projection. Before enlarging upon the pattern theory outlined in the preceding chapter we must consider this problem of systematic projection into two dimensions, for projections will be extensively used in our discussions. Unfortunately there is no simple solution which can combine both the linear and angular properties that must be represented, and they are usually treated separately. We saw in Chapter 1 how atomic positions within a unit cell are conveniently projected on to one of the cell faces with fractional co-ordinates giving indications of their heights. This simple method is quite adequate for a general impression of the crystal structure, and permits the accurate representation of linear features such as atom positions, interatomic distances, etc.; however, such projections do not readily provide a rapid method of calculating and expressing angular relations (bond angles, etc.) within the atomic pattern. More importantly, we saw in the last chapter that details of the atomic pattern can be replaced by the equivalent points generated by the lattice and symmetry operators; we need to represent the spatial distributions of rows and planes of lattice points and the general angular relationships between them. Furthermore, in Chapter 6 we shall see that among the most important characteristics of crystals with well-developed external shapes are the angles between their plane faces. In fact, many of the most significant crystallographic properties (like symmetry) are concerned with angular dispositions just as much as linear relationships.

Many projections which try to preserve angular relations are based on an initial projection on to a sphere. The subject for projection is imagined to be surrounded by a central sphere (often called the *circumscribing sphere*); for our purposes the subject might be two inclined lattice rows, a plane of lattice points, or a geometrical figure bounded by plane faces. Since we are concerned with angular features only, the directions in all three cases are shown as radii of the circumscribing sphere (Fig. 3.1); in (a) the lattice rows themselves are the radii, in (b) and (c) the radii are those normals to the planes which pass through the centre of the sphere. Each radius meets the surface of the circumscribing sphere at a point known as a *spherical pole*; in each example the distribution of poles on the surface of the sphere faithfully retains the angular

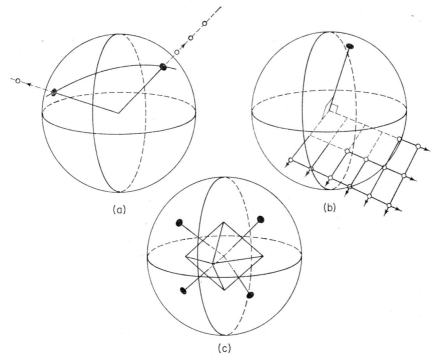

(a)

(b)

(c)

Fig. 3.1. Projection on to a circumscribing sphere. (a) Two lattice rows; the angle between the spherical poles is that between the lattice rows. (b) A plane of lattice points; the spherical pole is at the end of the radius normal to the plane. (c) A geometrical figure; each face is represented by a spherical pole at the end of a radius normal to the face (for clarity, the rear faces of the bipyramid and their spherical poles are omitted).

dispositions of the projection subject. For (a), the spherical arc between the poles is a measure of the angle between the lattice rows; while for (b), the position of the spherical pole fixes the orientation of the lattice plane with respect to a reference direction such as the N-S axis of the sphere, and in (c) the shape of the original figure could be constructed, for its planar faces are normal to the radii defined by the poles. So far, the representation is still three-dimensional with the subject replaced by a number of poles on the surface of a sphere. Before completing a projection into two dimensions, it is valuable to consider two important circular loci on the sphere.

Firstly, *great circles* on the surface of a sphere are the intersections with the spherical surface of planes passing through its centre; they have the same radius as the sphere and must pass through opposite ends of one of its diameters (Fig. 3.2(a)). Lines of longitude on the Earth are great circles passing through the North and South poles. A great circle may be drawn through any two points on the surface of the sphere, such as the poles representing the lattice rows in Fig. 3.1(a) (if the points are at opposite ends of a diameter, clearly

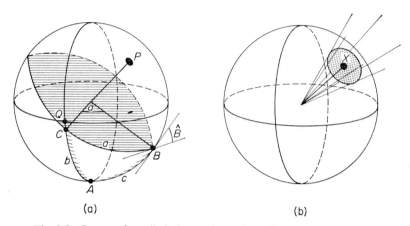

(a)   (b)

Fig. 3.2. Great and small circles on the surface of a sphere. (a) The inclined great circle is cut by the shaded plane through the centre of the sphere. The pole of this great circle is at P. The area ABC on the surface of the sphere bounded by the arcs of three great circles is a spherical triangle with six angular elements (sides a, b, c and included angles $\widehat{A}$, $\widehat{B}$, $\widehat{C}$). The length of a side a is given by the angle subtended by the arc BC at the centre of the sphere. An included angle, $\widehat{B}$, is measured by the angles between the tangents to the two great circles at their point of intersection; alternatively, it is the angular distance between the poles of the two great circles ($\widehat{PQ}$). (b) The small circle is cut by the shaded plane which does not pass through the centre of the sphere. Radii through points on the small circle form a cone so that they are equally inclined to the centre of the circle at X.

a third point is necessary to specify a particular great circle). The distance between two points on a great circle is the angle subtended by that arc of the circle at the centre of the sphere. The radius which is normal to the plane of the great circle gives a point on the sphere called the *pole of the great circle*, which is therefore at 90° to all points on the circle. The angle between two circles is measured between the tangents at the intersection of the circles, or, alternatively, as the angular distance between the poles of the great circles. Intersections between three great circles form a *spherical triangle* with six angular elements (three sides and three included angles); the relationships between these elements is the subject of spherical trigonometry (see Appendix

B). Secondly, *small circles* on the surface of the sphere are the intersections with the spherical surface of planes which do not pass through its centre; their geometrical radius is less than that of the sphere (Fig. 3.2(b)). The lines of latitude on the Earth are small circles with different angular radii drawn about the North and South poles; the equator is a small circle of radius 90°, i.e. a great circle. Since a small circle is the locus of points on the surface which are at an equal angular distance from a spherical pole, its use is mostly in locating a pole whose angular distances from two given positions are known. This enables great circles to be drawn through the three points to define a spherical triangle; arcs of small circles, however, cannot be used as elements of such triangles.

We shall return in the next section to the problem of projection, remembering that the subject has been replaced by a number of poles on a sphere; the relationships between these poles can be described in terms of their locations on great and small circles. In the final projection into two dimensions, experience and familiarity will soon dispense with the need to visualise the intermediate circumscribing sphere.

3.2. The method of projection and its properties

Whilst there are a number of ways in which the spherical poles could be projected into two dimensions, that most commonly employed in crystallography is the stereographic method, an adaptation of an old classical projection which was probably first used by the early Greek astronomers. The main virtues of *stereographic projection* are the following. (a) Subject only to the limits of scale, all directions in space can be accurately represented. (b) The angular truth of the sphere is preserved so that angular measurements between directions are unaltered and easily determinable with accuracy. (c) Projections of both great and small circles on the sphere appear as arcs of true geometrical circles, thus making their construction simple. The validity of these assertions is demonstrated in Appendix A, where the detailed properties of the projection are discussed.

To make a stereographic projection from the circumscribing sphere, we first choose points at the opposite ends of a diameter to be the north and south poles of the sphere; the *plane of projection* is then the equatorial plane of the sphere (Fig. 3.3). From each spherical pole in the northern hemisphere a line is drawn to the south pole of the sphere; the point at which this line cuts the equatorial section is marked as a dot on the projection plane. Such dots denote the stereographic poles representing directions in the projection subject; clearly these poles lie within the area of the projection bounded by the trace of the equatorial circle (known as the *primitive*) at distances from the centre of the projection given by $r \tan \psi/2$, where r is the radius of the projection (i.e. the circumscribing sphere), and ψ is the angular distance between the corresponding spherical pole and the north pole of the circumscribing sphere. If we continued in this way with any spherical poles in the southern hemisphere, the projected stereographic poles would lie outside the primitive; this becomes more inconvenient as a spherical pole gets closer to the south pole of the sphere. Projection becomes more manageable when those poles in in the southern hemisphere are projected by joining each of them to the north pole of the sphere. These lines will intersect the projection plane within the primitive, and stereographic poles so obtained are marked with open circles

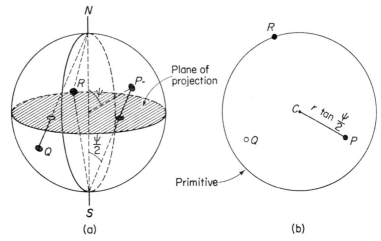

Fig. 3.3. Stereographic projection. (a) Projection of three spherical poles, P, Q and R in the northern and southern hemispheres and on the equatorial section respectively; R may be projected from either the N or the S pole, but conventionally is considered to be in the northern hemisphere. (b) The stereographic projection showing the location of the three poles P, Q and R; the N–S axis is at the centre of the projection at C.

to distinguish them from poles from the northern hemisphere. Thus a stereographic projection is restricted to poles (dots and circles) lying within the primitive. The particular arrangement of dots and circles depends both on the nature of the projection subject and the choice of projection plane, but with experience they may be interpreted in terms of the important crystalline features (symmetry, crystal shape, etc.) that they represent.

We have already mentioned the importance of great and small circles on the circumscribing sphere in relating spherical poles; these relationships are maintained on the corresponding stereographic projection (or *stereogram*), and so it is often necessary to construct the projections of great and small circles. On the stereogram these are the arcs of true circles, and various simple geometrical constructions are described in Appendix A. In practice, however, most angular manipulations on the stereogram are more easily carried out using a *stereographic net* (Fig. 3.4). This shows a series of great circles

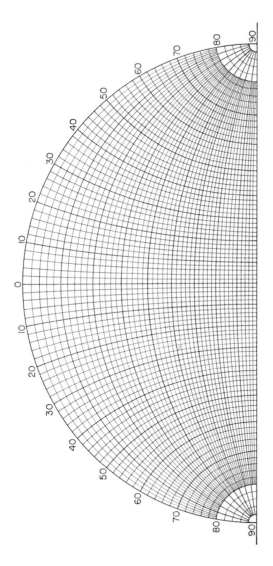

Fig. 3.4. A stereographic net.

(inclined at suitable angular intervals) drawn to pass through the two ends of a diameter of the primitive circle; superimposed on these is a series of small circles (with radii increasing at the same angular intervals) drawn about these diametrically opposite points. Some stereographic nets enclose the whole primitive, while others (as in the figure) overlap only half the projection. Naturally, the size of the net varies with the scale of the projection; for most elementary work a projection radius of $2\frac{1}{2}$ in. with the circles inscribed at angular intervals of 2° is adequate, though larger nets are available for more accurate work. Smaller nets are usually printed on transparent paper, though the larger ones (often as complete circles) are engraved and permanently mounted so that the actual projections may be carried out on tracing paper placed on top of them. Whatever its form and size, the stereographic net is an indispensable aid to constructions on the stereogram, for by rotation about the centre we can obtain the projections of all possible great circles and all possible small circles drawn about points on the primitive. We now describe briefly the use of the net in some of the more important stereographic constructions.

FIRSTLY, CONCERNING GREAT CIRCLES

(*i*) *To draw a great circle through any two points that are not diametrically opposite*

When both points are in the same hemisphere, the net is rotated about the centre of the stereogram until they are both on the same great circle locus. Should it be necessary to mark this great circle permanently on the projection, the true circle of which it is an arc is easily drawn. Its geometric centre must lie on the line perpendicular to the base of the net through the centre of the projection (Fig. 3.5(a)); its position on this line can be found by bisecting a chord. (Great circles of slight inclination are arcs of circles of very large radii; these can be most conveniently drawn using a thin metal blade flexed to the appropriate radius and placed so that it passes through the points on the circle.) The lower half of the great circle in the southern hemisphere (usually dotted or dashed) is the mirror image of the upper circular arc in the diameter along the base of the net.

If the points are in opposite hemispheres, the projection of the other end of a spherical diameter through one of them is found; this point, called the *opposite* of the original pole, will be on the same diameter of the primitive at an equal and opposite distance from the centre of the projection. The construction may then proceed as before. When one of the original points lies on the primitive or at the centre of the projection, the use of the net is obvious; the net can also be used to find a great circle of a given inclination to the primitive.

(*ii*) *To draw a great circle, given its pole*

The diameter through the pole of the great circle (P) and the centre of the projection is drawn; we now need to find the point at which the required great circle cuts this diameter, i.e. a point which is 90° away from P (Fig. 3.5(b)).

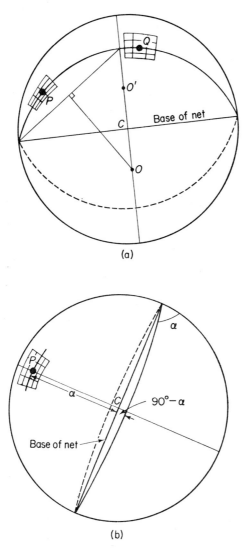

(a)

(b)

Fig. 3.5. Great circle constructions using a stereographic net. (a) A great circle through any two points within the primitive P and Q. (O, O′ are the geometric centres of the arcs of the two halves of the great circle). (b) A great circle of which P is the pole. For clarity only small areas of the net in the relevant regions are shown.

With the base of the net perpendicular to this diameter, the angular distance (α) of P from the centre of the projection is measured from the inclination of the great circle passing through it; the required great circle has inclination $90° - \alpha$ so that its intersection with the original diameter can be found after rotating the semi-circle of the net through 180°. It is then drawn to pass through this point and the ends of a diameter perpendicular to the original.

This construction can be reversed to find the pole of a given great circle.

NEXT, THE CONSTRUCTION OF SMALL CIRCLES

(*iii*) *To draw a small circle of given angular radius about a pole within the primitive*

At the top of Fig. 3.6, we wish to draw a small circle of angular radius (α) about a point P in a general position on the stereogram; this will be a circle

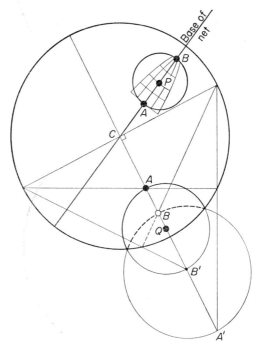

Fig. 3.6. Small circle construction using a stereographic net. At the top, a small circle of angular radius α ($=20°$) is drawn about P; note that the geometric centre of the circle is not at P. At the bottom, for a small circle of the same angular radius about Q, the points A and B located by the net are in opposite hemispheres. It is necessary to re-project these points outside the primitive by the construction shown to give A' and B' on the diameter produced. The upper and lower halves of the small circle pass through AB' and BA' respectively.

whose geometric centre is at a different point on the same diameter. If the diameter through P is drawn, points A and B on it must be located so that $\widehat{AP} = \widehat{PB} = \alpha$; this can be done quite simply using the engraved small circles of the net placed with its base on this diameter. The geometric centre of the circle is then found by bisecting AB, and the small circle is drawn to pass through A and B. The converse construction can be used to find the angular centre of a given small circle.

When the given pole is on the primitive, the small circle is already drawn on the net; if the given pole is at the centre of the projection, the small circle has its geometric centre at the same point. Arcs of a small circle can lie in both the northern and southern hemispheres of the circumscribing sphere. If the given pole lies on the primitive, the projections of the parts in the two hemispheres are superposed. If the given pole is in a more general position, the upper and lower halves are projected as circles of different radii both of which must be constructed. When the use of the net gives the points A and B in different hemispheres, and not superposed (the bottom of Fig. 3.6), a further construction step is necessary. Each in turn must be re-projected outside the primitive on to the same diameter to A' and B' by the construction shown in the figure; the justification of this construction is given in Appendix A. The geometric centres of the two circles representing the upper and lower halves of the small circle are found by bisecting AB' and BA'. It will be noticed that if A happens to coincide with the centre of the projection, the lower half of the small circle is a straight line, the chord joining points on the primitive intersected by the upper half, and vice versa when B coincides with the south pole of the projection.

NEXT, FOR COMMON ANGULAR MEASUREMENTS ON THE STEREOGRAM

(iv) To measure the angular separation of two given poles

This requires the location of the great circle on which both poles lie, as in (i) above. Once this is done by rotation of the net, the angle between them can be read off as the difference in the angular radii of the small circles on which they lie (Fig. 3.7(a)). With two poles in opposite hemispheres, both the upper and lower parts of the great circle must be measured, i.e. one must count to the primitive along both arcs and add to get the total angle.

(v) To measure the angle between two great circles

This may be carried out in a variety of ways (Fig. 3.7(b)). Without a net, the tangents to the circles at their point of intersection can be drawn and the angle between these lines measured by an ordinary protractor. One may use a net to plot the poles of the two great circles, and measure the angle between them as described earlier. Alternatively, the net allows us to draw the great circle of

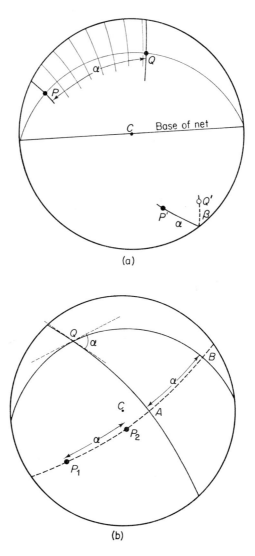

Fig. 3.7. Angular measurements on the stereogram. (a) The angular separation of two poles. In the upper half of the diagram the two poles P and Q are in the same hemisphere; the angle (α) between them is read off from the small circles on the net. In the lower half P′ and Q′ are in opposite hemispheres and the angle between them is $\alpha + \beta$. (b) The angle between two great circles. This angle (α) may be measured (i) as the angle between tangents at the point of intersection Q; (ii) as the angle between the poles P_1 and P_2 of the two great circles; (iii) as the angle between the intersections A and B of the given great circles with the great circle of which Q is the pole.

which the point of intersection Q is the pole; this will pass through the poles of the original great circles. The required angle may be obtained by measuring the arc of this great circle intercepted by the original great circles, for the two points of intersection A and B are necessarily at 90° to the original inter-section of the two great circles (imagine the point Q to be at the 'north pole' of a circumscribing sphere, when A and B will lie on the 'equator' of the sphere).

FINALLY, ROTATION OF THE STEREOGRAM

The distribution of points (dots and circles) for a given subject depends on the orientation of the plane of projection, i.e. the choice of the N-S axis of the circumscribing sphere; rotation of the projection may be desirable to bring the distribution of poles into a more favourable orientation. So far as spheri-cal poles on the circumscribing sphere are concerned, this amounts to the rotation of the sphere about a particular diameter so as to bring another point into the north pole position, normal to the plane of the new stereographic projection. As the sphere undergoes the necessary angular rotation about the given diameter, poles on its surface will be displaced. They will in general move through the appropriate angular distance around the loci of small circles whose angular centres are at the ends of the diameter of rotation. In terms of the same movements on the stereogram, we can quite simply relocate the poles after rotation of the projection as follows.

(vi) Rotation about the centre of the projection

Since the small circles are centred at the north and south poles of the pro-jection, the whole stereogram is rotated about the given angle measured on the primitive (Fig. 3.8(a)).

(vii) Rotation about a point on the primitive

All points on the stereogram move around small circles centred at the ends of the axis of revolution. This is conveniently done by moving the poles around the small circles on the net passing through them by the appropriate angular displacement (Fig. 3.8(a)). Care must be taken to ensure that rotation is in the same sense for all points.

(viii) Rotation about a point within the primitive

This is a much more tedious process, and starts with the construction of a small circle through each stereographic pole of radius appropriate to its angular distance from the axis of revolution. For each point two great circles must be found which intersect at the rotation axis inclined at the given angular rotation, in such a way that one passes through the undisplaced pole and the other gives its new position after rotation at the intersection with the appro-priate small circle (Fig. 3.8(b)); this is probably most simply achieved by

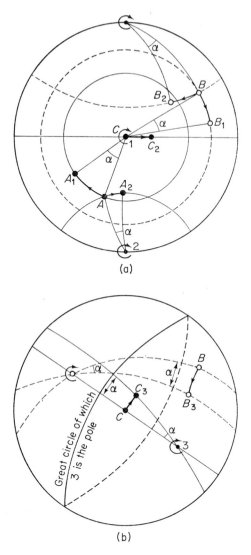

Fig. 3.8. Rotation of the stereogram. (a) Rotation through an angle α about the centre of the projection (axis 1) and a point on the primitive (axis 2) for three poles A, B and C. (b) Rotation through an angle α about a general direction (axis 3) for two poles B and C; the pairs of poles C, C_3 and B, B_3 lie on great circles through the axis of revolution which cut the great circle of which the rotation axis is the pole at points separated by the angle α; each of the pairs of related poles is then at the same angular distance from the axis of rotation (for clarity only arcs of the small circles are shown).

using intersections with the great circle for which the rotation axis is the pole; this adaptation of one of the constructions in (v) is shown in the figure. This involves a multiple use of the net for each pole, and in practice it can be simpler to use three separate rotations. In the first of these the rotation axis is brought to the centre of the stereogram; in the second the poles are further rotated through the given angle by (vi), whilst in the last stage the rotation axis is restored to its original position. However general rotation of this kind is carried out it is clearly a complex operation, and it is avoided if possible by careful consideration of the initial projection plane.

3.3. The use of stereographic projection

In crystallography this kind of projection is used whenever there are angles between directions to be specified; its applications are too numerous to be listed, and range from the geometry of diffraction to the co-ordination or environment of different atoms in a structure. Sometimes the stereogram has to be drawn accurately, sometimes a sketch of the distribution of the poles will suffice. Its value and utility in particular problems will be appreciated with experience, but to prepare for the use of the stereograms in later chapters, we shall consider in this section three fields where the use of this projection is invaluable; these are (i) angular relationships in lattices, (ii) the geometrical shapes of crystals, and (iii) the operation and combination of symmetry elements. Each of these will be explored in more detail in the appropriate chapter; the intention here is only to demonstrate the relevance of stereographic projection in these later discussions.

Angular values in lattices depend both on the shape and the dimensions of the unit cell; in Chapter 5, the crystallographic importance of lattice rows and planes is described and their relations for particular cells examined. The stereographic projection provides a simple way of deducing all the angular relations of the lattice rows and planes for a given cell; or conversely, from a knowledge of angular data it allows some conclusions to be drawn about the lattice cell. Let us give a simple illustration; Fig. 3.9 shows a simple cell of a lattice, in which the three lattice rows defining the edges of the cell are perpendicular but each has a different lattice translation. When a projection is made with c lattice translation along the N-S axis of the circumscribing sphere, the points on the stereogram a, b, c represent the directions of the three lattice rows of the cell edges; if any of these directions had been defined in the opposite sense, the opposite pole would be shown on the stereogram. Now considering a lattice row in some other direction, say d, both the magnitude and direction of its translation are fixed by the translations a and b; the same is true for any other row (e.g. e) in this plane. Given a and b, the directions of d and e may be calculated and plotted on the stereogram on the primitive circle, i.e. on the same great circle as a and b. This illustrates the general principle that directions lying within a plane will plot as stereographic poles on the same great circle; the row f will therefore lie on the great circle through b and c at an inclination determined by their magnitudes. We can now easily locate the direction of g, a more general lattice row, for it must lie both within the plane defined by c and d, and within the plane defined by a and f; it is therefore at the intersection of the great circles containing these pairs of poles.

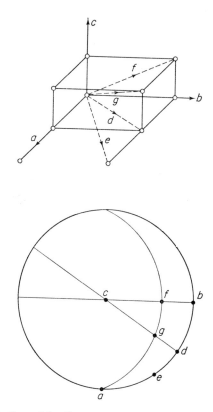

Fig. 3.9. Projection of the directions of lattice rows on the stereogram.

This method of locating a direction at the intersection of two great circles representing two planes to which it is common is extremely useful, and enables the position of any lattice row to be plotted once the unit cell translations are defined; it will be discussed at length in Chapter 5.1. The representation of the orientation of planes of lattice points by their normals is undertaken in a similar way. For example, a plane of points defined by the lattice translations c and b has a normal at the pole a; planes of lattice points do not always have normals which coincide with lattice rows, (e.g. the plane defined by c and d)

and the positions of their poles on the stereogram must be considered carefully. However, once a few such poles have been plotted on the stereogram,
others may be added using great circle relationships.

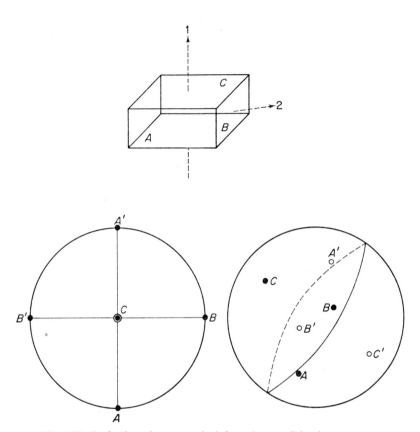

Fig. 3.10. Projection of a geometrical shape (a crystal) by the normals to
faces on the stereogram. The left-hand stereogram is constructed with
direction 1 as the N–S axis of the circumscribing sphere; the right-hand
stereogram is constructed with direction 2 at the centre of the projection.
The great circle on this latter projection shows the orientation of the
actual planes C and C'.

In Chapter 6 the external shapes of crystals developing under ideal growth
conditions are discussed as regular geometric polyhedra, which are conveniently represented on the stereogram where the characteristic angles between their plane faces can be displayed. In the simple illustration of Fig. 3.10,
the crystal is in the form of a rectangular block, similar to the unit cell discussed above. When a projection is made with the normal to the face C along
the N–S axis of the circumscribing sphere, the six poles formed by the normals

to the faces are disposed so that the shape and orthogonality of the crystal are evident. However, the stereogram would be much less easy to interpret if some other plane of projection had been chosen. Suppose that a body diagonal of the crystal (like g in Fig. 3.9) was placed at the centre of the projection. This would be equivalent to rotating the first projection to bring g to the centre; the effect is shown on the right-hand stereogram in the figure, where the essential orthogonality of the crystal shape is not so obvious to a casual inspection. Occasionally it is desirable to represent the orientation of faces (or planes) themselves rather than that of their normals; this is simply done by drawing on the projection the great circle of which the face normal is the pole.

This brief excursion into the projection of crystal shapes illustrates that stereograms are best drawn with due attention to the most convenient orientation; such orientations are usually related to any symmetry shown by the projection subject. Furthermore, the next chapter will show how the combination of symmetry elements into point groups is readily developed using this kind of projection. To conclude the present introduction to the use of stereograms, we shall look at the principles of operation and combination of symmetry elements in projection. In Chapter 2 it was shown that symmetry operators fall broadly into two categories, (a) those with rotational repetition, and (b) those which cause an inversion on repetition. Extension into three dimensions leads to a blurring of this distinction, with symmetry elements different in kind from those to be found in two dimensions (Chapter 4.1). For rotational symmetry, we shall omit any discussion of new operators for the present and deal with their stereographic representation when they are introduced in the next chapter; for inversion operations, we shall only consider the effects of simple inversion through a point and across a plane, a distinction which is also discussed in the next chapter. In principle the effect of rotation operators upon the stereogram is similar to rotation of the projection; for a particular symmetry axis a given pole will be repeated at the appropriate angular intervals as it moves around a small circle centred on the rotation axis. With conventional orientations of projections, rotation axes, with a few exceptions, are either at the centre of the projection or on the primitive (Fig. 3.11). These exceptions occur in some highly symmetrical combinations where diad axes can be found at 45° to the centre of the projection (Fig. 3.12(a)) to which triad axes are inclined at 54° 44'; this latter, apparently unusual, angle ($= \cos^{-1} 1/\sqrt{3}$) implies the triad axis is equally inclined to the projection centre and two perpendicular poles on the primitive (Fig. 3.12(b)). Inversion through a point will merely produce the opposite of the original pole (Fig. 3.13(a)). Inversion across a plane requires the construction of a great circle to cut that of the symmetry element at right angles; the given pole is then carried along this great circle to the intersection, and then repeated at an equal angular distance on the other side. In usual projection orientations symmetry planes are mostly in the plane of the primitive or vertical planes passing through the

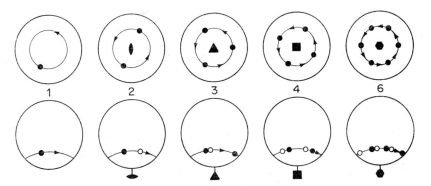

Fig. 3.11. The operation of rotation axes by symmetry on the stereogram. The repetition of a general pole by the various axes is shown in the upper row with the axis at the centre of the projection, and in the lower row with the axis on the primitive.

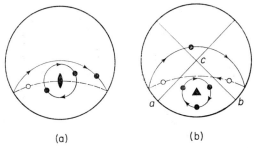

(a)   (b)

Fig. 3.12. The operation of inclined axes. (a) A diad axis inclined at 45° to the centre of the projection. (b). A triad axis inclined at 54° 44′ to the centre of the projection, i.e. equally inclined to the perpendicular directions represented by *a*, *b* and *c* so that it would repeat poles at these points. In both cases the repetition of a pole in each hemisphere is shown.

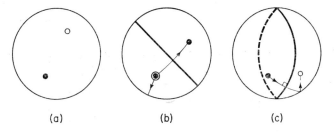

(a)   (b)   (c)

Fig. 3.13. Inversion operations on a stereogram. (a) Repetition of a general pole by inversion through a point (b) Repetition of a general pole by inversion across a plane through the centre of the projection or in the plane of the primitive (to the point in the southern hemisphere). (c) Repetition of a general pole by inversion across a plane inclined at 45° to the centre of the projection.

centre of the projection (Fig. 3.13(b)). Again in some highly symmetrical combinations, symmetry planes inclined at 45° to the centre of the projection are found (Fig. 3.13(c)).

Stereographic projections are also well suited to the study of symmetry element combinations. Although the full examination of individual point groups is made in the next chapter, we can illustrate now the value of stereograms in such studies by an example. Let us suppose that we wish to look at two simple combinations, with two perpendicular diad axes in the first, and one diad contained in a symmetry plane in the second. When such combinations are made, one can ask what is the effect of combination upon the total symmetry, i.e. in each case whether any extra symmetry elements have been developed, and whether the two combinations lead to total symmetry groupings different from each other. Both questions are easily answered by examining the effect

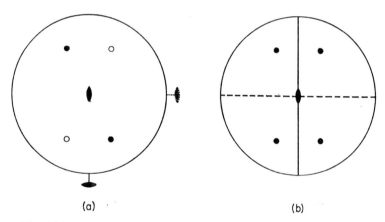

(a)   (b)

Fig. 3.14. The combination of symmetry elements on the stereogram. (a) Two perpendicular diad axes. (b) One diad contained in a symmetry plane. Note the different distribution of general poles and the extra symmetry elements (dotted and broken lines) that have been developed.

of the given symmetry elements of each combination upon a pole placed in a general position on a stereogram (Fig. 3.14); (a) shows the repetition of this pole by the first group, (b) by the second. In both cases new symmetry elements have been developed ; in (a) there is a third mutually perpendicular diad axis, whilst in (b) there is a second plane (also containing the diad axis) perpendicular to the first. Clearly the final symmetry groupings of the two combinations are quite distinct, a fact that is already reflected in the different distributions of general poles upon the stereogram. The number and arrangement of general poles is characteristic of each combination, and we shall use this in the systematic development of crystallographic point groups shortly.

3.4. Summary

A full and clear understanding of stereographic projection is essential to what follows, and it is worthwhile re-stating the main features of this chapter at this point; moreover, familiarity with the projection comes only with experience, and exercises provided at the end of the chapter are designed to help the reader to gain this essential experience.

In stereographic projection, the subject is first replaced by poles on the surface of the circumscribing sphere; each spherical pole is at the end of a radius of the sphere representing a direction in the projection subject. This arrangement of poles on the surface of the sphere is projected into two dimensions by choosing one diameter of the sphere to be the N-S axis, with the plane of projection as the corresponding equatorial section of the sphere. Poles in the northern hemisphere are joined to the S pole of the sphere, while poles in the southern hemisphere are joined to the N pole of the sphere; the intersections of these lines with the equatorial section give the stereographic poles, those from the northern hemisphere being indicated by dots, those from the southern hemisphere by open circles. Thus all points on the stereogram are contained within the equatorial, or primitive, circle.

This projection is most generally used in crystallography because it can represent all directions from 0° to 360°, it preserves angular truth, and its geometry is simple with both great and small circles on the sphere projecting as arcs of true circles. Constructions on the stereogram are readily carried out using stereographic nets, which are available to fit various projection scales. It can, in particular, represent the spatial angular distributions demanded by symmetry elements and their combinations in crystallographic point groups, as well as lending itself to angular calculations by spherical trigonometry. Probably its only real disadvantage in crystallographic work is the closer spacing of angular intervals near the centre; in certain circumstances this can lead to a crowding together of poles in this region of the stereogram, when it may be preferable to use an alternative method, the gnomonic projection, described briefly in Appendix A.

3.5. Exercises and problems

1. Familiarise yourself with the principles of projection by drawing sketches of the distribution of poles on a stereogram appropriate to certain aspects of simple objects. For example, using the rectangular parallelipiped of a match box (or book), make a sketch stereogram of the normals to its faces with the N-S axis of the circumscribing sphere (a) along the normal to the largest faces, (b) along the normals to other faces, (c) along other directions, such as face or body diagonals. Draw corresponding stereograms to

project the directions of the edges [of the parallelipiped. Repeat these exercises with other objects of simple geometrical shape.

2. Basic constructions on the stereogram can be carried out with a stereographic net; use these in the following exercise:
   (i) On your projection draw two vertical great circles inclined at 30°.
   (ii) On one of these locate a pole P at 30° from the centre of the projection.
   (iii) Find two poles Q and R on the other vertical great circle such that $\widehat{PQ} = \widehat{PR} = 20°$.
   (iv) Draw great circles which pass through P and Q, and P and R; determine the angle between these great circles.
   (v) Locate the positions of Q and R, when P is brought to the centre of the projection.
   (vi) Repeat (ii)–(v) with the pole P at 80° from the centre of projection and $\widehat{PQ} = \widehat{PR} = 50°$.

3. Show that a rotation tetrad axis combined with (a) a perpendicular symmetry plane, (b) a perpendicular diad symmetry axis, and (c) a parallel symmetry plane, are distinctive symmetry combinations although all three develop the same number of general poles. List the total number of symmetry elements in each of the combinations. Show that the addition of a further symmetry plane normal to the tetrad axis in (b) and (c) produces identical combinations; what symmetry elements does this new group contain?

4. (i) An air service is to be operated between London (52°N, 0°E) and Bombay (19°N, 73°E) with possible intermediate stops at Beirut (36°N, 36°E) and Teheran (36°N, 52°E). Show that, if the aircraft has a maximum range of only 2500 miles, refuelling stops must be made at both Beirut and Teheran. What is the minimum range for an aircraft that can eliminate either of the two intermediate stops?
   (ii) A man sets out to row a boat from Lisbon (39°N, 9°W) to Trinidad (11°N, 61°W); he hopes to average 20 miles per day. Determine the time he may expect to take and the bearing on which he should depart from Lisbon.
   (iii) On 21st June the sun's rays are inclined at an angle of 67° to the earth's north polar axis. At noon G.M.T. on this day has the sun risen in (a) Denver (40°N, 105°W), (b) Los Angeles (33°N, 118°W)? Has it set in (c) Peking (40°N, 116°E), (d) Sydney (34°S, 151°E)?
   (iv) A communications satellite is placed in orbit 2200 miles above the Earth's surface. Show that as it passes over London (52°N, 0°E) its transmissions can be received in New York (41°N, 74°W) and Teheran (36°N, 52°E) but not in Denver (40°N, 105°W) or Bombay (19°N, 73°E).
   (N.B. In all of these problems assume that the Earth is a perfect sphere of radius 4000 miles.)

SELECTED BIBLIOGRAPHY

General methods of projection

DEETZ, C. H. and ADAMS, O. S. 1934. *Elements of map projection.* Special publication No. 68, U.S. Dept. of Commerce, Coast and Geodetic Survey.
MAINWARING, J. 1942. *An introduction to the study of map projection.* Macmillan.

Stereographic projection

SOHON, F. W. 1941. The stereographic projection. Wiley.
TERPSTRA, P. and CODD, L. W. 1961. *Crystallometry.* Longmans.
TERTSCH, H. 1954. *Stereographische projektion in der kristallkunde.* Verl. für Angewandte Wiss.

4

CRYSTAL CLASSES AND SYSTEMS

4.1. Symmetry elements

The next three chapters are devoted mainly to those features of crystalline matter related to point group symmetry, i.e. the possible combinations of symmetry elements when lattice translations are neglected. As with planar point groups in Chapter 2.3, application of our discussions to the three-dimensional repetitive patterns of crystalline materials must follow, and so the point groups that are our concern are limited to the three-dimensional crystal point groups each displaying a set of symmetry elements to be found in one or other of the sub-divisions of crystalline matter known as the *crystal classes*. Every crystal can be assigned to a class which has its own assemblage of symmetry elements which pass through a point, so that an object is repeated in a characteristic way about the point. However, before looking at these combinations we must establish the nature of the symmetry elements that they can contain.

In Chapter 2, symmetry operations were found to be essentially of two types, those which produce rotational repetition and those which require an inversion. The concept of pure rotational symmetry can be taken over unchanged into three dimensions, save that the operation will be about the line of an axis (rather than a point). In the crystalline state there are the same limitations on the degrees of rotation axes of symmetry ($n = 1, 2, 3, 4$ or 6) for there are still repetitive lattice translations; in Chapter 7.1 the preservation of rotational symmetry in crystal lattice types is demonstrated. The origins of inversion between right- and left-handed enantiomorphous pairs needs further exploration. In planar patterns such pairs were formed by reflection lines; in crystalline matter they can arise either by inversion across a plane or by inversion through a point, and the two operations produce different spatial distributions of the enantiomorphous objects (already mentioned in Chapter 3.3). In Fig. 4.1, one member of the pair has co-ordinates xyz with respect to a set of reference axes; in (a) inversion through the origin point gives the co-ordinates of the other member as $\bar{x}\bar{y}\bar{z}$, whilst in (b) inversion by a plane containing the X and Z axes requires the enantiomorphous object to be at $x\bar{y}z$. We can regard the symmetry plane of (b) to be the three-dimensional equivalent of the symmetry line, and it is written symbolically as m. Inversion through a point, as in (a), does not have a two-dimensional analogy, for it will not lead to an enantiomorphous pair in a planar pattern; any symbolic representation will become apparent shortly.

Each of the two kinds of inversion can be combined with rotational repetition so as to express all symmetry elements in the crystal classes as forms of rotational operators. These would include pure rotation axes, and other types of symmetry axes (sometimes called improper axes) for which an inversion of hand in the object takes place at each repetition. The alternating rotational repetition of improper axes could be formed either as a combination of a rotation and a reflection or as a combination of a rotation and inversion through a point; the degrees of all axes must have the usual limitations on n.

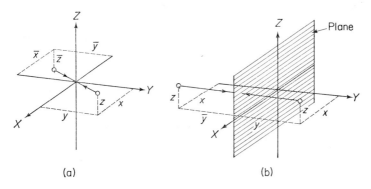

Fig. 4.1. Inversion operations in three dimensions. (a) Through a point. (b) Across a plane containing the X and Z axes.

In terms of these three kinds of rotational symmetry operator the permissible spatial combinations when the axes are constrained to pass through a point allow the symmetry of the crystal classes to be established. In practice this is unnecessarily cumbersome, for many of the improper axes are equivalent to one another. Furthermore, improper axes combining rotation and reflection are redundant in modern crystallographic usage; instead the symmetry plane (m) is retained as a separate symmetry element. Improper axes combining rotation and point inversion are used, and are known as *inversion axes of symmetry* (symbolically $\bar{1}$, $\bar{2}$, $\bar{3}$, $\bar{4}$, $\bar{6}$). The symmetry of any crystal class is described by means of planes of symmetry, inversion axes of symmetry, and rotation axes of symmetry.*

* Commonly in describing symmetry axes a prefix is used only to describe inversion axes; thus 'a tetrad' is taken to imply a rotation axis (4), whilst its improper analogue ($\bar{4}$) requires 'an inversion tetrad'.

To examine the combination of these elements we shall make extensive use of the stereographic projection; in the preceding chapter (Figs. 3.11–13), the operation of rotation axes and symmetry (or mirror) planes in projection was set out. To complete this review of symmetry elements, we shall look at inversion axes in the same way, both to familiarise their operation in projection, and to comment on certain features. Fig. 4.2 shows the distribution of general poles produced by each kind of inversion axis; those which require particular attention are $\bar{1}$, $\bar{2}$ and $\bar{6}$. The inversion monad ($\bar{1}$) has the opposite poles of a simple point inversion; when any crystal class has an arrangement of general

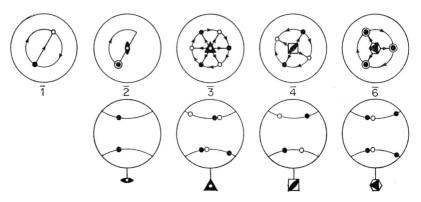

Fig. 4.2. The operation of inversion axes of symmetry on the stereogram. The repetition of a general pole by the various axes is shown in the upper row with the axis at the centre of the projection, and in the lower row with the axis on the primitive. Arrows in the upper row show the repetitive movements for clockwise rotation (cf. Fig. 3.11). Note the possibility of alternative symmetry descriptions for the $\bar{2}$ and $\bar{6}$ axes.

poles such that the opposite of every pole is present, there must be an inversion centre in the point group. As will be seen in later chapters, an inversion centre is of some physical importance, and it is usual to regard the inversion monad ($\bar{1}$) as representing a separate symmetry element, the *centre of symmetry*, to be listed as present or absent for each class. Next the inversion diad ($\bar{2}$) has in its operation an equivalence with a symmetry plane (m); the same distribution of general poles would occur if the axis were replaced by a mirror plane in a perpendicular direction. This alternative description of the same symmetry operation is usually preferred. Finally, the operation of the inversion hexad ($\bar{6}$) gives a distribution of general poles equivalent to that obtained if it is replaced by rotation triad (3) in the same direction together with a perpendicular symmetry plane. The dual nature of the degree of this axis is an illustration of the special relationships which exist between point groups with triads or hexads as their principal axes. For the present we will note this equivalence, and consider later which of the alternative symmetry descriptions is to be preferred.

4.2. The derivation of the crystal classes

(a) *The method of derivation.* In attempting to establish the crystal classes (or point groups of crystal symmetry), a systematic method can be adopted in which the possible self-consistent arrangements of rotation and all kinds of improper axes are found. Permissible angular inclinations of the various axes can be determined by Euler's construction, as illustrated in Appendix B. Certain of the improper axes are then decomposed into their equivalents in conventional symmetry elements to give the various crystal classes. We shall approach the problem in a different way, less rigorous but making use of the place of point groups in pattern theory, as described in Chapter 2 for planar patterns. We recall that each crystal class is a sub-division of a crystal system, which in turn is linked with a characteristic cell shape. For each cell shape there is a maximum number of symmetry directions with associated elements corresponding to the most symmetric (or *holosymmetric*) crystal class of each system. Within each system, however, there are minimum symmetry criteria which demand the particular cell shape; different crystal classes within each system arise as the essential symmetry is increased to the maximum permissible symmetry. We can in this way derive the possible crystal classes system by system.

For planar point groups the process is simple; lattice cell shapes are self-evident, and the restricted number and arrangement of symmetry elements ensures that the crystal classes of each system are easily recognised. In three dimensions, whilst the procedure is the same, the discussion is a little more complex. Firstly we must anticipate a full investigation of lattice types in Chapter 7 so as to start with the seven different cell shapes. Then the maximum symmetry compatible with each shape must be established together with any essential symmetry criteria common to all point groups of each system. Finally for each system, extra elements must be added step by step to the essential symmetry to build up the maximum symmetry of the holosymmetric class; as these elements are added we must decide whether a possible new combination (or crystal class) has been formed. This step by step progression to the holosymmetric class is not as formidable as it might appear for the symmetry elements of each point group must be self-consistent, i.e. they must act upon each other to give a restricted finite grouping; in more symmetrical classes this means there are often sets of similar elements which must all be present or absent. Recognition of each new crystal class as it evolves is made from its characteristic distribution of general poles upon a stereogram (see Chapter 3.3). It should also be realised that since all lattice types must be centro-symmetric by their nature, a symmetry axis in a particular direction may be rotation or inversion; when such a substitution is made the distribution of general poles will show what modification (if any) to the point group symmetry has occurred.

This method of investigating crystal classes is best illustrated in application, and in the remainder of this section the symmetry of classes within each of the systems will be briefly explored.

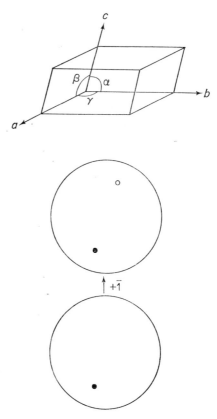

Fig. 4.3. The triclinic system. The upper stereogram is of the holosymmetric class, the lower the other class with no symmetry.

(b) *The triclinic system.* In this there is the most general cell shape (Fig. 4.3), a parallelogram-based oblique prism; no special relationships exist between the cell edges or the cell angles ($a \neq b \neq c$, $\alpha \neq \beta \neq \gamma \neq 90°$). Elements other than a centre of symmetry ($\bar{1}$) would demand a more regular cell shape, and so there are just two classes in this system, the holosymmetric class with the centre and another class which has no symmetry.*

* The stereogram of the holosymmetric class might appear to have a horizontal diad axis (cf. the poles of Fig. 3.11). To dispel this illusion insert a second set of general poles,

(c) *The monoclinic system.* This, too, has a non-orthogonal cell (Fig. 4.4), a parallelogram-based right prism; although no special relationships exist between the cell edges, the direction of one of them is perpendicular to the

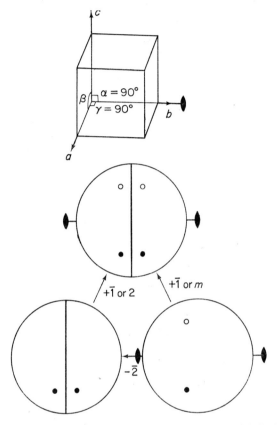

Fig. 4.4. The monoclinic system. The cell shows the essential diad. The upper stereogram is of the holosymmetric class; arrows show the effects on the lower symmetry class of adding (+) centres ($\bar{1}$), planes (*m*) or axes (2) or substituting (−) inversion axes ($\bar{2}$).

plane containing the other two ($a \neq b \neq c, \alpha = \gamma = 90°, \beta \neq 90°$; the choice of α and γ as the right angles is conventional). The maximum symmetry associated with this cell shape has three elements (a diad axis, a symmetry plane and a centre) all present in the holosymmetric class. The essential symmetry is the diad axis (conventionally along the direction of the *b* lattice translation). All three symmetry elements of the holosymmetric class are independent, so

when the only symmetry remaining will be the centre. This may be used in all classes to verify the presence of any symmetry element.

that the plane and the centre may be added separately to the lowest symmetry class, in which the rotation diad may be replaced by an inversion diad. These possibilities and their inter-relations are shown in the figure, from which we see that there are three crystal classes in this system.

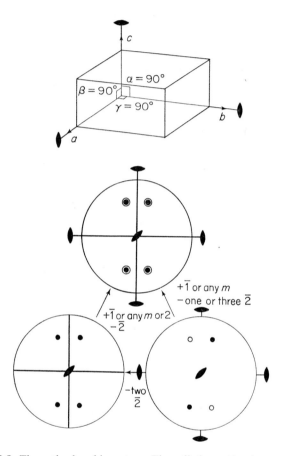

Fig. 4.5. The orthorhombic system. The cell shows the three essential perpendicular diads. The upper stereogram is of the holosymmetric class; arrows show the effects on the lower symmetry classes of adding (+) centres ($\bar{1}$), planes (m) or axes (2) or substituting (−) inversion axes ($\bar{2}$).

(d) *The orthorhombic system.* The most general orthogonal cell is the rectangular parallelipiped (Fig. 4.5) or a rectangle-based right prism; although there is no identity between the lattice translations, the cell angles are all right angles ($a \neq b \neq c, \alpha = \beta = \gamma = 90°$). Such a shape can be associated with a maximum of seven symmetry elements (three diads, three planes and a centre) all present in the holosymmetric class. The essential symmetry elements needed to maintain the orthogonality of the cell are the three perpendicular diads. All the symmetry elements are independent, and the effects of adding planes or a centre to this class together with the use of replacement inversion diads are set out in the figure; again there are three classes in this system.

(e) *The tetragonal system.* In this an orthogonal cell has the more regular shape of a square-based right prism (Fig. 4.6); now two of the lattice transla-

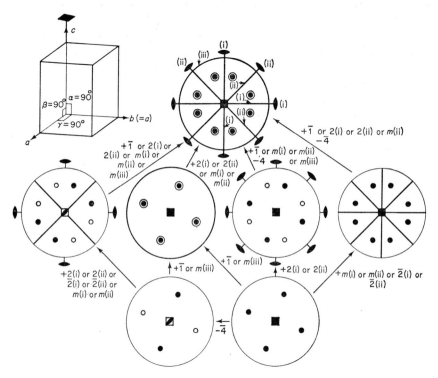

Fig. 4.6. The tetragonal system. The cell shows the essential tetrad axis. The upper stereogram is of the holosymmetric class; the elements of this class are divided into sets of which diads (i) and (ii), planes (i), (ii) and (iii) are marked. Arrows show the effects on the lower symmetry classes of adding (+) centres ($\bar{1}$), sets of planes (m(i) etc.) and diads (2(i) etc.) and substituting (−) inversion axes ($\bar{4}$).

tions are identical ($a = b \neq c$, $\alpha = \beta = \gamma = 90°$; the choice of a and b as identical is conventional). The maximum symmetry of the holosymmetric class has eleven elements (one tetrad, four diads, five planes and a centre). The only essential symmetry element is the tetrad axis normal to the square base of the cell, and the possible addition of the ten further symmetry elements leads to more complexities than have so far been encountered. The necessary presence of the tetrad axis implies that the extra elements of the holosymmetric group form six symmetry related sets; there are two sets of diads (i) and (ii), three sets of planes (i), (ii) and (iii) and the centre of symmetry (Fig. 4.6). Each of these sets may be added independently, with changes to inversion axes where appropriate. In the figure the effects of these changes are indicated schematically; from this we see that there are seven different crystal classes in this system.

(*f*) *The cubic system.* In this system (sometimes called the *isometric* system) there is the most symmetrical orthogonal cell with a cube shape (Fig. 4.7) in which the perpendicular lattice translations are all identical ($a = b = c$, $\alpha = \beta = \gamma = 90°$). The maximum symmetry associated with this shape has twenty-three elements (three tetrads, four triads, six diads, nine planes and a centre); their spatial arrangement is shown on the stereogram of the holosymmetric class, a complex distribution which the reader can verify by examining a model cube. The essential symmetry is the four equally inclined triad axes in the directions of the body diagonals of the cube; these ensure the identity of the lattice translations forming the cell edges. These essential triad axes divide the additional elements of the holosymmetric class into five symmetry related sets; there is the tetrad set, the set of six diads, two sets of planes (three in the set marked (i), six in that marked (ii)) and the centre of symmetry (Fig. 4.7). In this system, the lowest symmetry class is a little unusual in that the distribution of general poles given by the operation of the four triads implies that the point group must also have three perpendicular diad axes along the cell edges; symmetry axes are always present in cubic classes in these directions, diads in the lower symmetry classes and tetrads in the more symmetrical groupings. In view of this the best procedure is to add systematically the other four symmetry sets allowing the appropriate axes in the cell edge directions to develop by implication; this is illustrated in the figure, which shows that there are five distinct classes in this system.

(*g*) *The trigonal and hexagonal systems.* We have left these systems until last, as the special relationship between triad and hexad axes already mentioned implies similarities and overlap in their cell shapes. A full discussion of the cells of these systems is given in Chapter 7.1, and for the present it is only important to recognise that there are two further lattice cell shapes found in crystalline matter.

The first of these has the form of a rhombohedron (Fig. 4.8) which can be thought of as a cube elongated (or compressed) along one of its body diagonals; lattice translations are equally inclined to this direction (and to one another) and they are identical ($a = b = c$, $\alpha = \beta = \gamma < 120° \neq 90°$). Maximum symmetry for this cell has eight elements (one triad, three diads, three planes and a centre), in which the essential element is the triad axis

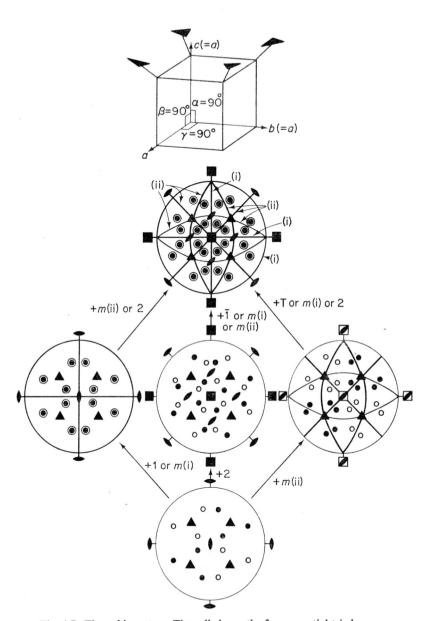

Fig. 4.7. The cubic system. The cell shows the four essential triad axes. The upper stereogram is of holosymmetric class; the elements of this class are divided into sets of which planes (i) and (ii) are marked. Arrows show the effects on the lower symmetry classes of adding (+) centres (Ī), and sets of planes (m(i), etc.) and diads (2).

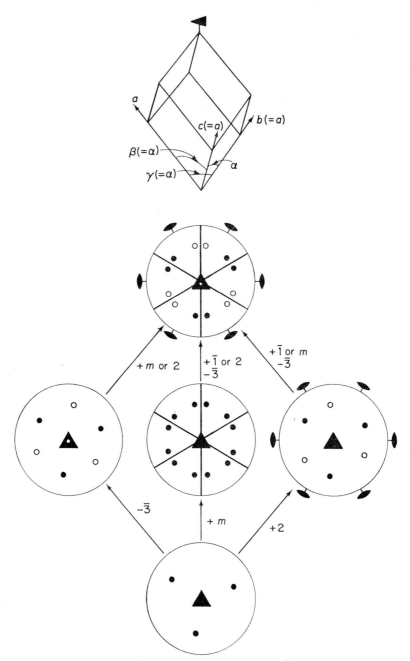

Fig. 4.8. The trigonal system. The cell shows the essential triad axis. The upper stereogram is of the holosymmetric class; the elements of this class are divided into sets. Arrows show the effects on the lower symmetry classes of adding (+) centres ($\bar{1}$) and sets of planes (m) and diads (2) and substituting (−) inversion axes ($\bar{3}$).

along the body diagonal. This axis divides the additional elements into three symmetry related sets, and following the usual procedure there are five classes associated with this cell shape.

The second cell (Fig. 4.9) is a 60° rhombus-based right prism in which two of the lattice translations are identical but different from the third ($a = b \neq c$, $\alpha = \beta = 90°$, $\gamma = 120°$). The maximum symmetry for this cell has fifteen elements (one hexad, seven planes, six diads and a centre). The essential symmetry, however, is only that of a triad axis needed to relate the identical lattice translations; in this lies a complication, for the symmetry of the five

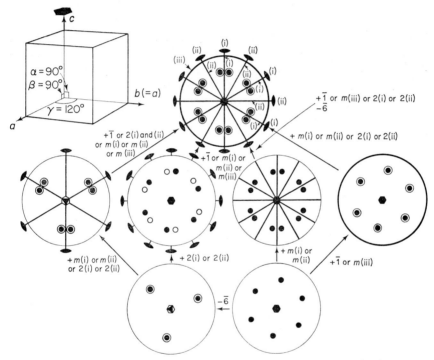

Fig. 4.9. The hexagonal system. The cell shows the essential hexad axis. The upper stereogram is of the holosymmetric class; the elements of this class are divided into sets of which diads (i) and (ii), and planes (i), (ii) and (iii) are marked. Arrows show the effects on lower symmetry classes of adding (+) centres ($\bar{1}$), sets of planes (m(i), etc.) and diads (2(i), etc.) and substituting (−) inversion axes ($\bar{6}$).

classes established above for a rhombohedral cell is such that they are equally consistent with this cell shape. Nevertheless, although the existence of these point groups is already known, there must be further classes uniquely associated with a hexagonal cell shape. These can be derived by imposing the minimum symmetry criterion of a hexad axis normal to the rhombus base, but in doing this we must remember that any sets of additional symmetry elements need only be related by the basic threefold symmetry. Strictly, therefore, there are six sets, three sets of planes (i), (ii) and (iii), two sets of diads (i) and (ii) and a centre to be considered, as well as any replacement of rotation by inversion axes. Fig. 4.9 shows the development of the seven extra classes for this cell.

With these two cell shapes, twelve new crystal classes have been formed; all twelve can be associated with the hexagonal cell, but the rhombohedral cell is consistent only with five. The division of these classes into separate systems is a vexed question. In most American usage, they are not usually divided and are said to belong to the sixth crystal system, the hexagonal system; this is often taken to have a trigonal sub-division to include the five rhombohedral classes. Other crystallographers recognise a division to give a total of seven crystal systems. When this is done, the essential criteria for inclusion in the trigonal system are the existence of a triad symmetry axis and point group symmetry in accord with a possible rhombohedral cell; on this basis in point groups where axes with alternative descriptions occur (viz. the problem of inversion hexad mentioned in 4.1), the class must be allocated to the hexagonal system because it requires a hexagonal cell. Indeed all classes which can only have this cell shape are put into the seventh system, the hexagonal system, for which the essential symmetry requirement is a hexad

Table 4.1. Crystal classes, systems and unit cell shapes

System	Cell shape	Essential symmetry	No. of classes
Triclinic	General parallelipiped $(a \neq b \neq c, \alpha \neq \beta \neq \gamma \neq 90°)$	None	2
Monoclinic	Oblique parallelipiped $(a \neq b \neq c, \alpha = \gamma = 90°, \beta \neq 90°)$	A diad axis	3
Orthorhombic	Rectangular parallelipiped $(a \neq b \neq c, \alpha = \beta = \gamma = 90°)$	Three mutually perpendicular diad axes	3
Tetragonal	Square-based right prism $(a = b \neq c, \alpha = \beta = \gamma = 90°)$	A tetrad axis	7
Cubic	Cube $(a = b = c, \alpha = \beta = \gamma = 90°)$	Four equally inclined triad axes	5
Trigonal	Rhombohedron $(a = b = c, \alpha = \beta = \gamma < 120° \neq 90°)$ or 60° Rhombus-based right prism $(a = b \neq c, \alpha = \beta = 90°, \gamma = 120°)$	A triad axis	5
Hexagonal	60° Rhombus-based right prism $(a = b \neq c, \alpha = \beta = 90°, \gamma = 120°)$	A hexad axis	7

axis. We shall adopt this latter convention with seven crystal systems, remembering that it implies that crystals of the trigonal system can have rhombohedral or hexagonal cells, an important point in space group considerations.

(*h*) *Summary*. We can now summarise quite briefly the important results of our investigations of crystal systems and classes in this section. In crystalline matter there are thirty-two crystal classes, or sub-divisions, in each of which there is a characteristic self-consistent arrangement of symmetry elements all passing through a point. Although the establishment of these crystallographic point groups is an important stage in the analysis of the three-dimensional regularity of atomic structures, it has also an importance in its own right as will be seen in subsequent chapters. Certain crystal classes may be grouped together to give seven crystal systems; the basis of this grouping is the presence of certain symmetry elements necessary to maintain a particular cell shape. The relationships between systems, classes, etc. are set out in Table 4.1.

4.3. A symbolic notation for the crystal classes

Each of the thirty-two crystal classes has a characteristic distribution of general poles on the stereogram from which the point group symmetry may be deduced. However, the importance of these sub-divisions of crystalline matter is so great that a less cumbersome method of denoting the symmetry of a particular class is essential, and various forms of symbolic nomenclature have been devised. The commonest notation is that shown in Fig. 4.10, where all the crystal classes are collected together and arranged in a conventional grid format to show the relationships between the systems and the classes within them; beneath each point group is a symbol which expresses the distinctive symmetry which it contains. This symbol is constructed from the notation for symmetry elements (rotation axes, X, inversion axes, $\bar{X}$ and symmetry planes m) so as to show their mutual orientations in the combination; the rules which govern the formation of compound symbols are discussed presently, but there is a general observation to be made first. As in planar patterns we wish to extend the use of this symbolism to the description of crystalline space groups, and this is an important consideration when deciding the conventional form of the symbol for a particular class. If we were concerned with point group symmetry alone, the conventional form for each class need only show those elements from whose presence any additional symmetry could be deduced; but since each class may be linked to several distinct space groups, its conventional symbol must have sufficient flexibility to allow the display of the various symmetries of the space groups. We shall comment on this again, although the symbolic notation for classes shown in the figure is relatively simple and soon becomes familiar.

The rules of the notation concern the mutual orientations of elements, and in the twenty-seven non-cubic crystal classes these are essentially orthogonal

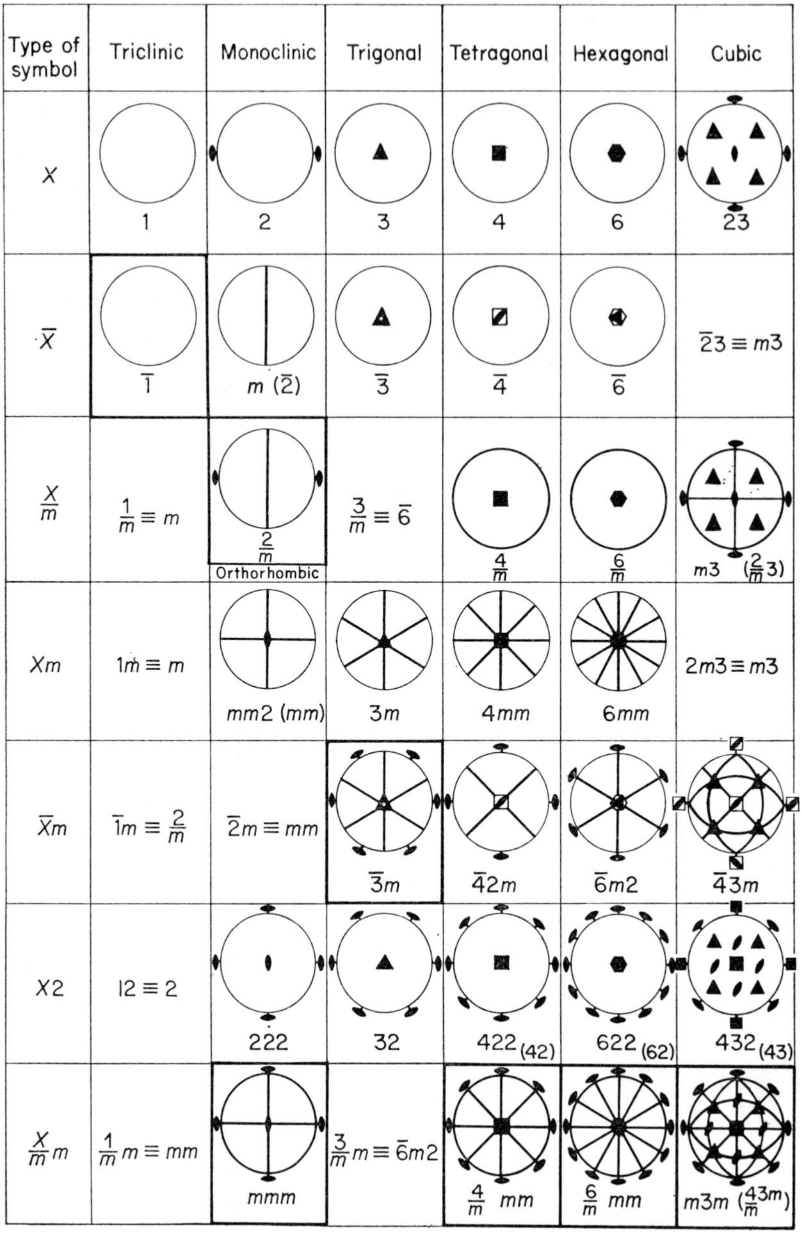

Fig. 4.10. The 32 crystal classes and their symbolic notation. Equivalences are shown in unoccupied sectors of the grid; alternative forms of symbols in common use are in brackets. The seven sectors of the holosymmetric classes are heavily outlined.

or parallel. In these classes the important relationships can be specified by a few simple conventions:

(i) The symbol $\dfrac{X}{m}$ (often written or printed X/m) represents a symmetry plane normal to an axis of degree X;

(ii) The symbol Xm represents a symmetry plane containing an axis of degree X;

(iii) The symbol $X2$ represents a diad axis normal to an axis of degree X.

Whilst these conventions alone are sufficient to devise a suitable symbol for each of the non-cubic classes, some further explanation of the conventional forms is necessary. For example, it might be expected that any symbol for the holosymmetric class of the orthorhombic system should contain some reference to the essential perpendicular diads, whatever the desirability of mentioning the three perpendicular planes; in terms of the conventions above a suitable class symbol might be $\dfrac{2}{m}\dfrac{2}{m}\dfrac{2}{m}$. However, in the related space groups, the nature of the planes (which may be mirror or glide) is more important than the axes; in a sense, the diads are redundant, for they will be implied once the three planes are specified, and the conventional class symbol is abbreviated to mmm (Fig. 4.10). This kind of manipulation and abbreviation means that it becomes convenient to remember the interpretation of certain groupings of symbols, although they arise only by adaptations of the original conventions. They are:

(a) The grouping mm, which represents two perpendicular planes; these on their own imply that their line of intersection is a diad axis, which is sometimes included as $mm2$ or $2mm$ (the order of symbols has a significance which will be explained elsewhere).

(b) The grouping mmm representing three mutually perpendicular planes in the orthorhombic holosymmetric class.

(c) Arising from (a), we can have X/mm $\left(\text{or } \dfrac{X}{m}m\right)$ representing an axis of degree X lying in one of the two planes with its direction perpendicular to the other.*

In this way, interpretation of the conventional symbols for the twenty-seven non-cubic classes is readily made; the compound symbol starts with the degree of the axis indicative of the crystal system except in the monoclinic

---

* In Fig. 4.10 it will be noticed that X/mm does not appear in the conventional symbols. For example, the holosymmetric classes of the tetragonal and hexagonal systems are described as $4/mmm$ and $6/mmm$ whereas the symbols $4/mm$ and $6/mm$ would appear to be adequate. In these cases the third m symbol refers to the extra set of vertical planes (i.e. (ii) in Chapter 4.2(e) and Chapter 4.2(g)) bisecting those containing lattice translations. The designation of these planes is necessary in the development of associated space groups.

and orthorhombic systems. It is perhaps necessary to comment briefly on the classes $\bar{3}m$, $\bar{4}2m$ and $\bar{6}m2$ in the line of the grid $\bar{X}m$; in these classes there can be some ambiguity in the orientation of the symmetry elements (other than the $\bar{X}$ axis) with respect to the edges of the smallest unit cell. If, when this cell is selected for a material of the tetragonal class, the diad axes are perpendicular to the lattice translations of the square base (and, due to the orthogonality of the cell edges, necessarily parallel to these directions) the class symbol is written $\bar{4}2m$; if the smallest lattice cell has planes of symmetry in these directions the symbol is $\bar{4}m2$. For the hexagonal class, the diads and planes are within the same plane and the question now is whether the normal to this plane is along the lattice translations of the rhombus base of the cell or perpendicular to them; by analogy with the tetragonal class, the former leads to the symbol $\bar{6}2m$ and the latter to the symbol $\bar{6}m2$. Both variants are needed to describe alternative space groups associated with this class, but there is a morphological convention to take the planes perpendicular to the crystallographic axes so that these are not parallel to the diads; for this reason, only the symbol $\bar{6}m2$ is shown in Fig. 4.10. The trigonal class presents the most difficulties in that it can be associated with either a rhombohedral or a hexagonal cell. With a rhombohedral cell no difficulties can occur, and the cell edges must lie within the symmetry planes; with a hexagonal cell ambiguities similar to those for tetragonal crystals can arise, though these are of real importance only in space groups.

Finally we turn to the remaining five cubic classes with their characteristic groupings of four triad axes. The presence of these axes must be indicated by the figure 3 in the symbol but there is a danger of confusion with trigonal classes. To overcome this it is necessary to adopt a fourth convention:

(iv) The symbol 3 appearing in a subsidiary position denotes a cubic class, and refers to the four equally inclined triad axes that such classes must possess.

A symbol $X3$, for example, denotes firstly cubic symmetry, and also, that this class has axes of degree X along the cell edge directions to which the triads are all equally inclined. In this way the notation for the non-cubic classes may be simply extended; for example, the holosymmetric class $4/m3m$ (often abbreviated to $m3m$) is cubic (the subsidiary position of the figure 3), but additionally has three tetrad axes along the cell edges with planes perpendicular to them, and six inclined planes containing the triad axes. For $\bar{4}3m$, there are no ambiguities in the choice of cell edges to be considered.

4.4. Other sub-divisions within the crystal classes

Division of the thirty-two crystal classes into seven crystal systems is of cardinal importance in any development of crystalline pattern theory. Nevertheless, it is valuable in other contexts, notably physical properties, to use different criteria for grouping together certain classes; within a system this can lead to fewer sub-divisions than the number of classes (as in X-ray diffraction, discussed in Chapter 9) or it can develop larger sub-divisions which embrace several systems (as in some of the physical properties described in Chapter 11). These criteria depend on the presence or absence of certain kinds of symmetry elements, and while we shall be examining their detailed origins at a later stage, it is appropriate to make some preliminary observations about them here.

The simplest symmetry element is the centre of symmetry; division on the basis of the presence or absence of a centre shows that there are 21 *acentric* (or non-centro-symmetric) classes and 11 *centric* (or centro-symmetric) classes ($\bar{1}$, $2/m$, mmm, $\bar{3}$, $\bar{3}m$, $4/m$, $4/mmm$, $6/m$, $6/mmm$, $m3$ and $m3m$). We can also select classes in which right- and left-handed motifs may be distinguished. These are called the *enantiomorphous* classes, and are those whose symmetry groupings do not contain a centre of symmetry, a plane of symmetry or an inversion axis; there are 11 enantiomorphic classes (1, 2, 222, 3, 32, 4, 42, 6, 62, 23 and 432). One can also recognise that directions in certain point groups have distinct positive and negative senses. For example, compare the diad axis directions in the trigonal classes $\bar{3}m$ and 32; in $\bar{3}m$ opposite ends of a diad are identical, whereas in 32 they are dissimilar. A direction with distinguishable senses is said to be *polar*, and a symmetry axis along such a direction is said to be *uniterminal*; the diads in 32 are uniterminal, whereas those in $\bar{3}m$ are biterminal. (Attempts are sometimes made to indicate this distinction on stereograms and space group diagrams by showing uniterminal axes with the axis symbol at one end only, and biterminal axes with symbols at both ends; whilst this is satisfactory for horizontal directions, it can lead to confusion for inclined or vertical directions and is not used here.) A simple method to determine whether an axis is uniterminal or not involves isolating those symmetry operators which demand identity at the two ends of the axis; these are (i) a centre of symmetry, (ii) a rotation axis of even degree perpendicular to the axis, (iii) a plane of symmetry normal to the axis, and (iv) an inversion operation in the direction of the axis. If none of these is present, the axis is uniterminal. From this discussion, it follows that some directions in all 21 acentric classes must be polar, but uniterminal symmetry axes occur in only 13 of these: 1 (any direction), 2, *mm* (the diad), 3, 3*m* (the triad), 32 (the diads), 4, 4*mm* (the tetrad), 6, 6*mm* (the hexad), $\bar{6}m2$ (the diads), 23 (the triads) and $\bar{4}3m$ (the triads). There is sometimes confusion between polar directions and the unique directions to which they are related. A *unique direction* is a polar direction which is not repeated by the symmetry of the class. Thus a unique direction must be able to represent the direction of a finite vector, and this permits such directions to be recognised. If we take each of the general poles produced by the symmetry elements of a class to represent a possible vector, the vector set can be summed; when this sum is zero there can be no unique direction in the class, but when it is finite, a unique direction exists in the direction of the resultant vector. Unique directions are found in 10 classes (1, 2, *m*, *mm*, 3, 3*m*, 4, 4*mm*, 6 and 6*mm*) often known, rather perversely, as the *polar classes*; for, while these ten classes belong to the 21 acentric classes in which polar directions can be found, many acentric classes do not possess unique directions (e.g. class 32 has no unique directions; the diads are polar but not unique). In general all unique directions are polar, but not all polar directions are unique.

4.5. Exercises and problems

1. (i) Many objects to be found in our surroundings are symmetrical. Familiarise yourself with the operations of symmetry elements by identifying the axes, planes and centres that they possess. (Remember that the degrees of axes are not necessarily restricted to those found in crystalline matter.)

   (ii) Consider the combinations of symmetry elements that these objects display, and, where possible, use the notation of Chapter 4.3 to devise suitable point group symbols. (Remember that point groups are not necessarily restricted to those found in crystalline matter.)

2. In the discussion of cubic point groups, the possibility of the four essential triad axes as inversion axes was not considered. By examining the effects of four inversion triads on a general pole on a stereogram show that this combination is already included in the conventional list of point groups.

3. $2\bar{2}2$, $\bar{3}/m2$ and $43\bar{2}$ are unusual descriptions of the symmetry of three crystallographic point groups. What are the conventional symbols? What do these classes have in common?

4. Which of the five classes of the trigonal system are enantiomorphous and have uniterminal symmetry axes? Which of these classes contain unique directions?

SELECTED BIBLIOGRAPHY

General

BUERGER, M. J. 1956. *Elementary crystallography*, Wiley.
INTERNATIONAL UNION OF CRYSTALLOGRAPHY. 1965. *International tables for X-ray crystallography*, vol. I. Kynoch Press, Birmingham.
PHILLIPS, F. C. 1963. *An introduction to crystallography*. Longmans.

Mathematical

HILTON, H. 1963. *Mathematical crystallography*. Dover Publications.
JASWON, M. A. 1965. *An introduction to mathematical crystallography*. Longmans.
KOSTER, G. F., DUNMODE, J. O., WHEELER, R. G. and STATZ, H. 1963. *Properties of the thirty-two point groups*. M.I.T. Press.

5

AXIAL SYSTEMS AND INDEXING

5.1. Plane indices and zone axis symbols

In the symbolism for the crystal classes we have the nucleus of a crystallographic shorthand, and in this chapter we will extend this with a simple nomenclature to specify directions in the lattices. Important lattice directions are those which (a) are defined by a row of lattice points, and (b) those which are the normals to families of planes that contain lattice points; in either case the direction indicated by a particular symbol depends on the form of the unit cell of a lattice. Since some classes with similar cell shapes are grouped to form a crystal system certain general features of the symbolism will be common to all classes within a system, although the exact angular location of a given direction may vary from class to class (or from material to material within the same class) due to the different dimensions of the unit cell; we can therefore conveniently consider the notation system by system for each distinctive cell shape. However, before embarking on this there are more general considerations of the nature of any symbolism, the choice of unit cell to which it refers, etc., to be discussed; in this section we shall examine only the widest aspects of the notation without reference to the choice of cell, axial system, etc., which can be advantageous in particular systems.

For the moment, then, it will be assumed that the unit cell of the lattice has been chosen, i.e. the most suitable lattice translations a, b and c have been selected. The first important direction to be described is that of any *lattice row*; this is defined by the vector joining two lattice points on the row. Any vector of this kind can be written as a linear combination of the lattice translations $Ua + Vb + Wc$, where U, V and W are positive or negative integers. The values of U, V and W define the direction of this particular lattice row from a chosen origin, but by the repetitive nature of infinite lattices there must be similar rows joining all lattice points; the integers specify the common direction of all such rows for the given cell. This is illustrated in Fig. 5.1 for a planar lattice with an oblique cell outlined by the lattice translations a and b. When the points A and B are chosen to form a row in a general direction,

$$\vec{AB} = 2a + 1b.$$

Due to the identity of all lattice points, $\overrightarrow{AB}$ is only one member of the family of identical rows joining all points in a similar manner. The integers $U = 2$, $V = 1$ are used to signify the common direction of all these rows in the lattice. They are known as *direction indices*, or more commonly as *zone indices* (or *zone axis symbols*) for reasons which will become apparent shortly, and are written symbolically with the integers in square brackets. Thus [21] indicates the rows of which $\overrightarrow{AB}$ is a member; if negative integers are used, e.g. $\overrightarrow{CD} = -1a + 1b$, the notation is [$\bar{1}$1], and is read 'bar one, one'. A reversal of the

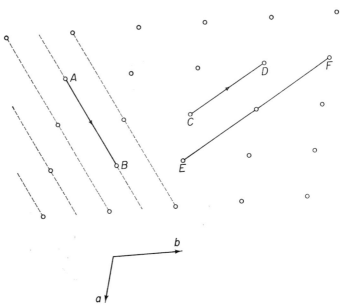

Fig. 5.1. Symbolism for lattice rows. AB is a member of a family of lattice rows; other adjacent rows of this family are shown as dashed lines. The common direction is [21] for the lattice translations a and b. CD (and EF) are in the direction [$\bar{1}$1].

sense of direction implies a change in sign for each of the integers in the symbol; thus $\overrightarrow{BA}$ is [$\bar{2}\bar{1}$] and $\overrightarrow{DC}$ is [1$\bar{1}$]. It is also customary to use a symbol which denotes the shortest vector between two points on the row, so that any common factor in the indices is removed; $\overrightarrow{EF} = -2a + 2b$ is in the same direction as $\overrightarrow{CD}$ and is also written [$\bar{1}$1]. Extension of this example into three dimensions to give zone axis symbols [UVW] is obvious; the integers of the symbol define a particular direction in a lattice (and crystal structure) in terms of a chosen unit cell. This direction is independent of whether the lattice is primitive or not, for this only determines the number of lattice points per unit length of row or the separation of the rows. For example, in the planar rectangular c-lattice the direction of a cell diagonal is $[\frac{1}{2}\ \frac{1}{2}]$, which becomes [11] when fractions are cleared; in a p-lattice cell of identical dimensions [11] denotes the same direction, although the spacing of lattice points is twice as great.

The second important requirement is for a notation to describe the *plane* defined by any three lattice points; again, by the nature of the lattices, such a plane is only a member of a family of planes obtained by joining all lattice points in a similar manner. Now, whereas lattice rows are primarily used to indicate directions, families of planes are important not only for their orientation but also, as we shall see in Chapter 8, for the spacing between the planes of a set; both orientation and spacing depend on the lattice cell. In this chapter we are mainly concerned with orientation, and the dependence of interplanar spacings on cell size and shape is considered in Chapter 7.2. The orientation of planes is usually described by the inclination of their common normal to the directions of the lattice translations, although the symbol representing the planes is constructed from the intercepts made on these directions. With a lattice point chosen as origin, we can write down the intercepts of each member of a family in terms of the appropriate cell edge. Thus for the very simple family of Fig. 5.2(a), the first plane has intercepts a, b and c on the lattice translation directions, the second has intercepts $2a$, $2b$ and $2c$, and so on. In other families, every member need not contain lattice points on the axial rows (see, for example, the intercepts on the c-lattice translation for the planes of Fig. 5.2(b)), but in every case the intercepts may be expressed as simple fractions of the appropriate lattice translation. The equation for any family of planes may therefore be written in the intercept form as

$$x \left/ \frac{a}{h} \right. + y \left/ \frac{b}{k} \right. + z \left/ \frac{c}{l} \right. = p$$

where h, k and l are integers and p is the position of the plane in the set. In this way, $p = 0$ defines the plane through the origin; $p = 1$ is the first plane from the origin with intercepts a/h, b/k, c/l, on axes xyz in the direction of the lattice translations; $p = 2$, the second plane with intercepts $2a/h$, $2b/k$, $2c/l$, and so on. For a given choice of cell, the orientation of a set of planes (and the inclination of their common normal) depends on the integers h, k and l. These are known as the *plane* (or *face*) *indices*, and they are written symbolically with the integers in round brackets. Thus the planes in Fig. 5.2(a) and (b) are (111) and (112) respectively; the directions of the common normals are specified by these symbols; though, as with direction symbols, a reversal of sense requires a change in sign for all the integers, i.e. ($\bar{1}\bar{1}\bar{1}$), read 'bar one, bar one, bar one', is the same family of planes as (111) but with the normal drawn in the opposite sense. A zero index shows that the planes are parallel to a lattice translation (i.e. they have infinite intercept); thus (110) represents a family which is parallel to the c-lattice translation (or z-axis).

Our definition of (hkl) to represent planes that divide the a-lattice translation into h parts, the b-lattice translation into k parts and the c-lattice translation into l parts, emphasises the repetitive nature of the set. When we are concerned only with orientation, any common factors can be removed to leave a set of indivisible integers; but in dealing with interplanar spacings it is

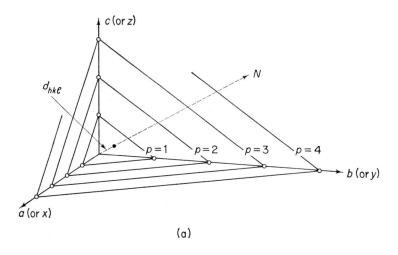

(a)

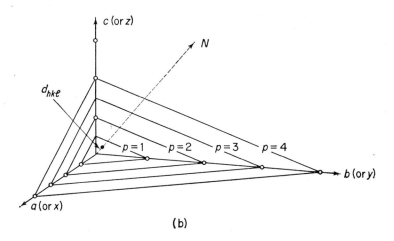

(b)

Fig. 5.2. Symbolism for lattice planes. (a) The (111) family of planes. (b) The (112) family of planes. N shows the direction of the common normal through the origin, and d_{hkl} the interplanar spacing. For clarity, only lattice points on the translation directions through the origin have been shown.

convenient (for reasons explained in Chapter 8) to distinguish between planes with a common normal direction such as (111) and (222), even though some planes of the latter set will not pass through any lattice points at all. The *interplanar spacing* (d_{hkl}) is the length of the normal from the origin to the first plane of the set ($p = 1$), so that

$$d_{hkl} = \frac{a}{h}\cos\widehat{Nx} = \frac{b}{k}\cos\widehat{Ny} = \frac{c}{l}\cos\widehat{Nz}$$

where N is the direction of the normal to the planes (hkl) and x, y and z are the directions of the lattice translations. This expression is often known as the *equations to the normal*, and it is mentioned here for its angular implications; d_{hkl} values are important only in diffraction studies.

So far we have established two types of symbol, one representing the directions of lattice rows and the other showing the orientation of normals to planes of lattice points. It is important to realise that these indicate two different kinds of direction in the lattice; in general, normals to lattice planes need not coincide with the directions of lattice rows, so that (111) and [111] represent essentially different directions, although some symmetrical cell shapes can allow general and particular coincidences as we shall see when considering indexing in various crystal systems. However, there can be some universal relationships between (hkl) and $[UVW]$ which are appropriate here. Families of planes $(h_1k_1l_1)$ and $(h_2k_2l_2)$ will intersect in a series of lines parallel to lattice rows, for the equations of those planes through the origin in each set are

$$x\left/\frac{a}{h_1}\right. + y\left/\frac{b}{k_1}\right. + z\left/\frac{c}{l_1}\right. = 0$$

and

$$x\left/\frac{a}{h_2}\right. + y\left/\frac{b}{k_2}\right. + z\left/\frac{c}{l_2}\right. = 0$$

with a line of intersection passing through the origin whose equation is

$$\frac{x/a}{k_1l_2 - l_1k_2} = \frac{y/b}{l_1h_2 - h_1l_2} = \frac{z/c}{h_1k_2 - k_1h_2}$$

This line will pass through another lattice point whose position relative to the origin is given by the vector

$$(k_1l_2 - l_1k_2)a + (l_1h_2 - h_1l_2)b + (h_1k_2 - k_1h_2)c$$

The coefficients of each lattice translation in this expression must be integral, so that it can be written in the form

$$U_{12}a + V_{12}b + W_{12}c$$

or symbolically as a zone index $[U_{12}V_{12}W_{12}]$ where

$$U_{12} = k_1 l_2 - l_1 k_2$$
$$V_{12} = l_1 h_2 - h_1 l_2$$
$$W_{12} = h_1 k_2 - k_1 h_2$$

The intersections of any two lattice planes must therefore be in the directions of lattice rows given by a zone axis symbol formed from the plane indices in this way. A convenient method of memorising this relationship is to form the zone index by a process known as 'cross multiplication' of the plane integers. These are written down twice, with corresponding integers in the same columns; using the planes (111) and (112) of Fig. 5.2 we get

$$[1 \times 2 - 1 \times 1, 1 \times 1 - 2 \times 1, 1 \times 1 - 1 \times 1]$$

i.e. $[1 \ \bar{1} \ 0]$

$$[k_1 l_2 - k_2 l_1, l_1 h_2 - l_2 h_1, h_1 k_2 - h_2 k_1]$$

i.e. $[U_{12} \ V_{12} \ W_{12}]$

The first and last columns are deleted and cross-multiplication in pairs in the direction of the arrows takes place in separate stages; each stage gives one integer of the zone symbol when the second product is subtracted from the first.

Since any pair of planes has intersections given by $[U_{12}V_{12}W_{12}]$, one can ask what other pairs of planes would also have intersections parallel to this direction. We would then have a number of planes of different indices whose mutual intersections are always in the same direction $[U_{12}V_{12}W_{12}]$. An assemblage of planes which intersect one another in parallel lines is said to form a *zone*, and the common direction of their intersections is called the *zone axis*; the term zone (or zone axis) symbol, commonly employed as a synonym for direction index, is derived from this. An alternative form of the original question is to enquire what relationship must exist between plane indices (*hkl*) and zone indices [*UVW*] in order that this set of planes is a member of the zone. Geometrically to satisfy this condition the normal to the planes and the direction of the lattice row of the zone symbol must be perpendicular, or alternatively, the lattice row must lie in the plane that passes through the origin (Fig. 5.3(a)). The equation of the row is

$$\frac{x}{Ua} = \frac{y}{Vb} = \frac{z}{Wc},$$

whilst that of the plane is

$$x \Big/ \frac{a}{h} + y \Big/ \frac{b}{k} + z \Big/ \frac{c}{l} = 0.$$

Thus the geometrical condition that ensures that (hkl) is in the zone represented by $[UVW]$ is met if

$$Uh + Vk + Wl = 0.$$

The expression is usually known as the *Weiss zone equation* or *law*, and is used in a variety of ways. For example, to find other planes (hkl) which lie in the zone $[1\bar{1}0]$, substitution in the zone equation shows that their indices must be such that $1 \times h - 1 \times k + 0 \times l = 0$, i.e. $h = k$; any pairs of planes fulfilling this condition (113, 221, $\bar{1}\bar{1}4$, 223, etc.) will have mutual intersections which

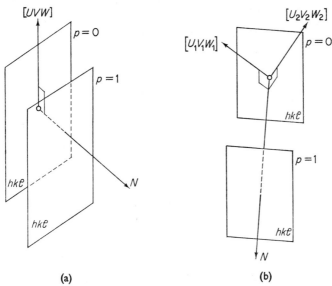

Fig. 5.3. Geometrical conditions relating plane and zone indices. (a) The condition that (hkl) is in the zone $[UVW]$. (b) The condition that (hkl) is at the intersection of the zones $[U_1 V_1 W_1]$ and $[U_2 V_2 W_2]$.

are parallel to this lattice row. Moreover, the zone law allows us to find the indices $(h_{12}k_{12}l_{12})$ which define the normal to planes which are at the intersection of the zones $[U_1 V_1 W_1]$ and $[U_2 V_2 W_2]$ (Fig. 5.3(b)).

Clearly these planes must satisfy both zone equations

$$h_{12}U_1 + k_{12}V_1 + l_{12}W_1 = 0$$

and

$$h_{12}U_2 + k_{12}V_2 + l_{12}W_2 = 0,$$

so that

$$h_{12} = V_1 W_2 - W_1 V_2$$
$$k_{12} = W_1 U_2 - U_1 W_2$$
$$l_{12} = U_1 V_2 - V_1 U_2.$$

As before, this solution can be obtained by 'cross multiplication'. For example, using $[1\bar{1}0]$ and $[211]$ we get

$$
\begin{array}{cccccc}
1 & \bar{1} & 0 & 1 & \bar{1} & 0 \\
& & & & & \\
2 & 1 & 1 & 2 & 1 & 1
\end{array}
$$

$$(\bar{1} \times 1 - 1 \times 0, 0 \times 2 - 1 \times 1, 1 \times 1 - 2 \times \bar{1})$$

i.e. $(\bar{1}\ \bar{1}\ 3)$

as the indices of the planes at the intersection of these two zones. The formal relationship between indices of planes within a zone expressed by the Weiss zone law can be used in an extremely valuable practical way. Since any two planes $(h_1k_1l_1)$ and $(h_2k_2l_2)$ form a zone $[UVW]$ for which

$$Uh_1 + Vk_1 + Wl_1 = 0$$

and

$$Uh_2 + Vk_2 + Wl_2 = 0,$$

we may also write

$$U(mh_1 + nh_2) + V(mk_1 + nk_2) + W(ml_1 + nl_2) = 0,$$

in which the zone equations are added after multiplication by the integers m and n (positive or negative). Comparing with the zone equation for any other planes (hkl) in this zone, we obtain

$$h = mh_1 + nh_2$$
$$k = mk_1 + nk_2$$
$$l = ml_1 + nl_2,$$

i.e. other plane indices in the zone formed by $(h_1k_1l_1)$ and $(h_2k_2l_2)$ are found by the addition of multiples of original indices. Thus in the zone defined by (111) and (112), there are planes (223) for $m = n = 1$, (001) for $m = -1$, $n = 1$, (334) for $m = 2$, $n = 1$, etc.; naturally the indices must satisfy the condition $h = k$ imposed by the zone index $[1\bar{1}0]$. This adaptation of the zone law gives rise to the commonest method of indexing planes at the inter-section of two zones, a method which is simpler than the formation of the zone symbols followed by cross-multiplication. Indices of the common planes (hkl) must be able to be expressed in the multiple form for other known planes in both zones, and a little mental dexterity enables the values of h, k and l to be obtained by the process known as 'cross-adding' in zones. Two simple examples are given in Fig. 5.4, where in (a), P is at the intersection of zones defined by (100), (111) and (110), (101), and in (b) Q is located by the zones (100), (021), and (110), (101). By cross-adding multiples of indices in zones, we can soon find the common set of integers in each case and so index P and Q. In (a) P must be (211), i.e. $1 \times (100) + 1 \times (111)$ in one zone and

$1 \times (110) + 1 \times (101)$ in the second. In (b) Q is (321), for these are the only common set of integers which can be found by cross-adding; for the first zone above $m = 3$, $n = 1$ gives (321), whilst for the second $m = 2$, $n = 1$ also gives (321). Indexing by using zonal relationships and the 'cross-adding' technique is extremely convenient, and an essential familiarity with its use is soon obtained.

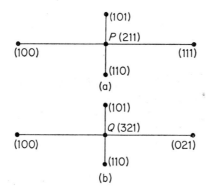

Fig. 5.4. Indexing by 'cross-adding' in zones.

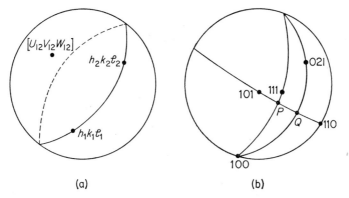

Fig. 5.5. Normals to planes and zone axis directions on the stereogram. (a) The zone axis $[U_{12}V_{12}W_{12}]$ for the zone defined by $(h_1k_1l_1)$ and $(h_2k_2l_2)$ is the pole of the great circle passing through them. (b) The location of P and Q in Fig. 5.4 on the projection.

Up to this point, these general relationships between plane indices and zone symbols have been treated geometrically and analytically. Since they will be used in conjunction with the stereograms described in the last chapter, we will conclude this section by transposing them into projection. Firstly, the directions of the normals to planes (hkl) and zone axis symbols $[UVW]$ are represented by poles on the stereogram; conventionally the poles (dots or circles) are identified by writing the appropriate integers by them, the round brackets being omitted for planes though the square brackets are usually retained for zone axes. Since the indices $(h_1 k_1 l_1)$ and $(h_2 k_2 l_2)$ define a zone (represented by $[U_{12} V_{12} W_{12}]$), this zone will be shown on the stereogram as a great circle passing through these two poles; moreover the direction $[U_{12} V_{12} W_{12}]$ of the zone axis will be the pole of this great circle (Fig. 5.5(a)). The indexing of a pole (hkl) by 'cross-adding' in zones means that the projection of the pole is at the intersection of the great circles representing the two zones (Fig. 5.5(b)).

5.2. The choice of axial systems and transformations between them

In the preceding section we saw how the directions of normals to planes and zone axes could be described in terms of the lattice translations defining a unit cell. The cell edge directions form a crystallographic axial system (xyz) for which the lengths of the edges (a, b, c) are the units of measurement along each axis; conventionally the axial systems of crystallography are right-handed (Fig. 5.6). In a crystalline material any three non-co-planar lattice

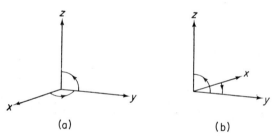

(a)   (b)

Fig. 5.6. Axial systems. (a) Right-handed axes; the senses of rotation from x to y and y to z are the same. (b) Left-handed axes; the senses of rotation from x to y and y to z are opposite.

repeats can be used as an axial system, but this would usually result in more complex and arbitrary descriptions than are necessary; in other words, just as any co-ordinate system is simplified by selecting reference axes advantageously, the simplest indexing occurs when natural axial systems are employed. To do this we must choose the unit cell (and hence the kind of axial system) in harmony with symmetry; crystallographic reference axes will then change with the crystal system according to the cell shape demanded by the symmetry, ranging from three perpendicular axes with identical units for cubic crystals to three generally inclined axes with different units of measurement for triclinic crystals. Once axes are chosen, the indexing of planes and zone axes can proceed; but this choice of cell can present problems that are discussed in this section.

These problems are broadly divided into two: (a) those which can still arise even when the fullest data on the crystal lattice is available, and (b) those which occur when only the most general information on the lattice is to hand. A situation such as (a) can occur when diffraction studies have established the full angular and linear distributions of lattice points; that of (b) can happen when only the symmetry of the crystal is known (as in morphological studies in Chapter 6) so that the form of the angular arrangement of points is inferred together with the characteristic cell shape but nothing is known about the linear dimensions.

It might be thought that given the arrangement of lattice points, as in (a), the choice of a suitably shaped unit cell upon a criterion of minimum volume is unambiguous. For highly symmetrical systems with regular cell shapes this is so, but for more general cells a variety of problems can arise. In an orthorhombic crystal, for example, the orthogonal cell of minimum volume with edges in the directions of the diad axes can be identified, but we must decide which of these translations shall be labelled a, b and c (or which directions are the x-, y- and z-axes); they differ only in their dimensions, and any choice must be arbitrary. In less symmetrical systems with non-orthogonal cell shapes the problem is further complicated by ambiguity in the choice of a minimum volume cell. For a monoclinic crystal, for example, the b-lattice translation (or the y-axis) is fixed by the direction of the essential diad axis; in the perpendicular plane there are innumerable pairs of generally inclined lattice rows which outline cells of the same minimum volume (as with planar oblique cells in Chapter 2.2). Each pair gives different directions to the x- and z-axes and has its own units a and c on these axes; once again any choice must be arbitrary. Any resolution of these arbitrary choices must be by convention, and we will discuss this in detail for each system later in this chapter.

But when indexing is carried out with only a knowledge of point group symmetry, as in (b), the ambiguities of choice are multiplied. Only the general shape of the cell is known, with no data on its dimensions or orientation at all. In order to index visible features, such as a crystal face, not only have axial reference directions to be chosen but unit translations a, b and c on these axes must also be assigned. Clearly the process becomes much more arbitrary, and is unambiguous only for the highly symmetrical classes which form the cubic system. In all other systems, there is some kind of choice in the axial directions to be made, and then a further step in the selection of the axial units. It should be realised that when angular aspects of plane and zone indices are solely concerned, a choice of axial units need only be given in terms of their ratios. Fig. 5.7 shows the intersection of the first member of the (111) planes on the reference axes; its intercepts are in the ratio $a:b:c$ on x, y and z respectively, as are those of any parallel plane; and so it is these ratios (rather than the absolute magnitudes of the translations) which determine the normal to the planes. In the form $\frac{a}{b}:1:\frac{c}{b}$, these are known as the *axial ratios*, and their numerical values are quoted to define the choice of axial units that has been made; for a specific axial system they fix the orientation of the (111) planes, often known as the *parametral planes*. Let us illustrate the problems of indexing under these circumstances by supposing that the point group symmetry is known to be 4/*mmm* (Fig. 5.8). The desirable tetragonal cell is a square-based

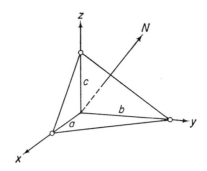

Fig. 5.7. The orientation of the parametral (111) planes.

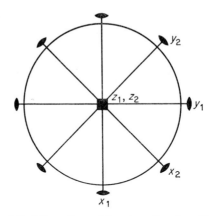

Fig. 5.8. Alternative axial systems for class 4/*mmm*.

right prism, and the direction of the tetrad axis can be taken as the z-axis. The other two axes will be symmetry related and there are obvious possibilities in the two sets of diad axes (x_1y_1 or x_2y_2); an arbitrary choice (say x_1y_1) must be made at this point. Next, the axial ratios must be fixed; because of the identity of the x- and y-axes, $\frac{a}{b} = 1$, leaving only a choice of $\frac{c}{b}\left(=\frac{c}{a}\right)$. Since $a = b$ the normal to the parametral planes (111) must lie on the diameter of the projection bisecting the x_1- and y_1-axes; it could be chosen at any point on this line in the lower right-hand quadrant, but is usually related to some feature of the particular crystal (often the position of an external face). With the choice of the parametral pole the value of the axial ratio has been determined, and indexing can proceed on this axial system. In this procedure the possibilities of divergence from an indexing scheme based on a full knowledge of the crystal lattice are obvious; even if the choice of x_1 and y_1 happens to correspond to the directions of the edges of the cell of minimum volume, there remains the arbitrary axial ratio which may not be

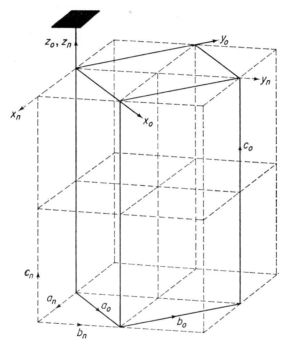

Fig. 5.9. Alternative descriptions of a tetragonal lattice. Lattice points are at the intersections of all broken lines. The eight new cells of minimum volume refer to an axial system $x_n y_n z_n$ defined by the lattice translations $\vec{a}_n$, $\vec{b}_n$ and $\vec{c}_n$; the old cell of larger volume is heavily outlined and refers to a different axial system $x_o y_o z_o$ implying the lattice translations $\vec{a}_o \vec{b}_o$ and $\vec{c}_o$. Symmetry demands that $\vec{a}_n = \vec{b}_n$ and $\vec{a}_o = \vec{b}_o$.

that given by the cell dimensions (although in this case it must be rationally related to the true value). Moreover, indexing of the same crystal by two different observers, although compatible, could be quite different in detail. In less symmetrical systems we shall face many more possibilities for the choice of axial directions and axial ratios when only the crystal class is known.

Later we shall see how the method of indexing may be best rationalised in each system, but the problems outlined above suggest that there will be from time to time a conflict in the representation of the crystallography of a particular material. In such cases each description is self-consistent and valid, but differs from other descriptions with alternative axial systems and unit translations. When this occurs, it is convenient to have a simple algebraic method of transforming one set of axes (and their indices) to another. Essentially this axial transformation reflects a change from one lattice cell to another defined by lattice translations which may differ in direction and dimensions. In general this may be expressed by vector equations in which the translations of one cell are given in terms of the other, so that

Old → new   New → old

$$\vec{a}_n = u_1\vec{a}_o + v_1\vec{b}_o + w_1\vec{c}_o \qquad \vec{a}_o = u_1'\vec{a}_n + v_1'\vec{b}_n + w_1'\vec{c}_n$$
$$\vec{b}_n = u_2\vec{a}_o + v_2\vec{b}_o + w_2\vec{c}_o \quad \text{or} \quad \vec{b}_o = u_2'\vec{a}_n + v_2'\vec{b}_n + w_2'\vec{c}_n$$
$$\vec{c}_n = u_3\vec{a}_o + v_3\vec{b}_o + w_3\vec{c}_o \qquad \vec{c}_o = u_3'\vec{a}_n + v_3'\vec{b}_n + w_3'\vec{c}_n$$

in which a_o, b_o, c_o and a_n, b_n, c_n are the lattice translations of the old and new cells respectively; the coefficients $u_1 \ldots$, $u_1', \ldots$, etc. define the relationships in each case. Fig. 5.9 shows two alternative cells describing a tetragonal lattice; one of these, labelled new, is the cell of minimum volume, whilst the other, labelled old, is much larger with a translation in the tetrad direction twice that of the new cell and rotation of the x- and y-axes through 45°. In this example the new cell might be that apparent after an X-ray examination of the internal crystal structure, whilst the old cell could be that implied in an earlier description based on external features alone. The equations relating the two cells are

Old → new   New → old

$$\vec{a}_n = \tfrac{1}{2}\vec{a}_o - \tfrac{1}{2}\vec{b}_o + 0\vec{c}_o \qquad \vec{a}_o = 1\vec{a}_n + 1\vec{b}_n + 0\vec{c}_n$$
$$\vec{b}_n = \tfrac{1}{2}\vec{a}_o + \tfrac{1}{2}\vec{b}_o + 0\vec{c}_o \qquad \vec{b}_o = -1\vec{a}_n + 1\vec{b}_n + 0\vec{c}_n$$
$$\vec{c}_n = 0\vec{a}_o + 0\vec{b}_o + \tfrac{1}{2}\vec{c}_o \qquad \vec{c}_o = 0\vec{a}_n + 0\vec{b}_n + 2\vec{c}_n$$

The coefficients in these vector equations can be written as matrices, so that generally

Old → new   New → old

$$\begin{vmatrix} u_1 & v_1 & w_1 \\ u_2 & v_2 & w_2 \\ u_3 & v_3 & w_3 \end{vmatrix} \quad \text{or} \quad \begin{vmatrix} u_1' & v_1' & w_1' \\ u_2' & v_2' & w_2' \\ u_3' & v_3' & w_3' \end{vmatrix}$$

which are sometimes printed $u_1v_1w_1/u_2v_2w_2/u_3v_3w_3$ and $u_1'v_1'w_1'/u_2'v_2'w_2'/u_3'v_3'w_3'$ respectively. These matrices have useful properties; in particular, they can be used to transform plane indices from one axial system to the other. To find what $(hkl)_o$ becomes on the new system, or conversely what indices $(hkl)_n$ had in the old system, we can use the appropriate matrix in the form,

$$
\begin{aligned}
h_n &= u_1 h_o + v_1 k_o + w_1 l_o & h_o &= u_1' h_n + v_1' k_n + w_1' l_n \\
k_n &= u_2 h_o + v_3 k_o + w_2 l_o \quad \text{or} & k_o &= u_2' h_n + v_2' k_n + w_2' l_n \\
l_n &= u_3 h_o + v_3 k_o + w_3 l_o & l_o &= u_3' h_n + v_3' k_n + w_3' l_n
\end{aligned}
$$

For the parametral planes $(111)_o$ and $(111)_n$ of the two cells of Fig. 5.9, these transform to become

$$
\begin{aligned}
h_n &= \tfrac{1}{2} \times 1 - \tfrac{1}{2} \times 1 + 0 \times 1 = 0 \\
k_n &= \tfrac{1}{2} \times 1 + \tfrac{1}{2} \times 1 + 0 \times 1 = 1 \\
l_n &= 0 \times 1 + 0 \times 1 + \tfrac{1}{2} \times 1 = \tfrac{1}{2}
\end{aligned}
$$

and

$$
\begin{aligned}
h_o &= 1 \times 1 + 1 \times 1 + 0 \times 1 = 2 \\
k_o &= -1 \times 1 + 1 \times 1 + 0 \times 1 = 0 \\
l_o &= 0 \times 1 + 0 \times 1 + 2 \times 1 = 2.
\end{aligned}
$$

Clearing fractions and multiples, $(111)_o \rightarrow (021)_n$ and $(111)_n \rightarrow (101)_o$, reflecting, of course, the changes in directions of the x- and y-axes and the unit intercept on the z-axis. We shall not attempt to justify this property of the matrices here, nor do more than mention certain others. For example, it will be noticed that the two matrices, old $\rightarrow$ new and new $\rightarrow$ old, are different. Mathematically they are related by inversion and their determinants give the numerical ratios of the cell volumes; a change in sign in the determinant denotes a change in hand of the reference system.

When using these matrices it is important to realise that whilst they relate plane indices (hkl), different matrices are needed to transform zone axis symbols $[UVW]$. These can be obtained by interchanging rows and columns in the inverse of the original matrix to give

Old $\rightarrow$ new   New $\rightarrow$ old

$$
\begin{vmatrix} u_1' & u_2' & u_3' \\ v_1' & v_2' & v_3' \\ w_1' & w_2' & w_3' \end{vmatrix} \quad \text{or} \quad \begin{vmatrix} u_1 & u_2 & u_3 \\ v_1 & v_2 & v_3 \\ w_1 & w_2 & w_3 \end{vmatrix}
$$

These new matrices are then used to transform zone axis symbols in the same way as before. With $[111]_o$ and $[111]_n$ of the two cells in Fig. 5.9, we shall get

$$
\begin{aligned}
U_n &= 1 \times 1 + -1 \times 1 + 0 \times 1 = 0 & U_o &= \tfrac{1}{2} \times 1 + \tfrac{1}{2} \times 1 + 0 \times 1 = 1 \\
V_n &= 1 \times 1 + 1 \times 1 + 0 \times 1 = 2 \quad \text{and} & V_o &= -\tfrac{1}{2} \times 1 + \tfrac{1}{2} \times 1 + 0 \times 1 = 0 \\
W_n &= 0 \times 1 + 0 \times 1 + 2 \times 1 = 2 & W_o &= 0 \times 1 + 0 \times 1 + \tfrac{1}{2} \times 1 = \tfrac{1}{2}
\end{aligned}
$$

Clearing fractions and multiples, $[111]_o \rightarrow [011]_n$ and $[111]_n \rightarrow [201]_o$. In practice, when only one or two zone symbols have to be transformed, it is often convenient to select two faces which lie in the zone and find the new index by cross-multiplying these after transformation with the original matrix. For example, the two faces $(1\bar{1}0)_o$ and $(11\bar{2})_o$ lie in the zone $[111]_o$ for they satisfy the zone equation; transforming these by the original matrix $(1\bar{1}0)_o \rightarrow (100)_n$ and $(11\bar{2})_o \rightarrow (01\bar{1})_n$, which on cross-multiplying give the direction $[011]_n$.

Transformation matrices can be important both in the reconciliation of older data with more recent investigations and in describing the relations between alternative choices of axial systems and cells in the lower symmetry systems. A fuller account of the properties of the various matrices and their uses will be found in the bibliography.

5.3. Determination of angular relationships

For a particular material, once the lattice translations are chosen, the indexing of planes and lattice rows embraces all angular and linear relationships of the lattice; indeed, as we saw in the previous section, the angular values alone determine the shape of the cell and the axial ratios. In this section we shall briefly review the methods by which angles between planes or zones can be evaluated from cell data; conversely these methods could permit conclusions about the cell from angular measurements. The most important angles are usually those between the normals to two sets of planes $(h_1 k_1 l_1)$ and $(h_2 k_2 l_2)$ and those between two zone axis directions $[U_1 V_1 W_1]$ and $[U_2 V_2 W_2]$; we shall refer to these as ρ and ϕ respectively.

Since all indexing is based on axial co-ordinate systems it is possible to set up algebraic expressions for ρ and ϕ in terms of the cell constants. With orthogonal axes one can adapt the results of co-ordinate solid geometry, in particular those which state that the sum of the squares of the direction cosines of a line is unity and which express the angle between two lines in terms of their direction cosines. For example, the equations to the normal for a family of planes (see Chapter 5.1) show that it has direction cosines

$$\cos \widehat{Nx}\left(\propto \frac{h}{a}\right), \cos \widehat{Ny}\left(\propto \frac{k}{b}\right) \quad \text{and} \quad \cos \widehat{Nz}\left(\propto \frac{l}{c}\right)$$

in such systems. When these are substituted in the standard relationship

$$\cos \rho = \frac{\dfrac{h_1 h_2}{a^2} + \dfrac{k_1 k_2}{b^2} + \dfrac{l_1 l_2}{c^2}}{\left[\left(\dfrac{h_1^2}{a^2} + \dfrac{k_1^2}{b^2} + \dfrac{l_1^2}{c^2}\right)\left(\dfrac{h_2^2}{a^2} + \dfrac{k_2^2}{b_2} + \dfrac{l_2^2}{c^2}\right)\right]^{1/2}}$$

and a comparable expression for $\cos \phi$ may be developed in the same way. These expressions are simplified when the crystal system has symmetry which requires one or both the axial ratios on the orthogonal axes to be unity. But, except for cubic crystals, they remain complex enough to encourage computational error and comparable expressions for a non-orthogonal axial system are even more unwieldy to use without computer facilities. When only a few angles are required, it is more convenient to obtain them from the stereogram.

Both ρ and ϕ correspond to angular distances on great circles on the projection and can be measured (by the methods described in Chapter 3.3) with reasonable accuracy on stereograms of a suitable scale from the locations of the appropriate poles; but as mentioned in this earlier section the stereogram is admirably suited to accurate calculation of spherical triangles by spherical trigonometry. This requires that three elements (of the six) in the triangle are known; in practice, crystallographic stereograms often contain an element of 90°, and the trigonometric solutions are simpler for such right-angled (or Napierian) triangles. Whenever possible the calculations are formulated in terms of Napierian triangles, which can be solved when two elements (other than the right angle) are known; the method of solution most suited to a particular problem is acquired by experience. We shall not attempt to illustrate these calculations on the stereogram here, but some examples are given for each of the crystal systems, together with the essentials of the spherical trigonometry, in Appendix B.

5.4. Axial systems and indexing for each crystal system

(a) Triclinic System

(i) *Choice of axial system.* The lattice cell is a general parallelipiped defining non-orthogonal crystallographic axes with different units on each. Even if the distribution of lattice points is known, there is no unique choice of a primitive cell of minimum volume; the decision to select one particular cell is arbitrary, and can only be governed by convention. Unfortunately, over many years there have been several different conventions and many ill-defined customs, and even now there are no universally recognised rules which lead to a standard form of description for this system. We shall refer to those used in the compilation of *Crystal Data*, a source book which lists the basic facts of symmetry and geometry for as many crystalline materials as possible. In this all triclinic crystals are described by:

(i) Choosing the primitive cell of minimum volume whose edges give the three shortest lattice translations; this is sometimes known as the *reduced cell*;

(ii) labelling the translations (and crystallographic axes) according to their lengths so that $c(\parallel z\text{-axis}) < a(\parallel x\text{-axis}) < b(\parallel y\text{-axis})$;

(iii) taking the positive directions of these translations to form a right-handed set of axes with the interaxial angles α and β obtuse; depending on the geometry of the particular cell, γ may be obtuse or acute.

If these rules, inevitably arbitrary, are followed, some consistency in the presentation of data for this system can be achieved; but it will soon be found that the records of triclinic crystals, even now, are by no means standardised. Apart from the use of other conventions, further factors that contribute to this non-uniformity include: (a) the choice of a cell (sometimes even non-primitive) whose axes are selected and labelled in a different manner in order to facilitate comparison of the structure with that of another crystal, often of higher symmetry; (b) maintenance of older descriptions based solely on external evidence and which implies an unconventional choice of cell. Whatever the origin of the divergence, the very general character of the triclinic axial system is common to all descriptions.

(*ii*) *The orientation of the parametral planes.* This can be calculated if the axial constants, the axial ratios $\frac{a}{b}:1:\frac{c}{b}$, and the interaxial angles α, β, γ are known, but is more often found by using zonal relationships on a stereogram as we shall see in (iii) below. In this connection the geometry of Fig. 5.10(a) is relevant; this is shown for (111) planes but it can be generalised for any family of planes (*hkl*). From the three triangles in the *xy*, *zy* and *zx* planes formed by the intersections of a (111) plane, we see that

$$\frac{a}{b} = \frac{\sin \phi_1}{\sin \phi_2}; \quad \frac{c}{b} = \frac{\sin \phi_3}{\sin \phi_4}; \quad \frac{c}{a} = \frac{\sin \phi_5}{\sin \phi_6},$$

and $\phi_1 + \phi_2 + \gamma = \phi_3 + \phi_4 + \alpha = \phi_5 + \phi_6 + \beta = 180°.$

Now ϕ_1, ϕ_2, etc. are angles between lattice rows, i.e. they are angles between zone axes; these may be identified with the angles between corresponding zones at their points of intersection in the stereogram below.

(*iii*) *Stereographic projection.* Conventionally the right-handed axial set is projected with the *z*-axis at the centre of the projection, and the plane of the *x*- and *z*-axes as the vertical diameter; since the (010) planes are parallel to the *xz* plane, the pole of their common normal direction must be on the primitive at the right-hand end of the horizontal radius (Fig. 5.10(b)). If the interaxial angles are obtuse, the poles of the *x*- and *y*-axes are in the southern hemisphere; the points (100) and (001), representing planes parallel to *yz* and *xy* respectively, will be the poles of the great circles passing through these pairs of axes. This means that the external angles of the spherical triangle formed by (100)(010)(001) will be the interaxial angles α, β and γ, for apexes of this triangle are the intersections of great circles whose poles are the *x*-, *y*- and *z*-axes. The point (111)—and any pole representing planes with positive integers—must lie within this triangle, and its position may be found by drawing the great circles joining it to the corners of the triangle. These are zones formed by the pairs of planes (111)(001), etc. for which the zone axis directions are given by their mutual intersections; in Fig. 5.10(a) these are the

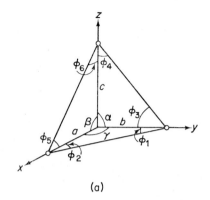

(a)

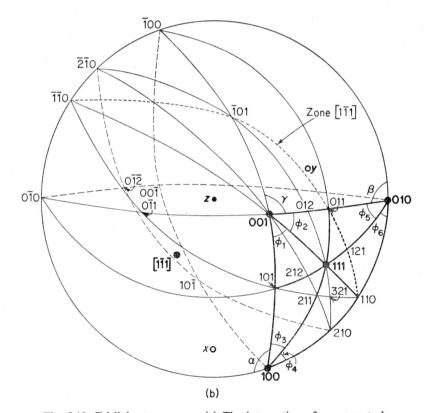

(b)

Fig. 5.10. Triclinic stereograms. (a) The intersection of a parametral (111) plane with the axes. (b) The corresponding stereogram. Heavy lines show the location of the (111) pole by zones within the spherical triangle (100) (010) (001). The indices (hkl) at the intersection of some other zones are shown. Any repetition by symmetry is neglected.

inclined lattice rows joining points on the x- and y-, z- and y-, and z- and x-axes. The angles ϕ_1, ϕ_2, etc. in this figure are enclosed between these rows and those along the axial directions, i.e. ϕ_1 is between the zones (001)(111) and (001)(100) at their point of intersection, and so on; they therefore locate the great circles which join (111) to the corners of the triangle (100)(010)(001). Values of ϕ_1, ϕ_2, etc. can be calculated from the axial ratios and interaxial angles by the equations in (ii), when these circles intersecting in (111) can be drawn. We shall adapt the application of this general geometry to the more symmetrical axial arrangements of the other crystal systems in succeeding sections. With these points on the stereogram all other poles (hkl) can be found by zonal relationships; a few simple examples are shown in the figure. In this system poles of planes $(hk\bar{l})$ intersecting the negative direction of the z-axis lie in the southern hemisphere; they will not be beneath the points (hkl) and must again be found by zonal relations.

The projection of $[UVW]$ does not coincide with the pole of planes with identical indices (viz. [100] (the x-axis) is in a different position to (100)), and each zone axis must be separately determined. It is usually best to find planes on the corresponding zone by the Weiss zone equation, and then plot the pole of this great circle. For example, the poles (110) and (011) must lie on the zone $[1\bar{1}1]$; the zone axis is plotted as the pole of the great circle passing through them (Fig. 5.10(b)).

(b) Monoclinic System

(i) *Choice of axial system.* The lattice cell chosen in accord with the essential symmetry of a diad axis is an oblique parallelipiped; conventionally the y-axis is identified with this diad, so that the x- and z-axes are generally inclined in the perpendicular plane, with different units on all three axes. In a monoclinic array of lattice points the y-axis is distinguishable, but there can be no unique cell of minimum volume and the choice of x- and z-axes must be governed by convention. Some authors (and *Crystal Data*) follow the rules set out for triclinic crystals, and choose the two shortest lattice translations in the lattice net normal to the y-axis to be the x- and z-axes, labelling them so that $c < a$ and the angular inclination β is obtuse; others have given priority to choosing a cell so that β, although obtuse, is as close to 90° as possible. There is a further complication in that some monoclinic crystals can have irreducible non-primitive cells (see Chapter 7.1), and in striving for a uniform description for such cells we can impose an axial system that contravenes other conventions. Taken with other aberrations due to structural comparisons and prior morphological descriptions, it is difficult to discern any uniformity of presentation in recorded data for this system. Nevertheless, the nature of the axial system with the diad as the y-axis, and the other two axes inclined at an obtuse angle in the perpendicular plane is generally preserved in most descriptions; occasionally in some early observations the diad is taken as the z-axis, but this is now discouraged.

(*ii*) *The orientation of the parametral planes.* As before, this requires a knowledge of the axial constants, the axial ratios $\frac{a}{b}:1:\frac{c}{b}$ and the interaxial angle β. The geometry of Fig. 5.10(a) is simplified, for α and γ are 90°, so that

$$\phi_1 + \phi_2 = \phi_3 + \phi_4 = 90°$$

to make the axial ratios become

$$\frac{a}{b} = \tan \phi_1; \quad \frac{c}{b} = \tan \phi_3; \quad \frac{c}{a} = \frac{\sin \phi_5}{\sin \phi_6}$$

which we can again use to plot the zones which locate (111). Alternatively these zones can be plotted from the angles $(100)\widehat{}(110)$ and $(001)\widehat{}(011)$ which are simply related to the axial constants (Fig. 5.11) as

$$(100)\widehat{}(001) = 180° - \beta$$

$$\tan (100)\widehat{}(110) = \frac{a}{b} \sin \beta$$

and

$$\tan (001)\widehat{}(011) = \frac{c}{b} \sin \beta.$$

The last two expressions can be generalised for planes ($hk0$) and ($0kl$) respectively.

(*iii*) *Stereographic projection.* There are two conventional orientations of projection. In the first (Fig. 5.12(a)), the z-axis is at the centre of the projection and the y-axis on the primitive at the right-hand end of the horizontal radius; this makes the xz plane the vertical diameter, with the pole of the x-axis in the southern hemisphere when β is obtuse. In the second (Fig. 5.12(b)) the y-axis is taken at the centre, with the z-axis on the primitive at the upper end of the vertical radius; for a right-handed set of axes and obtuse β this places the x-axis on the primitive in a general position in the lower left-hand quadrant. Both orientations have advantages and disadvantages; (a) maintains the location of the z-axis at the centre of the projection, but retains the feature that poles ($hk\bar{l}$) in the southern hemisphere are not immediately below the corresponding points (hkl); (b) eliminates this for points in the southern hemisphere are ($h\bar{k}l$), which are beneath poles (hkl), but it is now not uniform with the conventional orientation of projections in all other systems. A choice can only be made in terms of personal preference, and we shall mainly use the orientation of (a); but, whichever form is employed, the use of zonal relationships for indexing follows the same pattern.

As before, the poles (100), (010) and (001) are found as the poles of great circles containing the appropriate pairs of axes. (111) can then be located by zones to the vertices of the spherical triangle; these zones can be constructed either by use of the angles ϕ_1, ϕ_2, etc. or from the positions of the poles (110) and (011) determined on the zones (100)(010) and (001)(010) respectively by the relations above. Incidentally we notice from the stereogram that

$$\frac{c}{a} = \frac{\sin \phi_5}{\sin \phi_6} = \frac{\sin (00\overline{1})\widehat{(101)}}{\sin (100)\widehat{(101)}} = \frac{\sin (00\overline{1})\widehat{(101)}}{\sin [\beta - (00\overline{1})\widehat{(101)}]}$$

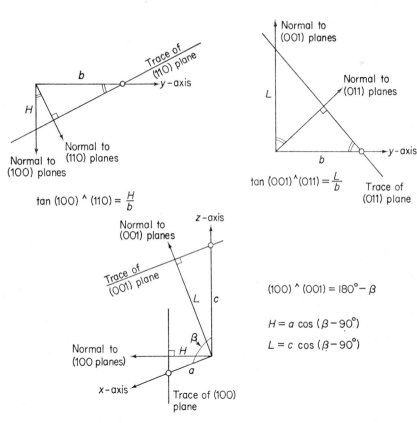

Fig. 5.11. Angles in principal zones of a monoclinic crystal. The three diagrams show planes perpendicular to the x, y and z-axes.

after substituting $(00\overline{1})\widehat{(100)} = 180° - \beta$; this enables the position of (101)—and any other poles $(h0l)$ in the zone (001)(100)—to be determined independently from the axial ratio $\frac{c}{a}$ and β. With these points on the stereogram, all other poles can be found by zonal relations; a few examples are shown in the figure.

With the exception of [010] (the y-axis) which coincides with (010), zone

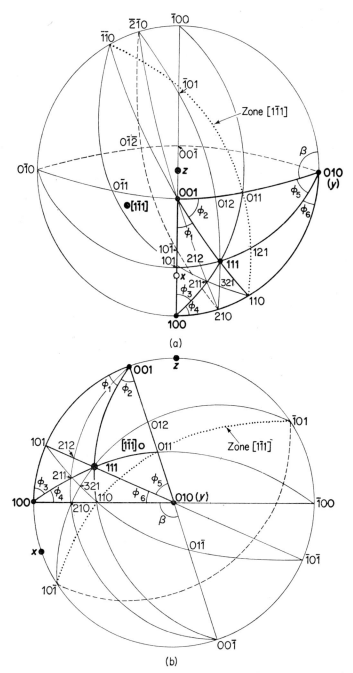

Fig. 5.12. Monoclinic stereograms. (a) With the z-axis at the centre of
the projection. (b) With the y-axis at the centre of the projection. Heavy
zones show the location of the (111) pole within the spherical triangle
(100)(010)(001). The indices (hkl) at the intersections of some other zones
are shown. Any repetition by symmetry is neglected.

indices $[UVW]$ are in different positions to the poles of corresponding plane indices, and should be found as poles of great circles containing the relevant plane indices (e.g. $[1\bar{1}1]$ in the figure).

(c) Orthorhombic System

(i) *Choice of axial system.* A lattice cell shaped as a rectangular parallelipiped is demanded by the essential symmetry of three perpendicular diad axes; the x, y and z crystallographic axes are along these diads with different units on each axis. From the array of lattice points there is no difficulty in recognising these directions and finding the cell of minimum volume, but the naming of the axes allows some freedom. One can follow the criteria of length of lattice translation ($c < a < b$), but this may conflict with the conventional nomenclature for the particular space group (see Chapter 7.4). Once again other priorities may be fixed according to the problem, and we must accept the possibility of an interchange of axial labelling in recorded data; but all descriptions of orthorhombic crystals are made with a right-handed orthogonal axial set.

(ii) *The orientation of the parametral planes.* With this axial system, there are no problems in calculating (as in Chapter 5.3) the positions of these planes from the axial constants, the axial ratios $\frac{a}{b} : 1 : \frac{c}{b}$. But pursuing the zonal approach of the two previous systems the increased symmetry with $\alpha = \beta = \gamma = 90°$ permits further simplication in the relations to axial ratios. Not only does $\phi_5 = (00\bar{1})\widehat{\ }(101)$ and $\phi_6 = (100)\widehat{\ }(101)$ as in the monoclinic system, but also $\phi_1 = (100)\widehat{\ }(110)$, $\phi_2 = (010)\widehat{\ }(110)$, $\phi_3 = (00\bar{1})\widehat{\ }(011)$ and $\phi_4 = (010)\widehat{\ }(011)$, so that

$$\frac{a}{b} = \tan \phi_1 = \tan (100)\widehat{\ }(110);$$

$$\frac{c}{b} = \tan \phi_3 = \tan (00\bar{1})\widehat{\ }(011);$$

$$\frac{c}{a} = \tan \phi_5 = \tan (00\bar{1})\widehat{\ }(101),$$

which can be respectively generalised for any planes of the type $(hk0)$, $(0kl)$ and $(h0l)$.

(iii) *Stereographic projection.* The conventional orientation has the z-axis at the centre and the y-axis on the primitive at the end of the right-hand horizontal radius; the x-axis is therefore also on the primitive at the lower end of the vertical diameter. The poles (001), (010) and (100) coincide with the poles of the axes, and the usual zones to plot (111) are shown on the stereogram in Fig. 5.13. The points $(hk\bar{l})$ in the southern hemisphere will project as circles immediately below the corresponding (hkl) poles; when both (hkl) and $(hk\bar{l})$ are present on the stereogram, it is usual to label only the upper one.

Although the axial directions [100], [010] and [001] coincide with the poles (100), (010) and (001), all other zone axes $[UVW]$ are in different directions from the poles of planes with corresponding indices (e.g. in the figure $[1\bar{1}1]$, plotted by the usual procedure, is not on the great circle $(0\bar{1}0)(101)$).

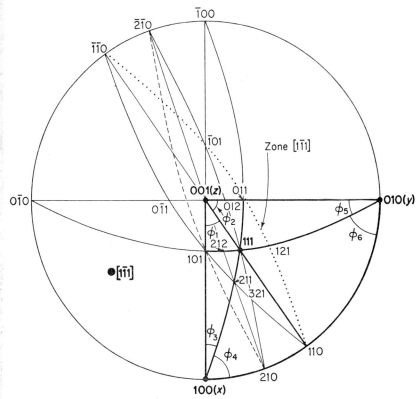

Fig. 5.13. Orthorhombic stereograms. Heavy zones show the location of the (111) pole within the spherical triangle (100)(010)(001). The indices (hkl) at the intersections of some other zones are shown. Any repetition by symmetry is neglected.

(d) Tetragonal System

(i) *Choice of axial system.* In the array of lattice points, a suitable square-based prism cell is identifiable, and the requirement of minimum volume is unambiguous; the essential tetrad axis perpendicular to the square base of the cell is the z-axis with its unit (c), whilst the other two symmetry related edges provide the x- and y-axes with identical units $(a = b)$. This axial system requires no arbitrary conventions; ambiguities can arise only if full lattice data are not available, as described in the examples of Fig. 5.9. In general, there are few problems in the description of tetragonal crystals; occasionally the choice of a different larger cell is advantageous for a particular purpose, but its relation to the true minimum volume cell is usually obvious.

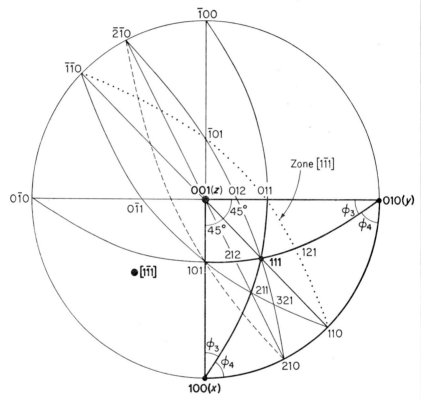

Fig. 5.14. Tetragonal stereograms. Heavy zones show the location of the (111) pole within the spherical triangle (100)(010)(001). The indices (*hkl*) at the intersections of some other zones are shown. Any repetition by symmetry is neglected.

(*ii*) *The orientation of the parametral planes.* The axial constant (an axial ratio usually quoted as *c/a*) determines the position of the (111) planes, which can be found by the usual zonal relations. The more regular shape of the cell ensures that

$$\frac{a}{b} = 1 = \tan \phi_1 = \tan (100)\widehat{}(110);$$

$$\frac{c}{b} = \frac{c}{a} = \tan \phi_3 = \tan \phi_5 = \tan (001)\widehat{}(101)$$

$$= \tan (001)\widehat{}(011)$$

so that $\phi_1 = \phi_2 = 45°; \quad \phi_3 = \phi_5; \quad \phi_4 = \phi_6.$

(*iii*) *Stereographic projection.* Conventional orientation is the same as for the orthorhombic system, with the *z*-axis at the centre and the *x*- and *y*-axes on the primitive; (001), (100) and (010) are at the poles of the respective axes. From the zonal relations above (110) must be at 45° to the *x*- and *y*-axes. The zone (110)(001) is the vertical great circle at 45° to the horizontal and vertical diameters; the position of (111) is found from the usual zones containing (101) and (011) which are equally inclined to the centre of the projection (Fig. 5.14). It will also be noticed that the positions of poles (*hk*0) on the primitive are independent of the axial ratio and must be the same for all tetragonal crystals. All other poles (*hkl*) depend on the axial ratio and can be inserted by zonal relationships, though (*hkl*) and (*hkl̄*) are coincident.

Due to the identity of units on the *x*- and *y*-axes, all zone axes [*UV*0], including [100] and [010], coincide with the poles of planes of corresponding indices, as do [001] and (001); more general zone axes (e.g. [1ī1]) are independently located.

(e) Cubic System

(*i*) *Choice of axial system.* The essential symmetry of four equally inclined triad axes requires a distribution of lattice points in which the cell must be a cube with the triads as the body diagonals; the edges of this cube are the *x*, *y* and *z* axes with identical units on all three axes. The high symmetry allows no ambiguity in this choice of a minimum volume cell.

(*ii*) *The orientation of the parametral planes.* These must be such that their normal is equally inclined to the three crystallographic axes at an angle of 54° 44′ ($= \cos^{-1} 1/\sqrt{3}$, for the sum of the squares of its direction cosines must be unity). The equations governing zonal relationships reduce to

$$\frac{a}{b} = \frac{c}{b} = \frac{c}{a} = 1 = \tan \phi_1 = \tan (100)\widehat{}(110), \text{ etc.}$$

so that $\qquad\qquad \phi_1 = \phi_2 = \phi_3 = \phi_4 = \phi_5 = \phi_6 = 45°.$

(*iii*) *The stereographic projection.* With the *x*, *y* and *z* axes projected as for the orthorhombic and tetragonal systems, (100), (010) and (001) are coincident with these poles. The position of (111) must be the same for all cubic crystals; the zonal relations above show that the poles (110), (011) and (101) are at 45° to the respective pairs of axes, so that (111) is at the intersection of three great circles inclined at 45° to the axial planes (Fig. 5.15). With the fixed position of (111) all other poles (*hkl*) are in identical positions for all cubic crystals.

Moreover the identity of units on all three axes requires all zone axes [*UVW*] to have poles coincident with those for corresponding plane indices.

(f) Trigonal and Hexagonal Systems

In the discussion of Chapter 4.2(g) point groups of these two systems are associated with rhombohedral and hexagonal cell shapes; trigonal crystals can have lattices described in terms of either cell, but hexagonal crystals can only have minimum volume cells of hexagonal form. This poses a problem in the indexing of trigonal crystals, for some should be described on the basis of

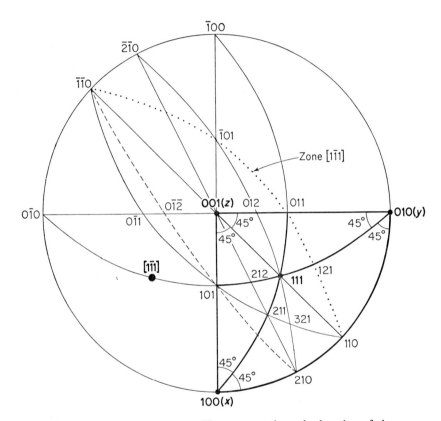

Fig. 5.15. Cubic stereograms. Heavy zones show the location of the (111) pole within the spherical triangle (100)(010)(001). The indices (hkl) at the intersections of some other zones are shown. Any repetition by symmetry is neglected.

rhombohedral axial geometry and others need a hexagonal axial system. The solution most commonly adopted when the material shows a rhombo-hedral distribution of lattice points is to choose to describe it by means of a larger cell of hexagonal shape; the relation of this cell to the rhombohedral lattice and its minimum volume cell is explained in Chapter 7.1(g). In practice, therefore, both trigonal and hexagonal crystals are usually indexed in the same way, though for completeness we shall describe first rhombohedral indexing which is occasionally used for trigonal crystals.

Rhombohedral cell

(i) *Choice of axial system.* A minimum volume cell with a body-diagonal along the essential triad symmetry can be chosen from a truly rhombohedral array of lattice points; the lattice translations of the cell edges are equally inclined to the triad axis, and they define the x-, y- and z-axes mutually inclined at angle α ($< 120° \neq 90°$) with identical units on each of the axes. Rhombohedral indexing on this axial system gives sets of three integers often known as *Miller indices*.

(ii) *The orientation of the parametral planes.* (111) planes must be equally inclined to the directions of the three axes, i.e. their normal is along the direction of the essential triad axis, and is defined by the axial constant, the interaxial angle α. From the geometry of Fig. 5.16(a) we see that this angle of inclination is given by

$$\rho = \sin^{-1}\left(\frac{2}{\sqrt{3}} \sin \frac{\alpha}{2}\right).$$

With this arrangement of axes the zonal equations become

$$\frac{a}{b} = \frac{c}{b} = \frac{c}{a} = 1 = \frac{\sin \phi_1}{\sin \phi_2} = \frac{\sin \phi_3}{\sin \phi_4} = \frac{\sin \phi_5}{\sin \phi_6}$$

so that $\qquad \phi_1 = \phi_2 = \phi_3 = \phi_4 = \phi_5 = \phi_6 = 90° - \dfrac{\alpha}{2}.$

(iii) *The stereographic projection.* Conventionally, the triad axis and pole of the (111) planes is placed at the centre of the projection, so the problem is turned into the location of the poles of the axes. These must lie on a small circle of radius ρ about the centre; the x-axis is taken at the lower intersection of this circle with the vertical diameter, and the y- and z-axes at similar intersections with diameters at 120° intervals (Fig. 5.16(b)). (100), (010) and (001) will be at the poles of great circles defined by the appropriate pairs of axes, i.e. on the same diameters as the poles of x, y and z but not coincident with them. All other poles (pqr) can then be inserted by zonal relationships.

With the exception of [111], zone axes [UVW] are at different positions from the poles of planes with corresponding indices, and are found in the usual way.

Hexagonal cell

(i) *Choice of axial system.* A minimum volume cell in the shape of a rhombus-based right prism can be chosen for the lattice point array for all hexagonal crystals and some trigonal crystals; as mentioned above, a cell of this kind

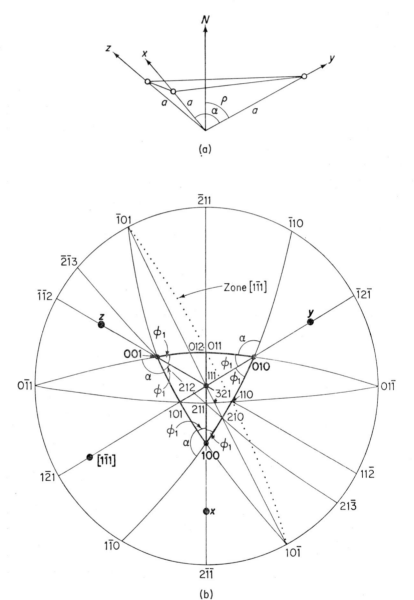

Fig. 5.16. Trigonal stereograms with rhombohedral axes. (a) The intersection of a parametral (111) plane with the axes. (b) The corresponding stereogram. Heavy zones show the location of the poles (100), (010) and (001). The indices (*pqr*) at the intersections of some other zones are shown. Any repetition by symmetry is neglected.

can also be selected for all other trigonal crystals, though it will not be of smallest volume. The natural axial system suggested by this cell shape has the z-axis along the triad or hexad direction with units (c) and the x- and y-axes in the direction of other cell edges inclined at 120° with identical units ($a = b$).

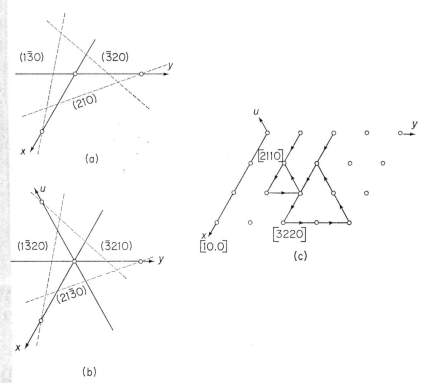

Fig. 5.17. Three- and four-axis systems for trigonal and hexagonal crystals. (a) A three-axis system for planes. The diagram shows the plane perpendicular to the z-axis and the traces of three symmetry related planes parallel to the z-axis. (b) A four-axis system for planes. The diagram is similar to that in (a). (c) An array of lattice points in a plane normal to the triad or hexad axis. Zone symbols on the three- and four-axis systems for the x-axis direction are shown.

Such an axial system can be—and is—commonly used in X-ray work, but it is not entirely satisfactory in other contexts, particularly for morphological studies; for example, where it is desirable that symmetry related planes should have related indices, this does not always happen (Fig. 5.17(a)). This kind of difficulty can be overcome by using an additional axis (called the u-axis) in the plane perpendicular to z so that it externally bisects x and y; it is identical to them and must have similar units. Hexagonal indexing on this axial system gives sets of four indices ($hkil$), often known as *Miller–Bravais indices*; in such indices the integers h, k and i cannot be independent, and in all cases

$$h + k + i = 0$$

though this does allow symmetry-related planes to have related indices (Fig. 5.17(b)). It is most common to use the Miller–Bravais system for the indices of planes; a simple reversion to the three-axis system can be made by omitting the redundant integer $(hk \cdot l)$; the dot is used to emphasise which index is missing, for it might be advantageous to drop either h, k or i on some occasions.

It is appropriate at this point to clarify the use of zone axis symbols, sometimes a source of confusion with hexagonal axes. With a three-axis system the nomenclature is straightforward and strictly comparable to that of all other systems; thus the direction of the x-axis (Fig. 5.17(c)) is described as [10 . 0] Using a four-axis system, there is an ambiguity in the choice of integers to define the direction of a lattice row in the xyu plane; in the figure the x-axis direction can be written [1000], [2110], [3220] and so on. To rationalise the infinite variations we choose the symbol $[U'V'J'W']$, which has no common factors and for which

$$U' + V' + J' = 0,$$

an analogous condition to that for plane indices; thus the direction of the x-axis is $[2\bar{1}\bar{1}0]$. This example shows that the conversion from four-axis zone symbols to their equivalent three-axis symbols, unlike that for plane indices, is not a simple matter of omitting a redundant index. The change from a three-axis symbol $[UV . W]$ (i.e. $[UV0W]$) to its four-axis equivalent $[U'V'J'W']$* can be made by subtracting $\frac{1}{3}(U + V)$ from U, V and 0 to satisfy the condition $U' + V' + J' = 0$; hence $U' = \frac{1}{3}(2U - V)$, $V' = \frac{1}{3}(2V - U)$, $J' = -\frac{1}{3}(U + V)$ and $W' = W$. For [10 . 0], we get $U' = \frac{1}{3}(2)$, $V' = \frac{1}{3}(-1)$, $J' = -\frac{1}{3}(1)$, $W' = 0$, to give $[2\bar{1}\bar{1}0]$ after fractions are cleared. Conversely, the conversion of four-axis symbol $[U'V'J'W']$ to its three-axis equivalent $[UV . W]$ involves the relations $U = U' - J'(= 2U' + V')$, $V = V' - J'$ $(= U' + 2V')$ and $W = W'$; so that for $[2\bar{1}\bar{1}0]$ we get $U = 2 - (-1)$, $V = -1 - (-1)$, $W = 0$ to give [10 . 0]. In considering the relations between plane indices in a zone, the Weiss zone law applies to all three-axis systems so that

$$Uh + Vk + Wl = 0$$

remains the condition for $(hkil)$ to lie in the zone $[UV . W]$. In four-axis systems it must be written in the form

$$U'h + V'k + J'i + W'l = 0.$$

---

* Some writers use general symbols $[UV . W]$ and $[uvtw]$, or $[uv . w]$ and $[UVJW]$ to denote corresponding directions on three- and four-axis systems.

The case for three- or four-axis zone symbols can be argued, and both methods have advantages and disadvantages. The four-axis system gives an identity of direction to the zone symbol $[U'V'J'0]$ and the normal to planes of corresponding indices; thus $(2\bar{1}\bar{1}0)$ and $[2\bar{1}\bar{1}0]$ are both in the direction of the x-axis, which has quite different plane and zone indices on a three-axis system.

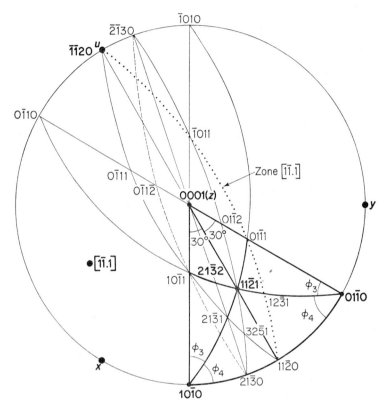

Fig. 5.18. Trigonal and hexagonal stereograms with hexagonal four-axis system. Heavy zone shows the location of the $(11\bar{2}1)$ pole in the spherical triangle $(10\bar{1}0)(01\bar{1}0)(0001)$. The indices $(hkil)$ at the intersections of some other zones are shown. Any repetition by symmetry is neglected.

But in more general directions this equivalence does not hold and the symbols are more cumbersome to interpret, e.g. $[2\bar{1}\bar{1}3]$ requires more thought than $[10 . 1]$. The four-axis system finds favour in work where directions $[U'V'J'0]$ are important, particularly in the metallurgical field; but we shall use the more popular alternative form $[UV . W]$, with plane indices written as $(hkil)$; contraction to $(hk . l)$ is usually made in diffraction studies.

(ii) *The orientation of the parametral planes.* With the Miller–Bravais axial system the parametral planes with indices $(11\bar{2}1)$ are in a direction determined by the axial constant (the axial ratio usually quoted as c/a); their normal, equally inclined to the x- and y-axes, must lie in the plane of the $z\bar{u}$ axes at an inclination (ρ) to the triad or hexad axis given by

$$\rho = \tan^{-1}\left(\frac{2c}{a}\right).$$

With these axes, the zonal relations become

$$\frac{a}{b} = 1 = \frac{\sin\phi_1}{\sin\phi_2}; \quad \frac{c}{b} = \frac{c}{a} = \tan\phi_3 = \tan\phi_5$$

so that $\phi_1 = \phi_2 = 30°; \quad \phi_3 = \phi_5$ and $\phi_4 = \phi_6.$

Alternatively, these zones can be found from the positions of poles such as $(10\bar{1}1)$; by examining a projection down the y-axis $\tan(000\widehat{1})(10\bar{1}1) =$ $\dfrac{c}{a\cos 30°}.$

(*iii*) *The stereographic projection.* Conventionally, the z-axis (with the essential triad or hexad) is at the centre of the projection, with the y-axis on the primitive at the end of the right-hand horizontal diameter; this also places the x- and u-axes on the primitive in the left-hand lower and upper quadrants respectively (Fig. 5.18). The points $(10\bar{1}0)$, $(01\bar{1}0)$, etc. are the poles of great circles containing the pairs of axes to which they are parallel, i.e. they are in positions on the primitive bisecting the axial directions; (0001) is at the centre of the projection. Once the position of $(11\bar{2}1)$ is found, all other poles $(hkil)$ can be inserted by zonal relationships and the usual cross-adding procedure; $(hki\bar{l})$ is immediately below $(hkil)$.

In the discussion above we saw that (with the exception of $[00\,.\,1]$ and (0001)) zone axes $[UV\,.\,W]$ do not lie in the directions of the normals to planes with corresponding indices; $[1\bar{1}\,.\,1]$ in the figure is the pole of the zone containing $(11\bar{2}0)$ and $(01\bar{1}1)$.

5.5. Exercises and problems

[*Note:* Many of these problems can be solved by applying the methods described in Appendix B, though it is also possible to use accurate constructions.]

1. (i) Draw out a number of unit cells of a planar lattice for which $a = 2$ cm, $b = 3$ cm and $\gamma = 110°$.
   (ii) Insert the directions [11] and [1$\bar{2}$], and measure the angle between them.
   (iii) Draw in several of the lines in the sets (11) and ($\bar{2}$3) and measure the spacings of each set.

(iv) Determine the angle $(11\widehat{)}[11]$.

(v) Choose a new cell so that a' is $[1\bar{2}]$ and b' is $[11]$; what are the values of a', b' and γ'?

(vi) Determine the indices of the two sets of lines of (iii) in terms of this new choice of cell.

2. (i) Find the zone symbols formed from the plane indices (a) (010) and $(3\bar{2}5)$, (b) (103) and (112).

(ii) Planes with indices (130), $(01\bar{1})$, $(\bar{3}35)$, (251) also lie in one or other of these two zones; decide which planes belong to which zones.

(iii) From the zone symbols of (i) find the indices of planes which are common to both zones; check your answer by cross-adding the plane indices of (a) and (b).

3. In a transformation from old to new axes, the following changes in plane indices were made:

$$(210)_o \rightarrow (212)_n$$
$$(101)_o \rightarrow (102)_n$$
$$(012)_o \rightarrow (012)_n$$

What do $(111)_o$ and $[111]_o$ become when indexed on the new axes?

4. For a material crystallising in class $\bar{1}$, $a:b:c = 0\cdot917:1:0\cdot720$, $\alpha = 90° \ 06'$, $\beta = 102° \ 02'$, $\gamma = 105° \ 45'$.

(i) On an accurate stereogram locate the poles (100), (010) and (001), and determine the angles between them.

(ii) Locate (111), and from this plot the positions of (110), (101) and (011). What are the angles $(100\widehat{)(}110)$ and $(100\widehat{)(}1\bar{1}0)$?

(iii) Determine the angle $(110\widehat{)}[110]$.

5. In a crystalline substance of class $2/m$, $(110\widehat{)(}1\bar{1}0) = 61° \ 13'$, $(110\widehat{)(}001) = 67° \ 47'$ and $(010\widehat{)(}\bar{1}11) = 63° \ 08'$. Determine the axial constants for the cell implied by these measurements.

6. For a crystal belonging to class mmm with axial ratios $a:b:c = 0\cdot813:1:1\cdot903$ a set of planes (hkl) is orientated so that $(001\widehat{)(}hkl) = 55° \ 30'$ and $(010\widehat{)(}hkl) = 77° \ 33'$. Determine the indices (hkl) and the value of the angle $[\bar{1}30][0\bar{5}1]$ for this crystal.

7. For a compound with point group symmetry $4/m$ and an axial ratio $c/a = 1\cdot575$, a set of planes (hkl) lie in the same zone as (201) and $(2\bar{1}0)$ at an inclination of $15° \ 15'$ to $(2\bar{1}0)$. What are the indices of (hkl)? Later examination of the same material showed that planes formerly indexed as (201) should have been chosen as the parametral planes in order to give a more conventional choice of cell. Re-index the other two sets of planes and determine the new axial ratio.

8. In a crystal of class 432 the angle between two sets of planes $(hlh)\widehat{(lhh)} =$ 37° 51′, where $h > l$. Determine the angle $(hh\bar{l})\widehat{(hh\bar{l})}$, and the values of the indices h and l.

9. In all trigonal crystals the planes $(10\bar{1}1)$, $(\bar{1}101)$, $(0\bar{1}11)$ must be related by symmetry. If for a particular crystalline substance with trigonal symmetry $(10\bar{1}1)\widehat{(\bar{1}101)} = 87° 23′$, determine the angles between the zone axes in which these three sets of planes intersect. What are the zone symbols for these directions (i) with a three-axis hexagonal system, (ii) with a four-axis hexagonal system?

SELECTED BIBLIOGRAPHY

General
BUERGER, M. J. 1956. *Elementary crystallography*. Wiley.
INTERNATIONAL UNION OF CRYSTALLOGRAPHY. 1965. *International tables for X-ray crystallography*, vol. I. Kynoch Press, Birmingham.
PHILLIPS, F. C. 1963. *An introduction to crystallography*. Longmans.

Analytical approach
HILTON, H. 1963. *Mathematical crystallography*. Dover Publications.
INTERNATIONAL UNION OF CRYSTALLOGRAPHY. 1966. *International tables for X-ray crystallography*, vol. II. Kynoch Press, Birmingham.

Axial transformations
AITKEN, A. C. 1939. *Determinants and matrices*. Oliver and Boyd.
BUERGER, M. J. 1942. *X-ray crystallography*. Wiley.

Source book for crystallographic data
DONNAY, J. D. H., DONNAY, G., COX, E. G., KENNARD, O. and KING, M. V. 1963. *Crystal data* (*determinative tables*). American Crystallographic Association.

6

MORPHOLOGICAL CRYSTALLOGRAPHY

6.1. Introduction

In this chapter we are concerned with the shapes of natural crystals; until the discovery of X-ray diffraction in 1912 these external regularities had provided the main experimental data for the development of scientific crystallography. Undoubtedly man has shown an interest in the crystalline state from prehistoric times, if only for the rare beauty and economic value of the many natural forms; but the real foundations of the present science were laid only in the late eighteenth century by Abbé Haüy. On his work later generations built up a mixture of experimental observation and mathematical theory into a kind of 'classical crystallography', self-coherent but restricted by an inability to examine any internal structure responsible for external regularities. In the past fifty or sixty years X-ray diffraction and other techniques have been used to discover internal atomic arrangements, and inevitably the importance of crystalline shape to the modern crystallographer has decreased as the spectrum of specimens that he can examine has broadened. Nevertheless, the classical experimental data, the arrangement and regularity of crystalline faces, whose study is often known as morphological crystallography, still have a place in any modern work.

At the outset it must be emphasised that crystal morphology is concerned only with those shapes formed by natural processes; artificial faces, such as those cut upon gem stones to display their qualities are of no fundamental scientific interest. Our passing familiarity with a crystal as a homogeneous polyhedral solid bounded by naturally developed plane faces is usually gained either from examples seen in museums and in natural rock formations, or from elementary growth experiments from saturated solutions in the laboratory; in the polyhedron there is a wealth of information to be gleaned. Well-developed crystals occur in different sizes and shapes. Common sizes vary from crystals which can be comfortably handled (of dimensions in centimetres) to those in which faces can only be resolved using a magnifying glass or a powerful microscope; in natural geological processes, larger crystals rarely have the opportunity to develop, although artificially controlled growth conditions can give very large synthetic crystals. Whatever the size, there are two important and distinguishable factors in the crystal shape; (i) the particular polyhedral faces which are developed and the symmetry which is displayed in their arrangement, and (ii) the relative importance of the different faces of the particular specimen. Early crystallographers soon realised that under ideal growth conditions (i) is of more fundamental significance than (ii); whilst similar faces are arranged in the same symmetrical way, crystals of a particular substance may appear quite dissimilar due to the accentuation of different faces. Fig. 6.1 illustrates two crystals of the same substance; in (a) the rectangular faces are well developed and dominant, whereas in (b) it is the

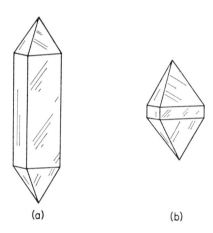

(a)   (b)

Fig. 6.1. Two crystals with the same symmetry and interfacial angles.

triangular faces which determine the general appearance. Despite this difference due to the second factor (ii), both crystals show the same polyhedral faces and the same symmetry; this is demonstrated by the identity of the measured angles between pairs of corresponding faces on the two specimens. Indeed by very extensive measurements of interfacial angles, Haüy proposed a *law of constancy of angle*, which stated that in all crystals of the same substance the angles between corresponding faces have constant values. This law is central to all developments in classical crystallography, though its real origins do not become clear until we have explored the relations between the faces developed by a particular crystal and its atomic structure. We shall discuss this in Chapter 6.3(a); some other factors important in the growth shapes of real crystals are considered in Appendix C. But first, before turning to the principles of morphological analysis and their application to real crystals, we shall describe the experimental measurement of interfacial angles.

6.2. Methods of measurement

Any morphological study is based on the accurate measurement of interfacial angles, followed by their representation on a projection; each pole on the stereogram corresponds to the orientation of the normal to a face of the crystal (as in Chapter 3.3), and the distribution of the poles will indicate any symmetry in the arrangement of the faces.

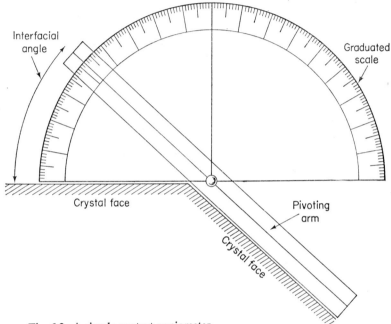

Fig. 6.2. A simple contact goniometer.

Methods of measuring interfacial angles depend on the nature of the specimen and the desirable accuracy. For very large crystals, or when the highest accuracy is not important, a simple device known as a *contact goniometer* can be used. This has a graduated semi-circle (like a protractor) with a pivoting flat arm (Fig. 6.2). The flat base of the semi-circular scale is placed on one face and the flat arm pivoted to lie along the other face; the angle between the normals to the faces is read off from the graduated scale. Contact goniometers made of perspex are suitable for elementary work, but for any accurate measurements the instruments have a metal scale and arm with vernier attachments. However, opportunities to use these goniometers in accurate work is limited by the rarity of crystals of suitable size and quality, and in the nineteenth century optical goniometers were developed in which measurements are made by reflecting light from the crystal faces.

During the last century, with the flowering of classical crystallography, a wide variety of optical goniometers were designed and constructed. It is not worthwhile here to recount their construction in detail; many of them are still

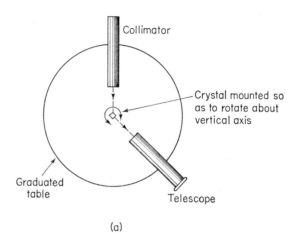

(a)

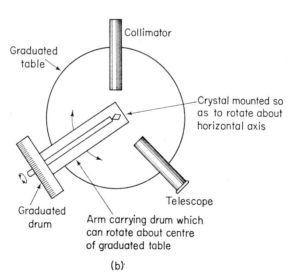

(b)

Fig. 6.3. Optical goniometers (schematic). (a) Single circle instrument. (b) Two circle instrument.

preserved and in use, and more modern goniometers still employ the same general principles as are found in the original designs. The commonest instrument is the *single-circle goniometer* shown schematically in Fig. 6.3(a). The crystal is mounted on an axis through the centre of a graduated circular scale;

as it is rotated about this axis its setting is read on the scale. A parallel light signal from the fixed collimator is reflected from the first face, and the crystal setting is adjusted so that this image is observed on the crosswires of a telescope clamped in a fixed position; the angular setting of the crystal is noted. The crystal is then rotated about the central axis until the reflection of the signal from a second face is on the telescope crosswires; the new setting is recorded, when the angular difference between the two positions of the crystal is the interfacial angle measured between the normals to the two planes. To ensure that the two reflected images are in the same plane as the collimator signal and telescope crosswires, the crystal must be mounted so that the edge or zone axis formed by the intersection of the planes of the two faces (extended if they are not adjacent) is parallel to the rotation axis; whilst this is roughly achieved by eye, the final orientation is performed with a pair of perpendicular arcs with tilts of up to 20–30° mounted on the spindle carrying the specimen. The symmetrical arrangement of faces into zones (with parallel mutual intersections) allows all the interfacial angles within one zone to be measured once the crystal has been adjusted. A preliminary visual examination identifies such zones, and an initial crystal mounting usually sets the most prominent zone axis along the rotation axis; measurements continue with re-mounting for each new zone, until every visible face is included in at least one zone. More sophisticated optical goniometers are designed to eliminate the re-mounting of the crystal necessary after each set of zonal measurements. In the *two-circle goniometer*, shown diagrammatically in Fig. 6.3(b), a second graduated drum is carried vertically so that it can rotate about an axis which intersects that of the horizontal circle. When the crystal is set with its most prominent zone axis parallel to the axis of the vertical circle, rotations on both circles make it possible to obtain reflections of the signal on the telescope crosswires from nearly all crystal faces; the recorded settings on the two circles allow the angular positions of the faces to be plotted.

In practice, all optical goniometry requires crystals of a suitable size whose well-developed faces are of good quality. When faces have irregularities or are imperfectly flat, the reflected signal is distorted (or even multiple) and the accuracy of measurement is reduced. Often, the best quality crystals are the smallest, but a lower limit of size (usually a few millimetres) is set by the necessity of handling the specimen and adjusting it on the goniometer. After the most suitable specimen of a batch has been selected a preliminary visual examination provides a sketch of its appearance on which faces are labelled for identification and probable zonal configurations established. In the subsequent measurement of interfacial angles on an optical goniometer, the accuracy depends on the experimental conditions, not least the sharpness of the reflected images; with crystals of good quality and modest equipment an accuracy of $\pm 1'$ is quite usual. After measurement, a stereogram showing the angular distribution of the faces is plotted, and on this projection the first task is to recognise any symmetry elements (whose presence may have been suspected in the preliminary visual examination) and, if possible, decide the crystal class; rotation of the stereogram into a conventional orientation may then be required. The detailed examination of the crystal morphology involving the selection of crystallographic reference axes and the analysis of the polyhedral shape is then completed; we shall consider the nature of this analysis in the next section, and a detailed example of a morphological study is set out in Chapter 6.5.

6.3. The morphological analysis of crystal shapes

(a) General Considerations

So far, we have referred to crystal shapes as kinds of geometrical polyhedra which often have a regularity and symmetry in the arrangement of their faces. The particular polyhedron that crystallises for a substance under given circumstances depends on all the factors that can influence growth; under real conditions these can change and distort the crystal faces in a complex way. But crystal shapes developed under ideal uniform growth conditions reflect the underlying atomic pattern, and the analysis of these ideal shapes is an essential part of the background to any morphological studies; in this section we shall describe only how the possible regular polyhedra may be discovered, and their relationship to actual crystals will be considered in the subsequent section.

A connection between crystal faces and the structural pattern was suspected from the earliest days of the subject; Haüy showed how various shapes could be built up by the stacking of fundamental units whose form was characteristic of the symmetry. Later these units were identified with the various lattice cells, though only in this century has it been possible to discern the atomic pattern within these cells. As these ideas evolved, it was natural to associate the faces of crystals with the families of lattice planes; since each face is parallel to a particular stack of lattice planes, there must be a constancy of angles between corresponding faces, because all angles are fixed by the cell constants of the particular substance. But, as we saw in the last chapter, there are an infinite number of different sets of lattice planes (hkl) depending on the integers h, k and l; only very few are represented in the finite number of faces of a crystal, and these crystal faces usually have low values of h, k and l. In a general way we can see that this limitation arises in the assembly of the constituent atoms in the growth process, when the lattice planes with the fastest growing surface areas are those with the largest number of atoms per unit area; faces parallel to these planes will eventually dominate the external shape of the growing crystal. This may be pursued to relate the actual observed indices h, k and l of faces to the particular structure, an aspect of growth theory mentioned in Appendix C; whatever the validity of the detailed deductions for individual materials, this theory provides a reasonable explanation for the low values of h, k and l generally associated with crystal faces, for such lattice planes must have the highest density of atomic material in any structure.

Individual faces of a crystal may therefore be identified by the indices (hkl) of the planes to which they are parallel; as we saw in the last chapter, conventional indexing relates to an axial system determined by any symmetry elements present, and this ensures that faces reproduced by symmetry have related indices. Our crystal polyhedron must display the symmetry of one of the thirty-two crystal classes; its allocation to one of these classes (or at least a system) enables indices to be assigned to the various faces, but we must also disentangle which faces are symmetry related to one another for this is important in analysing the nature of the particular polyhedron. It is best to do this as an exercise in which all polyhedral forms consistent with all point group symmetries are worked out; any ideal crystal belonging to a particular class must have a shape built up of one or more of these forms. As usual the stereogram is invaluable in this exercise, and we have already shown the projections of symmetry related general poles in Chapter 4.2; in each class the distributions represent a possible geometric polyhedron displaying the appropriate symmetry elements. In morphological terms the set of symmetry related planes constituting each polyhedron is known as a *(crystal) form*; symbolically a form is written by placing the integers of one of its faces in braces, so that $\{hkl\}$ represents the assemblage of faces reproduced by the operation of the point group symmetry on the face (hkl).* The shapes defined by these earlier stereograms are known as the *general forms* of each class, but in nearly every case there are other different shapes that are consistent with the same point group symmetry. These *special forms* are the assemblage of faces reproduced by the point group symmetry acting on a face in a position preferentially related to the symmetry elements; there may be several special forms in a class, each with a different shape, and all different from the general form for that class. Again we can use the stereogram to discover possible special forms as illustrated in Fig. 6.4 for class $4/m$. The upper stereogram shows the distribution of poles for the general form $\{hkl\}$; beside it is the polyhedral shape represented by these poles, known as a tetragonal bipyramid, the general form of this class. The representative shown is, of course, only one of the family of tetragonal bipyramids all with the same general appearance but having different interfacial angles; the particular member of the family is specified by the integers h, k and l. The middle stereogram shows the distribution obtained when an initial pole is placed on the symmetry plane around the primitive, i.e. it is the special form developed by the

---

* The analogous grouping of zone axes obtained by the operation of the point group symmetry on a direction $[UVW]$ is usually expressed symbolically $\langle UVW \rangle$, though some objections have been raised to this and an alternative version is $[[UVW]]$.

operation of the point group symmetry on a face $(hk0)$. The faces of the form $\{hk0\}$ for this class are shown on the right as a tetragonal prism. This has four perpendicular faces, and is not an enclosed polyhedron; such distributions are called *open forms* and cannot exist independently but must occur in combination with other forms. Although the faces of any $\{hk0\}$ always have the

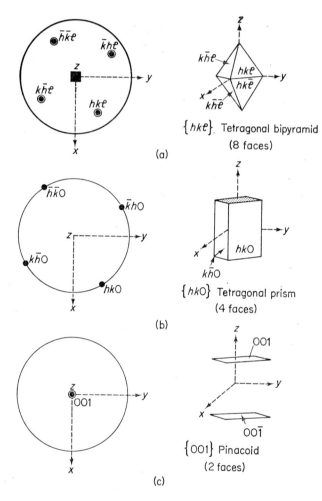

Fig. 6.4. Forms in the class $4/m$. (a) General form $\{hkl\}$. (b) Special form $\{hk0\}$. (c) Special form $\{001\}$.

same interfacial angles $(90°)$, particular values of the integers h and k are needed to define its orientation with respect to those other forms with which it is combined. The only other special form in this class is shown in the lowest stereogram of the figure; in this the initial pole is placed on the tetrad axis.

The form {001} is a pair of parallel planes, usually called a pinacoid, normal to the tetrad axis; again it is an open form, but it is unique in that the integers are specified. Ideal polyhedral crystals of any substance with symmetry $4/m$ must have shapes which are either any closed form (a tetragonal bipyramid) alone or combinations of forms, open and closed; the crystal of Fig. 6.1 combines a tetragonal bipyramid and a prism and could be an example of this class (though see the discussion in Chapter 6.4).

To summarise, the analysis of crystal shapes involves three stages: (i) the determination of crystal class, (ii) the recognition of the different forms, special and general, which may be present, and (iii) the choice of a suitable axial system and the assignation of indices to the faces of the various forms. We have dealt with the principles of (i) and (iii) in earlier chapters, and practical problems that can arise are discussed shortly; some account of (ii) has been given above, and an exhaustive approach to morphological analysis would now require the treatment exemplified by the discussion of $4/m$ to be repeated in the other thirty-one classes. This is tedious and less rewarding than in the past owing to the diminished importance of morphological studies in modern crystallography. Excellent detailed accounts are given in other books mentioned in the bibliography, and in the rest of this section we shall only derive the special and general forms for the holosymmetric classes of each system by way of illustration.

(b) Triclinic holosymmetric class ($\bar{1}$)

There are no symmetry elements other than the centre, and so there is no real distinction between special and general forms; all forms {hkl} are pinacoids, and crystals must show a combination of these sets of parallel planes (Fig. 6.5). (In terms of a chosen axial system, the pinacoids are sometimes given descriptive adjectives, e.g. {001} is often called the basal pinacoid; but this distinction is only of convenience and there is no real difference between the planes {001} and any other pinacoid {hkl}.)

(c) Monoclinic holosymmetric class ($2/m$)

Now a distinction between the special and general forms must be recognised; their stereograms and geometrical configurations are shown in Fig. 6.6. There are three types of form; the general form {hkl} is a prism, and the special forms {010} and {h0l} are both pinacoids. This is one of the commonest classes, and a large number of substances with this symmetry crystallise polyhedra which are a combination of these forms; an example is shown in the figure. (Again in terms of a chosen axial system, various descriptive adjectives are often employed (see Phillips, p. 76).)

(d) Orthorhombic holosymmetric class (mmm)

The closed general form {hkl}, a bipyramid, together with the special forms

$\{hk0\}$, $\{h0l\}$, $\{0kl\}$ as prisms and $\{100\}$, $\{010\}$, $\{001\}$ as pinacoids, are shown in Fig. 6.7. This class is almost as common as $2/m$, and most crystals show combinations of forms, as in the illustration. (Once again the descriptive terminology is often supplemented after an axial system has been chosen, even to the use of different names, e.g. 'domes' for the prisms $\{h0l\}$ and $\{0kl\}$ (see Phillips, p. 70).)

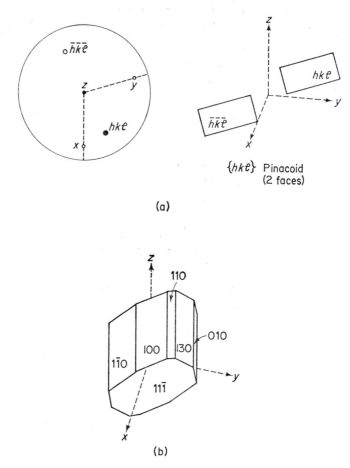

$\{hk\ell\}$ Pinacoid
(2 faces)

(a)

(b)

Fig. 6.5. Forms in the class $\bar{1}$. (a) General (and special) form $\{hkl\}$. (b) A crystal of $CuSO_4.5H_2O$ showing the forms $\{010\}$, $\{100\}$, $\{1\bar{1}0\}$, $\{110\}$, $\{130\}$, $\{11\bar{1}\}$.

(e) Trigonal holosymmetric class ($\bar{3}m$)

The special and general forms will be described by means of the Miller–Bravais four-axis indexing system (Chapter 5.4(g)). The general form $\{hkil\}$ is known as a ditrigonal scalenohedron; there are five distinctive special forms,

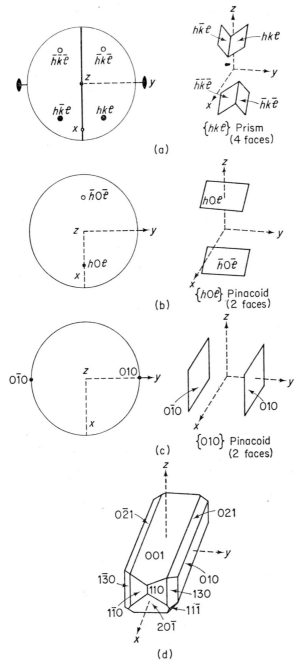

Fig. 6.6. Forms in the class $2/m$. (a) General form $\{hkl\}$. (b) Special form $\{h0l\}$. (c) Special form $\{010\}$. (d) A crystal of orthoclase ($KAlSi_3O_8$) showing general forms $\{110\}$, $\{130\}$, $\{021\}$, $\{\bar{1}11\}$ in combination with the special forms $\{010\}$, $\{20\bar{1}\}$, $\{001\}$.

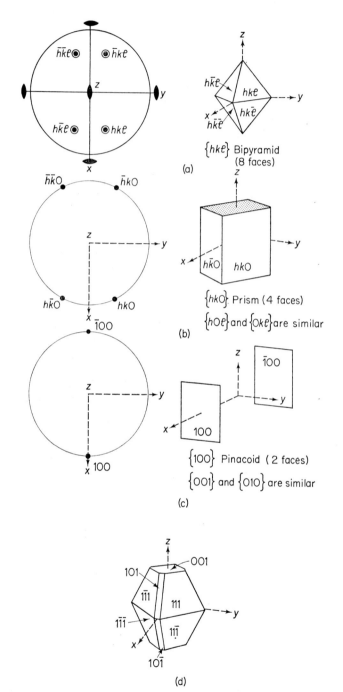

Fig. 6.7. Forms in the class *mmm*. (a) General form {*hkl*}. (b) Special form {*hk*0}, {*h*0*l*} and {0*kl*}. (c) Special forms {100}, {001} and {010}. (d) A crystal of PbSO$_4$ showing general form {111} in combination with special forms {101} and {001}.

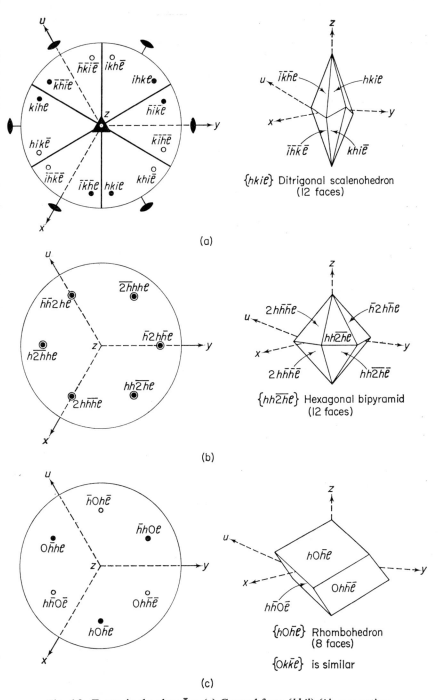

$\{hki\ell\}$ Ditrigonal scalenohedron
(12 faces)

(a)

$\{hh\overline{2}h\ell\}$ Hexagonal bipyramid
(12 faces)

(b)

$\{h0\overline{h}\ell\}$ Rhombohedron
(8 faces)

$\{0k\overline{k}\ell\}$ is similar

(c)

Fig. 6.8. Forms in the class $\overline{3}m$. (a) General form $\{hkil\}$ (i is a negative integer, $-(h + k)$). (b) Special form $\{hh\overline{2}hl\}$. (c) Special forms $\{h0\overline{h}l\}$ and $\{0k\overline{k}l\}$. (d) Special form $\{hki0\}$. (e) Special forms $\{11\overline{2}0\}$ and $\{10\overline{1}0\}$. (f) Special form $\{0001\}$. (g) A crystal of calcite, $CaCO_3$, showing the general form $\{21\overline{3}1\}$ in combination with the special forms $\{10\overline{1}1\}$ and $\{10\overline{1}0\}$.

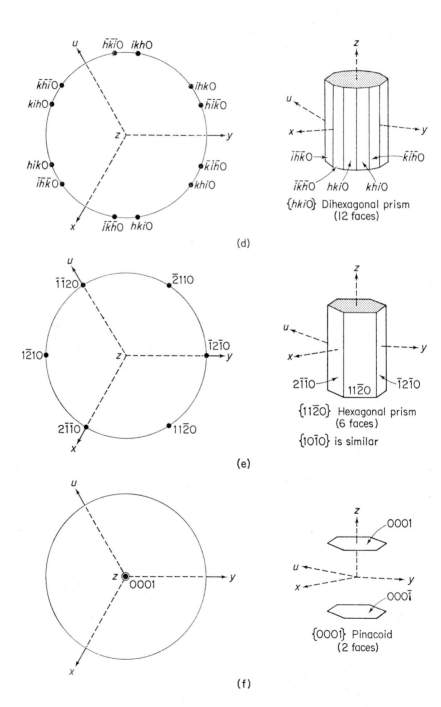

(d)

(e)

(f)

most of which emphasise geometrically the special relationships between the trigonal and hexagonal systems (Fig. 6.8); $\{hh2\bar{h}l\}$ is a hexagonal bipyramid, $\{hki0\}$ is a dihexagonal prism, $\{10\bar{1}0\}$, $\{11\bar{2}0\}$ are hexagonal prisms, $\{0001\}$ is a pinacoid, all shapes to be found in one or other of the hexagonal classes. The only specifically trigonal special forms are $\{h0\bar{h}l\}$, $\{0k\bar{k}l\}$, as rhombohedra, a shape to be found only in this and other trigonal classes. One of the commoner minerals, calcite, a form of $CaCO_3$, crystallises in this class with a wide variety of different forms; an example is shown in the figure.

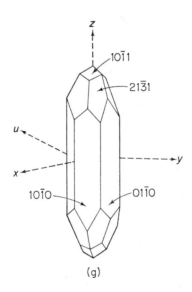

(g)

(f) Tetragonal holosymmetric class (4/mmm)

By comparison with *mmm*, the list of special forms must lengthen; for example, a distinction must be now made between the family $\{hk0\}$ and $\{110\}$, for the pole (110) is in a more symmetrical position than (hk0). The general form $\{hkl\}$, a ditetragonal bipyramid, and the various special forms, $\{h0l\}$, $\{hhl\}$ as tetragonal bipyramids, $\{hk0\}$, a ditetragonal prism, $\{100\}$, $\{110\}$ as tetragonal prisms and $\{001\}$ as a pinacoid are shown in Fig. 6.9; a zircon crystal with a combination of various forms is also illustrated.

(g) Hexagonal holosymmetric class (6/mmm)

The general form $\{hkil\}$ is a dihexagonal bipyramid; this is shown in Fig. 6.10, together with the special forms $\{h0\bar{h}l\}$, $\{hh\bar{2}hl\}$ as hexagonal bipyramids, $\{hki0\}$ as a dihexagonal prism, $\{10\bar{1}0\}$, $\{11\bar{2}0\}$ as hexagonal prisms, and $\{0001\}$ as a pinacoid; a combination of some of these forms found on a crystal of beryl, a semi-precious mineral, is also illustrated.

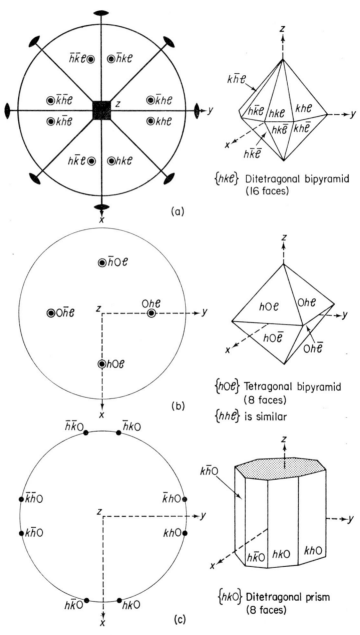

$\{hk\ell\}$ Ditetragonal bipyramid
(16 faces)

(a)

$\{hO\ell\}$ Tetragonal bipyramid
(8 faces)
$\{hh\ell\}$ is similar

(b)

$\{hkO\}$ Ditetragonal prism
(8 faces)

(c)

Fig. 6.9. Forms in the class $4/mmm$. (a) General form $\{hkl\}$. (b) Special forms $\{h0l\}$ and $\{hhl\}$. (c) Special form $\{hk0\}$. (d) Special forms $\{100\}$ and $\{110\}$. (e) Special form $\{001\}$. (f) A crystal of zircon $ZrSiO_4$ showing the general form $\{211\}$ in combination with the special forms $\{101\}$, $\{110\}$ and $\{100\}$.

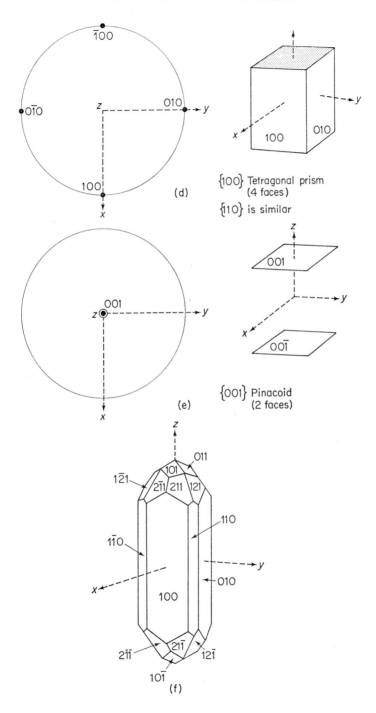

$\{100\}$ Tetragonal prism
(4 faces)

$\{110\}$ is similar

(d)

$\{001\}$ Pinacoid
(2 faces)

(e)

(f)

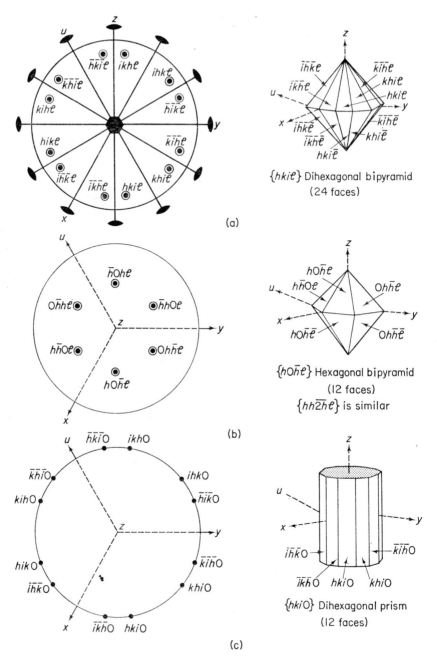

{$hki\ell$} Dihexagonal bipyramid
(24 faces)

{$h0\bar{h}\ell$} Hexagonal bipyramid
(12 faces)
{$hh\overline{2h}\ell$} is similar

{$hki0$} Dihexagonal prism
(12 faces)

(a)

(b)

(c)

Fig. 6.10. Forms in the class 6/*mmm*. (a) General form {*hkil*} (*i* is a negative integer, $-(h + k)$). (b) Special forms {$h0\bar{h}l$} and {$hh\overline{2h}l$}. (c) Special form {*hki*0}. (d) Special forms {$10\bar{1}0$} and {$11\bar{2}0$}. (e) Special form {0001}. (f) A crystal of beryl, $Be_3Al_2Si_6O_{18}$, showing the general form {$21\bar{3}1$} in combination with the special forms {$11\bar{2}1$}, {$10\bar{1}1$}, {$10\bar{1}0$} and {0001}.

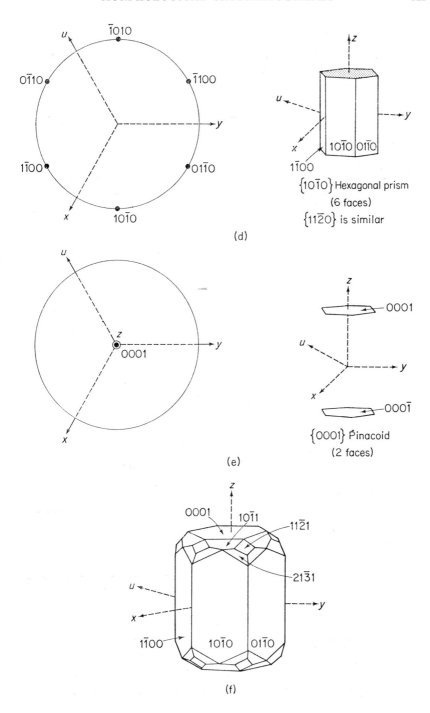

(d)

$\{10\bar{1}0\}$ Hexagonal prism
(6 faces)
$\{11\bar{2}0\}$ is similar

(e)

$\{0001\}$ Pinacoid
(2 faces)

(f)

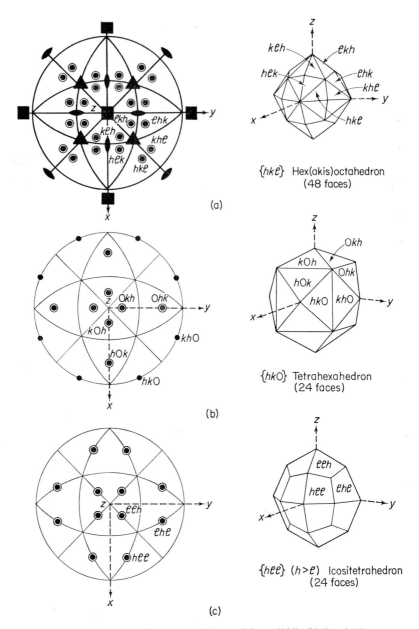

{hkℓ} Hex(akis)octahedron
(48 faces)

(a)

{hkO} Tetrahexahedron
(24 faces)

(b)

{hℓℓ} (h > ℓ) Icositetrahedron
(24 faces)

(c)

Fig. 6.11. Forms in class $m3m$. (a) General form {hkl}. (b) Special form {hk0}. (c) Special form {hll}($h > l$). (d) Special form {hhl}($h > l$). (e) Special form {111}. (f) Special form {110}. (g) Special form {100}. (h) A crystal of garnet, $X_3Y_2(SiO_4)_3$ with X = divalent metal, Y = trivalent metal, showing a combination of the special forms {211} and {110}.

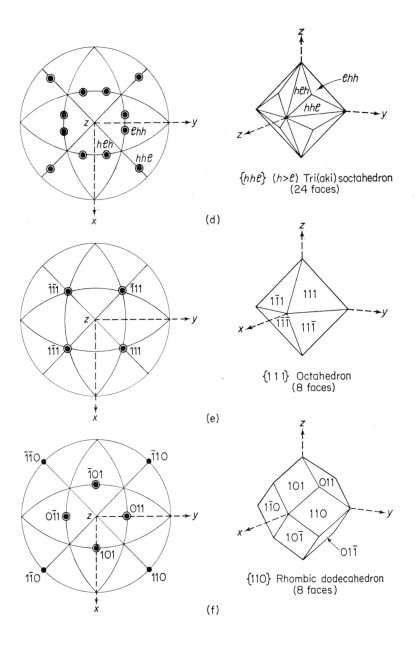

{hhℓ} (h>ℓ) Tri(aki)soctahedron
(24 faces)

(d)

{111} Octahedron
(8 faces)

(e)

{110} Rhombic dodecahedron
(8 faces)

(f)

(h) Cubic holosymmetric class (m3m)

This class has the largest number of symmetry elements, and as such has the greatest complexity in the varieties of form shapes; all special forms are closed, and can often be found alone for substances crystallising with this symmetry. The general form {*hkl*} is a hex(akis)octahedron (Fig. 6.11) and it is perhaps worthwhile to consider the generation of special forms shown in this figure a little more carefully than hitherto. Firstly, the initial pole to derive the special form may be placed in a general position on one of the symmetry planes; taking $h > l$, this leads to three different families, in which {*hk*0} is a tetra-hexahedron, {*hll*} is an icositetrahedron, and {*hhl*} is a tri(aki)soctahedron. Next we can explore initial locations in more symmetrical positions on symmetry axes; these lead to the unique special forms {110}, the rhombic dodecahedron for the diad, {111}, the octahedron for the triad, and {100}, the cube for the tetrad. Crystals belonging to this class are relatively common, particularly among metals and simple compounds such as metallic oxides and sulphides,

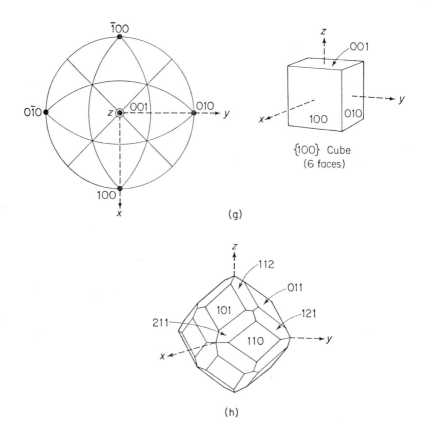

{100} Cube
(6 faces)

(g)

(h)

alkali halides, etc.; the illustration, however, shows a typical combination of forms found in crystals of garnet, another semi-precious mineral.

6.4. The external shapes of real crystals

In the earlier morphological analysis any ideal polyhedral crystal has a shape determined by the form or combination of forms that develop during growth; the description of the crystal is made in terms of its class and the indices of the forms present. Even under perfect growth conditions the physical appearance

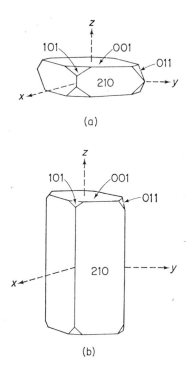

(a)

(b)

Fig. 6.12. Different habits in crystals of barytes, $BaSO_4$. Both crystals show the forms {210}, {101}, {011} and {001}; (a) is tabular due to the dominance of {001}, whilst (b) is prismatic due to the dominance of {210}.

can be profoundly changed by the relative developments of the forms in combination (as already shown in Fig. 6.1). As a further example, the two crystals of barytes, $BaSO_4$, in Fig. 6.12, are crystallographically identical (class *mmm*, $a:b:c = 1\cdot629:1:1\cdot312$, with the forms {210}{011}{101} and {001}) but are apparently quite different at first sight; in (a) {001} is the dominant form giving the flat appearance, whilst in (b) the crystal appears elongated due to the dominance of the prism {210} faces. Although these differences are not of any vital crystallographic significance, morphological description is incomplete until a statement about the relative importance of faces and forms is made; this usually refers to the *crystal habit*. Descriptions of habit are given in a variety of ways, sometimes, as above, by naming the dominant forms or sometimes by general descriptive terms; the two barytes crystals, for example, might be said to differ in that (a) has a tabular (or flat) habit, whilst the elongated crystals of (b) have a prismatic (or acicular) habit. Changes in the habit of crystals of the same material are usually due to different crystallisation conditions, e.g. concentration of solution, temperature gradients in the solution, impurities, etc. all of which can have a marked effect both upon the habit and the particular forms that crystallise. Moreover these factors can lead to distortion of the ideal regularity for they can cause preferential development of some but not all faces of a particular form; if during the growth of a cube they cause one pair of faces to develop faster than the others, the actual crystal will appear tabular rather than equi-dimensional. It should be emphasised that none of these changes which distort the habits of real crystals can affect the fundamental constancy of interfacial angles which will always be apparent after measurement and projection, but nevertheless they are confusing to the inexperienced eye.

Even when the complexities of habit shown by actual crystals are discounted, there is a much more important problem in any morphological study which relates to the determination of symmetry. This can be illustrated by the crystals of Fig. 6.1, which we said earlier might belong to class $4/m$ and be described as a combination of the general form $\{hkl\}$ with the special form $\{hk0\}$. Now these crystals have been drawn so that both the tetragonal bipyramid and prism have the integers h and k in common; in this case, the resulting crystals apparently have more symmetry than $4/m$ (it could be as high as $4/mmm$), and we could well be in doubt as to the true symmetry. This reflects a real problem in symmetry determination by morphological study alone, for some forms (particularly special forms) occur in several different classes; when only these are present, there must be some ambiguity in crystal class, and the crystals of Fig. 6.1 could belong to any one of the classes $4/m$, 42, $\bar{4}2m$ and $4/mmm$. Another obvious example is the cube which occurs as the special form $\{100\}$ in all cubic crystals, so that substances crystallising exclusively with this shape may belong to any one of the five different classes. However, there are usually particular diagnostic combinations of forms which give a certain indication of particular point group symmetry, and often the occurrence of a general form is sufficient. In systematic morphology, there must be a tabulation of all forms possible in the classes of the various systems with a subsequent consideration of their diagnostic significance in the determination of crystal class; such a tabulation and evaluation is given in Buerger, pp. 172–5. This demonstrates that, whilst most point groups can be recognised with certainty when critical forms are present, doubts must remain in the allocation of crystal class if the crystals under examination do not exhibit these forms. An additional hazard for the morphological crystallographer lies in the occurrence of twinned crystals (a form of composite crystal discussed in detail in Appendix D); in some crystals the growth process develops an imitative (or mimetic) twinning, sometimes on a sub-microscopic scale so that the composite crystal has a physical appearance consistent with a symmetry greater than that actually possessed by the structure.

In view of all the experimental difficulties and the subsequent interpretative pitfalls, it is a great tribute to the early morphological crystallographers that many of their results have suffered little or no later modification. These classical crystallographic studies provided the basis for all theories of the crystalline state until the advent of X-ray diffraction. So powerful was the influence they exerted that in the early years of this century a project was begun which aimed to recognise and characterise all compounds by their crystalline morphology; the resulting Barker index, although its original purposes were overtaken by events, remains an outstanding record of the achievements of the morphological crystallographer. Diffraction studies, begun in 1912, free the modern crystallographer from the necessary limitations of well-developed growth forms, and broaden the materials for study to the whole of the solid state; morphological work (where appropriate) provides only a small part of the total spectrum of data obtained mainly by other methods of examination. In general, these other techniques are physical in nature; as well as diffraction studies, which we shall consider at some length in later chapters, they involve other tests for symmetry described in Chapter 11, many of which were used by the morphologist to supplement the prime information contained in the crystal shape. As we shall see, many problems that defeated the earlier crystallographers can now be resolved but, as commonly happens, more refined difficulties appear; doubts as to the allocation of materials to a crystal class are far less common than when morphological studies were supreme, but in some cases symmetry determination can still prove a difficult and even intractable task.

6.5. An example of a morphological study

To conclude this chapter on crystal shapes, we shall describe a detailed morphological examination to illustrate this kind of work. The well-developed crystal X selected for study is sketched in Fig. 6.13; it appears to be highly symmetrical, and even from the sketch some symmetry elements are obviously present. Nevertheless *ab initio* these (and any other) elements must be deduced from a stereographic projection constructed from goniometric measurements; using a single-circle instrument, the measured values of interfacial angles in various zones are given in Table 6.1. Zone 1 appears to have the most important development of faces with some interfacial angles of 90°

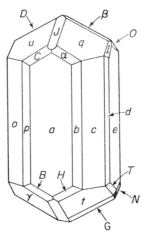

Fig. 6.13. Sketch of crystal X.

Table 6.1. Values of interfacial angles determined on a single-circle goniometer for the crystal X

	Zone 1		Zones 2	3	4	5	
a			a	a	e	e	
	26° 34′						39° 39′
b			α	C	I	O	
	18° 26′						28° 27′
c			q	u	q	v	
	18° 26′						21° 54′
d			β	D	J	P	
	26° 34′						21° 54′
e			v	r	u	r	
	26° 34′						28° 27′
f			A	E	K	Q	
	18° 26′						39° 39′
g			i	i	m	m	
	18° 26′						39° 39′
h			V	F	L	R	
	26° 34′						28° 27′
i			s	w	s	γ	
	26° 34′						21° 54′
j			U	G	M	S	
	18° 26′						21° 54′
k			γ	t	w	t	
	18° 26′						28° 27′
l			B	H	N	T	
	26° 34′						39° 39′
m			a	a	e	e	
	26° 34′						
n							
	18° 26′						
o							
	18° 26′						
p							
	26° 34′						
a							

(e.g. $\widehat{ae}$), suggesting that the zone axis is in the direction of a symmetry element and that there may well be some orthogonality in the axial system; we will start the projection by placing this zone around the primitive (Fig. 6.14), though a re-orientation of the stereogram could still be required after symmetry elements are recognised. Next we must locate the zones 2, 3, 4, 5 containing faces within the primitive; this may be done by first plotting a face common to two of these zones. For example, the measurements listed for

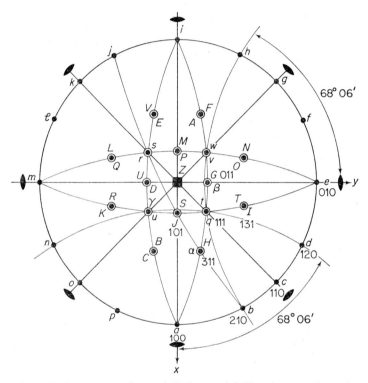

Fig. 6.14. Stereogram of crystal X. Letters labelling faces are beneath poles in upper hemisphere and above poles in lower hemisphere. Indices are shown only for positive integers.

zones 2 and 4 both contain the face q, with $\widehat{aq} = \widehat{eq} = 68° 06'$; small circles of this angular radius about the poles a and e on the primitive intersect at q, after which all poles on zones 2 and 4 can be plotted. Zones 3 and 5 are now drawn, for the poles u and v respectively in these zones are already on the projection.

The completed distribution of poles is consistent with the tetragonal symmetry $4/mmm$, for which the stereogram is conventionally orientated. For this class, the forms present are a general form $\{\alpha\}$, a ditetragonal bipyramid, and the special forms $\{J\}$ and $\{q\}$, tetragonal bipyramids, $\{b\}$ a ditetragonal prism, and {a} and {c}, tetragonal prisms. The presence of {α}, confirms the point group symmetry for the ditetragonal bipyramid is unique to this class (had it not been developed the remaining faces could also be found as forms in classes $\bar{4}2m$ and 42). In this system, the crystallographic axes are orthogonal; the z-axis must be chosen along the tetrad axis, but, as we have seen before, the selection of the x- and y-axes is arbitrary. From the crystal shape there is no evidence as to the choice which would correspond to a minimum volume cell, and we can only proceed in accordance with the expectation that the faces of any crystal have simple indices. There are two reasonable possibilities. (i) We can use the directions of the normals to the faces a and e as x- and y-axes respectively; or (ii) we can choose the normals to c and g as these reference axes. Any guidance on a final choice between (i) and (ii) could only come from experience in previous examinations of similar crystals, and lacking this, we will choose (i); the tetragonal prisms {a} and {c}, must therefore be indexed as {100} and {110} respectively. All other faces on the primitive belong to the ditetragonal prism {b} and may be immediately indexed before the choice of axial ratio; since

$$\widehat{ab} = 26° \, 34' = \tan^{-1}\left(\tfrac{1}{2}\right) = \tan^{-1}\left(\frac{k}{h}\right),$$

this form is {210}.

At this point, a choice of parametral plane cannot be avoided. In any tetragonal crystal, its pole must be on the radius containing c and q bisecting the x- and y-axes, but again the crystal shape cannot provide evidence as to the position required by the dimensions of the minimum volume cell; as before, we must proceed arbitrarily but in accordance with the desire to keep face indices small, so that the face q is a reasonable choice. This indexes one tetragonal bipyramid {q} as {111}, and requires all other faces within the primitive to be indexed in terms of this parametral plane. By zonal relationships on the stereogram J is (101), so identifying the second tetragonal bipyramid as {101}. The remaining faces are those of the general form one of which, α, is at the intersection of the zones (100)(111) and (210)(101), which cross-add to give its indices as (311); the particular bipyramid present on this crystal is {311}. The study is now complete except for the determination of the value of the axial ratio $\left(\frac{c}{a}\right)$ corresponding to the choice of parametral plane and a statement about the crystal habit. The formal record of crystal X is then:

Class 4/mmm   c/a = 0·439

Forms   {311}, {111}, {101}, {210} and {100}.

Habit   Elongated parallel to the prism zone.

6.6. Exercises and problems

(Note. Many of these problems can be solved by application of the methods of

Appendix B; it will also be necessary to refer to Appendix D for those involving twinning in crystals.)

1. Use sketch stereograms to find the number of distinct forms, special and general, in each of the following crystal classes (i) m, (ii) 222, (iii) $3m$, (iv) $\bar{4}2m$, (v) $\bar{6}$, (vi) $m3$. In each case list the number of faces possessed by the form and try to sketch its appearance.

2. (i) In which classes of the trigonal system is the rhombohedron to be found as a special or general form?

   (ii) In which classes of the tetragonal system is the tetragonal bipyramid to be found as a special or general form?

   (iii) In which class of the hexagonal system does the hexagonal prism not occur as a special or general form?

   (iv) In which classes of the cubic system does the trisoctahedron occur as a special form? If a crystal also showed the faces of a tetrahexahedron, which of these possible classes would be eliminated?

3. Many simpler crystals of a common mineral are described as: Class $2/m$, $a:b:c = 0.659:1:0.554, \beta = 116° 01'$, showing the forms $\{001\}, \{010\}, \{110\}$ and $\{\bar{1}01\}$; habit, either prismatic or elongated parallel to the x-axis.

   (i) Draw a stereogram showing all the faces of these forms.

   (ii) Determine the angles between faces in the zones [010] and [001].

   (iii) Make sketches of crystals with the two different habits.

   (iv) Crystals are often twinned with [001] as twin axis. Insert the twinned positions of faces on your stereogram, and determine the angle $\widehat{(001)(\bar{1}01)}$.

4. Goniometric measurements were made around zones on the upper half of a centrosymmetric crystal as follows:

Zone 1		Zones 2	3	4	5		Zone 6	
a		a	c	a'	c'		b	
	45° 00'					35° 10'		52° 46'
b		g	j	l	o		e	
	45° 00'					29° 30'		74° 28'
c		e	h	f	m		f	
	45° 00'					50° 40'		52° 46'
d		h	f	m	e		b'	
	45° 00'					29° 30'		
a'		i	k	n	p			
	45° 00'					35° 10'		
b'		a'	c'	a	c			
	45° 00'							
c'								
	45° 00'							
d'								
	45° 00'							
a								

(i) Determine the crystal class.

(ii) If g is indexed as (311), complete the description of this crystal.

5. A crystal of class 32 shows the forms $\{10\bar{1}0\}$ and $\{31\bar{4}1\}$, and measurement shows that $(10\bar{1}0)\widehat{}(31\bar{4}1) = 22° 41'$. Determine the axial ratio. Other crystals of the same substance are frequently twinned about the z-axis. Find the angle $(31\bar{4}1)\widehat{}(31\bar{4}1)$ for such crystals.

6. It is known that a substance belongs to the hexagonal system, and the external shapes of most of its crystals combine basal pinacoid, hexagonal prism and hexagonal bipyramid, which are indexed as $\{0001\}$, $\{10\bar{1}0\}$ and $\{11\bar{2}1\}$ respectively; measurement gives the angle $(0001)\widehat{}(11\bar{2}1) = 55° 46'$. What are the possible crystal classes? A rare crystal of this substance shows two further small faces in the zone $[0\bar{1}.1]$; goniometric measurement reveals that one of these faces is inclined at $30° 19'$ to $(10\bar{1}0)$, and that the second face is the opposite of the first. To what form do these faces belong? If its presence were confirmed, what is the probable crystal class?

7. A certain element shows crystals which are perfect regular hexoctahedra $\{321\}$. Draw accurate representations of central sections cut through these crystals (a) parallel to the plane (100), (b) parallel to the plane (110).

8. Common crystals of a particular mineral are described as class mmm, $a:b:c = 0·677:1:1·188$, showing the forms $\{hk0\}$ and $\{0kl\}$. If $(hk0)\widehat{}(h\bar{k}0) = 68° 13'$ and $(0kl)\widehat{}(0\bar{k}l) = 33° 05'$, index the forms of this morphological description. Later X-ray measurements were made in which the cell dimensions were chosen as $a = 9·51$, $b = 5·65$, $c = 6·42$ (all in Å). Show how this cell can be reconciled with the morphological axial ratios, and re-index the two common forms in terms of the X-ray cell.

SELECTED BIBLIOGRAPHY

Goniometry
TUTTON, A. E. H. 1922. *Crystallography and practical crystal measurement*, vol. I. Macmillan.
WOLFE, C. W. 1953. *Manual for geometrical crystallography*. Edwards, Ann. Arbor.

Morphology
BUERGER, M. J. 1956. *Elementary crystallography*. Wiley.
PHILLIPS, F. C. 1963. *An introduction to crystallography*. Longmans.
TERPSTRA, P. and CODD, L. W. 1961. *Crystallometry*. Longmans.

Source books for morphological data
GROTH, P. 1906–19. *Chemische krystallographie* (five volumes). Engleman, Leipzig.
PORTER, M. W. and SPILLER, R. C. 1951–6. *The Barker index of crystals* (two volumes). Heffer, Cambridge.

7

THREE-DIMENSIONAL LATTICE TYPES AND SPACE GROUPS

7.1. Lattice types

(a) Methods of Derivation

In the past few chapters the symmetries of point groups and their application to crystal classes and systems have been considered at some length; our discussions, although implicitly bounded by the existence of translational repetition, have not required a systematic development of the space lattices that can exist in crystalline matter, but in the further development of pattern theory this can no longer be delayed. The importance of planar lattices in two-dimensional patterns was described in Chapter 2, and before embarking on a discussion of their three-dimensional analogues it is worthwhile to re-state in the appropriate terms some of the principles set out in that chapter.

Any lattice associated with crystalline matter can be described as a regular repetitive pattern of points in space such that the environment of every point is identical; in a crystal structure each lattice point may be regarded as representing the pattern motif, a separate atom or a grouping of similar or dissimilar atoms. The array of lattice points can be produced by the repetitive operations of three non-coplanar lattice translations. The repeats of any three lattice rows can be chosen to outline a lattice cell, but there are conventional advantages in selecting this cell to be of minimum volume. Moreover, some arrangements of lattice points are related by symmetry operators, and any systematic nomenclature related to the choice of cell is simplified if the conventional selection is made in accordance with this symmetry. In effect this implies that for a given array the most suitable cell has a shape fixed by certain symmetry criteria, and that we must choose a cell of this shape with minimum volume. It follows that for some space lattices the most suitable cell contains more than one lattice point per cell and is multiply primitive, just as in others the cell is singly primitive. The symbolism of the various cell types is simple; P (to contrast with the planar analogue p) denotes a (singly) primitive lattice cell, and the other symbols in common use are listed in

Table 7.1. In this section we shall establish the total number of space lattices of each shape, singly or multiply primitive, that can be recognised in crystalline matter.

Table 7.1. Symbols for types of lattice cell

Symbol	Meaning	Lattice points per cell
A	Centring of cell faces defined by the b and c lattice translations; with these as the y and z crystallographic axes it could be said that this cell is centred on (100) planes.	2
B	Centring of cell faces defined by the a and c lattice translations, i.e. on (010) planes.	2
C^a	Centring of cell faces defined by the a and b lattice translations, i.e. on (001) planes.	2
F	Centring of all cell faces.	4
I	An extra lattice point at the centre of the cell.	2
R^a	Used to denote a primitive rhombohedral cell.	1

[a] The symbol (C) was also formerly used for a primitive hexagonal cell; this is now simply denoted by P. The extra primitive symbol (R) is necessary to distinguish trigonal crystals which have a minimum volume cell of rhombohedral rather than hexagonal shape.

In Chapter 4.2 we have already linked cell shape and crystal system and, in particular, set down the essential symmetry criteria for each of the seven crystal systems (Table 4.1); the problem of the possible space lattices can be approached by trying to build up distributions of lattice points which conform to these criteria expressed in terms of the presence of symmetry axes. Any spatial arrangement of lattice points can be regarded as the repetition of a planar array by a third non-co-planar lattice translation, and in Chapter 2.2 we have already determined the five distinctive planar lattices and the maximum symmetry with which each is consistent (Figs. 2.7 and 2.9). In stacking planar layers to build up a symmetrical space lattice we must use layers in which the points have the correct symmetrical array and ensure that their superposition preserves an appropriate regularity of points in the lattice as a whole; this will impose restrictions on the kinds of stacking pattern that are permissible. In particular, to build up a distribution of lattice points consistent with an n-fold symmetry axis the procedure should be:

(i) select the planar lattice which has an n-fold axis normal to its plane;
(ii) place the second layer of the stack parallel to the first so that the directions of the n-fold axes are in register. Since there are often several different n-fold axes intersecting the planar cell at different points, there may be several different placements for the second layer;
(iii) for each alternative stacking position, use the relative positions of the two layers to define stacking vectors whose repetitive operations re-

peat the planar lattice layers. Each stacking vector $\vec{D}$ can be specified by its components uvw parallel to the lattice translations of the planar layer and the normal to the layers; thus when the two layers are exactly above one another the stacking vector has components $00w$.

(iv) all possible distributions of lattice points are discovered by continuing for each different $\vec{D}$ to stack successive layers with the same relative displacements.

After this has been carried out for a particular system it remains to select suitable cells for each distribution; comparison of these cells will tell us whether or not all the lattices are different and so allows us to establish the number of distinctive lattice types. In the rest of this section a method of this kind is applied system by system.

(b) Triclinic System

This has no rotational symmetry, and, whilst any planar lattice could be used

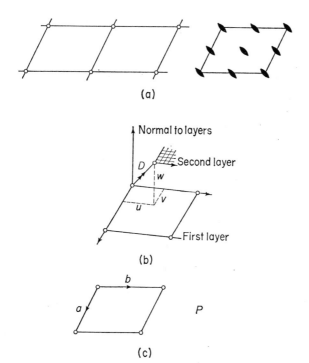

Fig. 7.1. Triclinic space lattices. (a) The oblique p-lattice and its maximum symmetry. (b) The stacking of the first two layers; the stacking vector $\vec{D}$ has components uvw. (c) The conventional projection of the cell of the triclinic P-lattice normal to (001); for simplicity the displacements of the upper lattice points due to the non-orthogonality of the lattice translations is not shown.

to form a general unrelated array of points in space, it is preferable that we should build this up from oblique p-lattices with generally inclined lattice translations. (In a way, this recognises that the least symmetrical cell shapes consistent with the symmetry are those which will be found in practice. In theory, there is no reason why a triclinic lattice should not have a cube cell provided that the atomic groupings associated with the lattice points destroy all the symmetry inherent in such a shape.) Such planar layers have diad symmetry (Fig. 7.1) but the stacking pattern is unrelated to this; the first two layers can have any relative displacement with a general stacking vector uvw. Repetition of this type must build up a space lattice without any symmetry in which there is an infinite number of ways of choosing a primitive cell of minimum volume. Conventions which govern the most suitable selection have already been described in Chapter 5.4(a), but there can be only one triclinic lattice type (P).

(c) Monoclinic System

Conventionally the direction of the essential diad is chosen as the y-axis, so that the lattice translations of the rows of the oblique p-lattice are a and c inclined at β. We must now preserve the symmetry normal to the layers, and this restricts the relative displacements of the first two layers. There are four alternative arrangements for which the stacking vectors are $00w$, $0\frac{c}{2}w$, $\frac{a}{2}0w$ and $\frac{a}{2}\frac{c}{2}w$ (Fig. 7.2); we notice generally that unless $u = v = 0$ the completion of a stacking sequence does not occur until further layers are added. In all four cases the cells have the characteristic monoclinic shape of an oblique parallelipiped, but there are only two different lattice types P and $C(\equiv A \equiv I)$; the equivalences are demonstrated in the figure, and arise because we can choose alternative lattice translations within a layer as the cell edges a and c (cf. the discussion of multiply primitive oblique planar lattices related to Fig. 2.8).

(d) Orthorhombic System

In Table 4.1 the minimum symmetry for this system is given as 222; in the present context it is convenient to re-state this as $2mm$ ($\equiv 2\bar{2}\bar{2}$), a legitimate modification for inversion and rotation axes may be interchanged in lattices (see Chapter 4.2(a)). This symmetry is consistent with planar rectangular lattices of both types, p and c. Again the preservation of this essential symmetry restricts relative displacements in the stacking of the layers; for both planar lattice types there are four different arrangements of the first two layers with stacking vectors $00w$, $0\frac{b}{2}w$, $\frac{a}{2}0w$ and $\frac{a}{2}\frac{b}{2}w$ (Fig. 7.3). The eight cells for the resultant space lattices all have the characteristic orthorhombic shape of a rectangular parallelipiped, but several of them prove to be identical when lattice translations are relabelled. In orthorhombic crystals there are four different lattice types P, $C(\equiv B \equiv A)$, I and F.

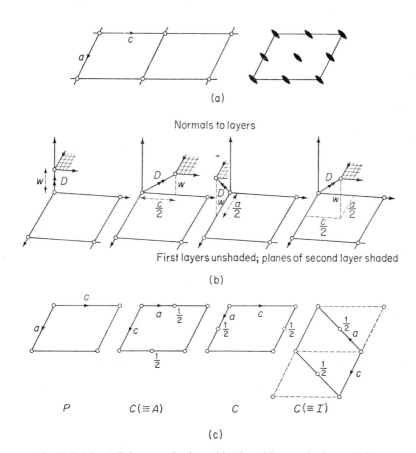

(a)

Normals to layers

First layers unshaded; planes of second layer shaded

(b)

(c)

Fig. 7.2. Monoclinic space lattices. (a) The oblique p-lattice and its monoclinic symmetry elements. (b) The alternative stacking of the first two layers; the stacking vector $\vec{D}$ has possible components $00w$, $0\frac{c}{2}w$, $\frac{a}{2}0w$ and $\frac{a}{2}\frac{c}{2}w$. (c) Projections of the cells of the four resultant space lattices normal to the layers.

(e) Trigonal System

In the definition of crystal systems we decided that point groups which have a triad symmetry axis and which can be consistent with a rhombohedral cell shape should be allocated to a separate trigonal system; in making this

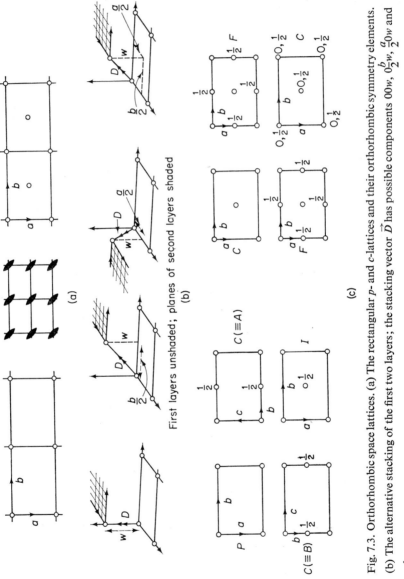

Fig. 7.3. Orthorhombic space lattices. (a) The rectangular *p*- and *c*-lattices and their orthorhombic symmetry elements. (b) The alternative stacking of the first two layers; the stacking vector $\vec{D}$ has possible components $00w$, $0\frac{b}{2}w$, $\frac{a}{2}0w$ and $\frac{a}{2}\frac{b}{2}w$. (c) Projections of cells of the eight resultant space lattices normal to the layers; the four on the left are derived from the *p*-lattice, whilst the four on the right come from the *c*-lattice.

distinction it was recognised that the same point group symmetries could also be found in crystals with hexagonal cells. We can now explore the origins of these alternative lattices and detailed relationships between the rhombohedral and hexagonal indexing of trigonal crystals described in Chapter 5.4(g).

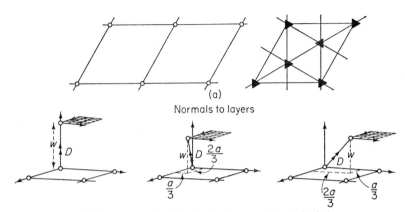

(a)
Normals to layers

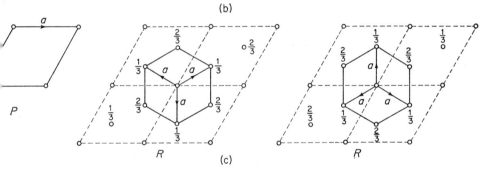

First layers unshaded, planes of second layer shaded

(b)

(c)

Fig. 7.4. Trigonal space lattices. (a) The hexagonal p-lattice and its trigonal symmetry elements. (b) The alternative stacking of the first two layers; the stacking vector $\vec{D}$ has possible components $00w$, $\frac{2a}{3}\frac{a}{3}w$ and $\frac{a}{3}\frac{2a}{3}w$. (c) Projections of the cells of the three resultant space lattices normal to the layers; note that the R cells require three layers to complete.

The alternative lattice arrays for trigonal crystals can be derived in the usual way by the stacking of hexagonal p-lattices. The maximum symmetry of hexagonal p-lattices is that shown in Fig. 2.7, but in trigonal crystals the hexads of this figure may be replaced by triads (whose repetition they include) to give the symmetry of Fig. 7.4. To preserve these essential triad axes, there are three different positions for the second layer which define stacking vectors $00w$, $\frac{2}{3}a\frac{1}{3}aw$ and $\frac{1}{3}a\frac{2}{3}aw$. The space lattice formed by stacking with $00w$ has a cell with the shape of a 60° rhombus-based prism, and is the hexagonal P lattice; the other two form lattices in which the minimum volume cell is a rhombohedron. These rhombohedral cells (shown in the figure) are both primitive, and are symbolised R to distinguish them from the hexagonal P cell. Apart from rotation through 180° about the normal to the layers the R cells are identical.

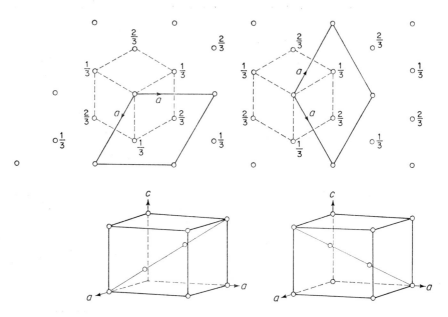

Fig. 7.5. Triply primitive hexagonal cells used to index rhombohedral crystals on the Miller-Bravais system. The alternative choices are shown in projection at the top and in perspective at the bottom; the obverse cell is on the left, the reverse cell on the right.

The second relevant problem is the nature of the multiply primitive hexagonal cell employed when crystals with an R lattice are indexed on the Miller–Bravais system. Fig. 7.5 shows a projection of planar hexagonal layers stacked to form a rhombohedral lattice; from these points we can choose a triply primitive hexagonal cell of the same height as the rhombohedral cell in two different relative positions (the obverse and reverse orientations). Either orientation of the triply primitive hexagonal cell can be used for indexing by the Miller–Bravais method; from the diagram we see that the matrices relating the axes of the cells are

$R \rightarrow$ Triple hexagonal Triple hexagonal $\rightarrow R$

$$\begin{vmatrix} 1 & \bar{1} & 0 \\ 0 & 1 & \bar{1} \\ 1 & 1 & 1 \end{vmatrix} \qquad \begin{vmatrix} \dfrac{2}{3} & \dfrac{1}{3} & \dfrac{1}{3} \\ \dfrac{\bar{1}}{3} & \dfrac{1}{3} & \dfrac{1}{3} \\ \dfrac{\bar{1}}{3} & \dfrac{\bar{2}}{3} & \dfrac{1}{3} \end{vmatrix} \begin{array}{l} \text{Obverse} \\ \text{orientation} \end{array}$$

$R \rightarrow$ Triple hexagonal Triple hexagonal $\rightarrow R$

$$\begin{vmatrix} 1 & 0 & \bar{1} \\ \bar{1} & 1 & 0 \\ 1 & 1 & 1 \end{vmatrix} \qquad \begin{vmatrix} \dfrac{1}{3} & \dfrac{\bar{1}}{3} & \dfrac{1}{3} \\ \dfrac{1}{3} & \dfrac{2}{3} & \dfrac{1}{3} \\ \dfrac{\bar{2}}{3} & \dfrac{\bar{1}}{3} & \dfrac{1}{3} \end{vmatrix} \begin{array}{l} \text{Reverse} \\ \text{orientation} \end{array}$$

These matrices may be used as described in Chapter 5.2 to relate rhombohedral plane indices (pqr) to Miller–Bravais plane indices ($hk \cdot l$); the symbol $i (= -(h + k))$ is omitted.

(*f*) Tetragonal System

With an essential tetrad axis, tetragonal space lattices must be built from layers of planar square lattices. There are two different arrangements of the first two layers, whose relative displacements are described by the stacking vectors $00w$ and $\frac{a}{2}\frac{a}{2}w$, (Fig. 7.6). In the resultant arrays of lattice points, the unit cells are square-based prisms of the two tetragonal lattice types P and I.

(*g*) Hexagonal System

If this system is restricted to those point groups with a hexad axis, the space lattices must be formed by stacking hexagonal p-lattices which are consistent with this symmetry (Fig. 7.7). The layers can only be stacked exactly in register if the hexads are to be preserved, so that the stacking vector $00w$ leads to an array with a primitive hexagonal cell (P).

(*h*) Cubic System

This system is slightly more difficult to discuss in that the lattice array must display four equally inclined triad axis directions. Although it is possible to examine lattice layers normal to these essential axes, it is simpler to consider layers normal to the three perpendicular directions to which the triads are equally inclined (see the operation of triad axes in Chapter 3.3 and 4.2(f)).

These layers have diad symmetry in some cubic classes but are normal to tetrad axes in the more symmetrical classes. In this way cubic lattices are constructed by stacking planar square lattices. However, the placements of successive layers within the stack is more critical than before, in that the spatial array

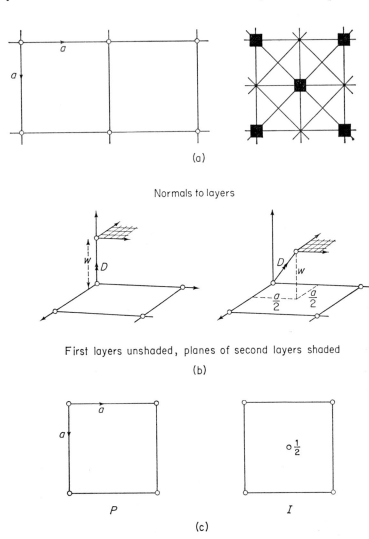

Fig. 7.6. Tetragonal space lattices. (a) The square p-lattice and its tetragonal symmetry elements. (b) The alternative stacking of the first two stacking layers; the stacking vector $\vec{D}$ has possible components $00w$ and $\frac{a}{2}\frac{a}{2}w$. (c) Projections of the cells of the two resultant space lattices normal to the layers.

of lattice points must be consistent with the triad axes inclined to the planes of the stack; this requires the component of the stacking vector w to have specific values to give three identical mutually perpendicular lattice translations related by the triads. This can be achieved in either of the two ways shown in Fig. 7.8; these show the triad axes contained in one or other of the two independent sets of symmetry planes that will develop as the planar nets are stacked. From the two placements of successive layers for each orientation of

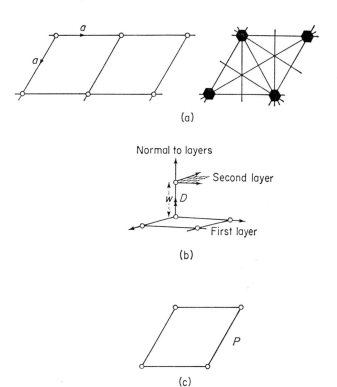

Fig. 7.7. Hexagonal space lattices. (a) The hexagonal p-lattice and its hexagonal symmetry elements. (b) The stacking of the first two layers; the stacking vector $\vec{D}$ has components $00w$. (c) Projection of the cell of the space lattice.

the triad axes, there are four possible stacking vectors $00a$, $\dfrac{a}{2}\dfrac{a}{2}\dfrac{a}{2}$, $00\sqrt{2}a$ and

$\dfrac{a}{2}\dfrac{a}{2}\dfrac{a}{\sqrt{2}}$. The four resultant arrays of lattice points have P, I, C and F cells, but we realise that the C lattice is lacking in the essential cubic symmetry; to satisfy this, the other two cell faces must be centred, and thus there are only three independent lattice types with P, I and F cells.

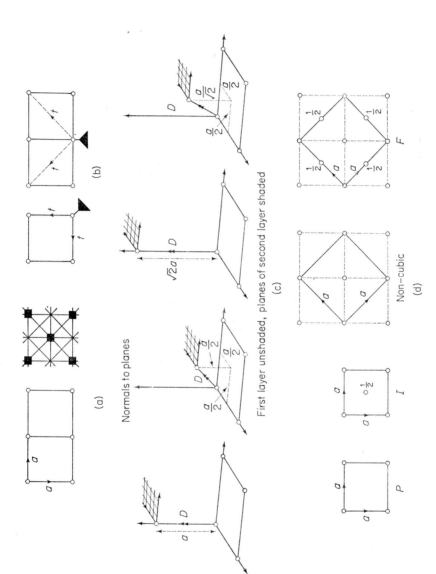

Fig. 7.8. Cubic space lattices. (a) The square p-lattice and its cubic symmetry elements. (b) The two orientations of the triad axis with respect to the square p-lattice; the symmetry of this axis requires the lattice translations t to be the same as that normal to the layers. (c) The alternative stacking of the first two layers; the stacking

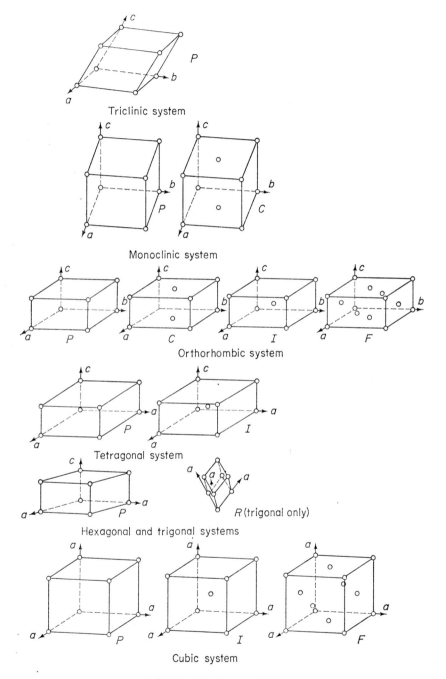

Triclinic system

Monoclinic system

Orthorhombic system

Tetragonal system

Hexagonal and trigonal systems

Cubic system

Fig. 7.9. The 14 space lattices.

(*i*) *Summary*

This brief investigation has demonstrated that there are fourteen different space lattices found in crystalline matter; these are sometimes known as the *Bravais lattices*. Fig. 7.9 groups the unit cells in accordance with the crystal system. Each space lattice within a system can show the symmetry operators of the holosymmetric class, so that with a suitable form of pattern motif this symmetry will be found in the arrangement of the atomic structure; but the introduction of the translational elements can generate new types of symmetry operators not found in point groups, and a full discussion of the permissible symmetry elements in crystal structures is given shortly in Chapter 7.3. Moreover, each space lattice within a system can also be consistent with all point groups of lower symmetry than the holosymmetric class (again with the possibility of new types of operators); this lower symmetry must arise from the nature of the pattern motif to be associated with each lattice point. The combination of repetition due to lattice type and all the different kinds of symmetry operators constitutes the different space groups, the ultimate subdivisions in crystalline pattern theory to be discussed in Chapter 7.4.

7.2. Dimensional relations in lattices

All linear and angular values in arrays of lattice points are related to the size and shape of the unit cell. In Chapter 5 we saw how a system of indexing based on cell shapes could describe different rows or planes of lattice points and how the values of the angles between the rows or planes of different indices could be calculated from cell constants expressed as axial ratios and interaxial angles. Any calculation of dimensional values in lattices is, of course, also made from cell constants, but we need to know the magnitudes of the lattice translations a, b and c, not just their ratios. Methods of computation vary with the geometry of the unit cell; it is not our intention here to perform such calculations system by system, but merely to draw attention to those linear values that are important in diffraction studies.

Firstly, we can consider the families of lattice rows $[UVW]$. Each family has two distinct dimensional characteristics, the distances between the rows and the separation of points along any lattice row of the set; the separation S_{UVW} of points on a row $[UVW]$ is of most immediate concern, for it is relevant to the discussion of diffraction by the Laue equations in Chapter 8.3(a). For P lattices the calculations are straightforward; for example in an orthorhombic crystal

$$S_{UVW} = (U^2a^2 + V^2b^2 + W^2c^2)^{\frac{1}{2}}$$

an expression which can be adapted for orthogonal cells in other systems. But some care must be taken with multiply primitive lattice types; depending on the integers U, V and W, the additional lattice points will either change the separation of the rows or the spacing of the points upon them. Fig. 7.10 shows two orthorhombic unit cells with identical dimensions, one primitive and the other C-face centred. For some directions S_{UVW} is the same for both cells (e.g. $S_{100} = a$ for both lattice types); the extra points of the centred lattice have created additional members of the family without changing the separation on a given row. For a direction such as [112], $S_{112} = (a^2 + b^2 + 4c^2)^{1/2}$ for the P-lattice, but is only $\frac{1}{2}(a^2 + b^2 + 4c^2)^{1/2}$ for the C-lattice; in such directions as this there are no extra members of the set, but the separation of points on each lattice row has been halved. In multiply primitive cells the effects of the extra lattice points on S_{UVW} must be considered for each set of integers U, V and W.

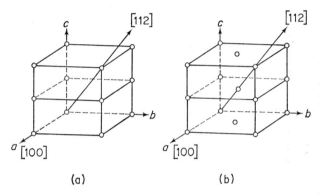

(a)   (b)

Fig. 7.10. The separation S_{uvw} of points on lattice rows. (a) Orthorhombic P-cell. (b) Orthorhombic C-cell. The zone axes [100] and [112] are marked for both cells.

The other dimensional values that are important concern the various families of planes (hkl); these have two linear characteristics, the distribution of lattice points on a particular member of the family and the interplanar spacing d_{hkl} separating the members of the set. Now it is the interplanar spacings which are most important to us, for they occur in the description of diffraction by the Bragg equation in Chapter 8.3(a); indeed their values for a particular substance can be used to characterise that material and aid its identification (see Chapter 10.4). Naturally, general expressions which relate d_{hkl} to the cell constants gain in complexity as the cell shape becomes less specialised in lower symmetry systems; Table 7.2 gives expressions for $1/d_{hkl}^2$, a form which is often useful in diffraction work. Fortunately, values of d_{hkl} can nowadays be calculated from a known set of cell constants by standard computer programmes. We should perhaps mention here that in the use of these spacings in the Bragg treatment of diffraction it is convenient to leave the integers h, k and l in multiple form, i.e. we do not divide out any common factors; this implies that d_{hkl} values relate only to cell constants and are independent of whether the cell is multiply primitive or not. In the orthorhombic

Table 7.2. Reciprocals of the square of interplanar spacings $(1/d^2_{hkl})$ for each cell shape

Cell shape	$1/d^2_{hkl}$	System
Cube of side a	$\dfrac{h^2 + k^2 + l^2}{a^2}$	Cubic
Right prism with square base a and height c	$\dfrac{h^2 + k^2}{a^2} + \dfrac{l^2}{c^2}$	Tetragonal
Rectangular parallelepiped of sides a, b and c	$\dfrac{h^2}{a^2} + \dfrac{k^2}{b^2} + \dfrac{l^2}{c^2}$	Orthorhombic
Rhombohedron with polar edges a inclined at angle α	$\dfrac{1}{a^2}\left(\dfrac{(h^2 + k^2 + l^2)\sin^2\alpha + 2(hk + hl + kl)(\cos^2\alpha - \cos\alpha)}{1 + 2\cos^3\alpha + 3\cos^2\alpha}\right)$	Trigonal
Right prism with 60° rhombus base a and height c	$\dfrac{4}{3a^2}(h^2 + hk + k^2) + \dfrac{l^2}{c^2}$	Trigonal and Hexagonal
Oblique parallelepiped with side b normal to a and c inclined at β	$\dfrac{1}{\sin^2\beta}\left(\dfrac{h^2}{a^2} + \dfrac{l^2}{c^2} - \dfrac{2hl}{ac}\cos\beta\right) + \dfrac{k^2}{b^2}$	Monoclinic
General parallelepiped with sides $a\,b\,c$ inclined at angles α, β and γ	$\dfrac{\begin{vmatrix} \frac{h}{a} & \cos\gamma & \cos\beta \\ \frac{k}{b} & 1 & \cos\alpha \\ \frac{l}{c} & \cos\alpha & 1 \end{vmatrix} + \begin{vmatrix} 1 & \frac{h}{a} & \cos\beta \\ \cos\gamma & \frac{k}{b} & \cos\alpha \\ \cos\beta & \frac{l}{c} & 1 \end{vmatrix} + \begin{vmatrix} 1 & \cos\gamma & \frac{h}{a} \\ \cos\gamma & 1 & \frac{k}{b} \\ \cos\beta & \cos\alpha & \frac{l}{c} \end{vmatrix}}{\begin{vmatrix} 1 & \cos\gamma & \cos\beta \\ \cos\gamma & 1 & \cos\alpha \\ \cos\beta & \cos\alpha & 1 \end{vmatrix}}$	Triclinic

example of Fig. 7.10 (100) and (200) planes are parallel and there are identical values of d_{100} ($= a$) and d_{200} $\left(= \dfrac{a}{2}\right)$ for both cells. It is irrelevant in this use of lattice planes that alternate (200) planes contain no lattice points for the P-cell, whereas all such planes are populated for the C-cell; moreover, the presence of an extra lattice point midway between the (100) planes for the C-lattice is taken to have no effect on the d_{100} value. No account of the siting of lattice points on planes is needed in this treatment to predict the expected positions of diffraction maxima; any difference in the arrangements of lattice points on the members of a family affects whether these maxima are observable or not. All these aspects of the use of interplanar spacings in diffraction will be clarified in the next chapter.

7.3. Symmetry operators in space groups

In the space groups which underlie the atomic patterns of crystalline matter we can expect to find rotation and inversion axes, mirror planes and centres combined as in the point groups; but since these elements are modified by the translational periodicity of the pattern we can also expect the development of new but related kinds of symmetry operators which can replace or supplement the elements of the point groups. In this section we examine the nature of these additional symmetry elements which may be found in space groups.

Centres of symmetry are repeated unchanged by lattice translations, but *glide planes* (analogous to the glide lines of Chapter 2.1) can be developed when translations are combined with mirror symmetry planes. In planar patterns the glide line successively repeats right- and left-handed forms of the pattern motif on either side of the line with a periodicity of half the lattice translation in the direction of glide; in crystal structures, although glide planes operate in much the same general manner, the extra dimension permits a greater variety of glide movement to be followed after reflection. When the plane is normal to one cell edge, say a, any glide translation can be either in the direction of b or c or both; thus there are three different kinds of glide plane which can occur in this orientation. Some examples of planes in various orientations and their repetition of lattice points are shown in Fig. 7.11; note particularly the symbols indicating elements when they are normal or parallel to the paper for these are used in the diagrammatic representation of space groups later.* The nature of the plane is described by the direction of glide;

---

* The symbol + is conventionally used to indicate a fractional co-ordinate z above the plane of the paper parallel to the third lattice translation; similarly the symbol − denotes a co-ordinate $\bar{z}$ below the plane of the paper. $\frac{1}{2}+$, $\frac{1}{3}+$, $\frac{1}{2}-$, etc. must be interpreted as heights $\frac{1}{2} + z$, $\frac{1}{3} + z$, $\frac{1}{2} - z$, etc. so that, for example, a point $\frac{1}{2}+$ is one half of the cell edge above a point $+$. Remember too that all points must be repeated in every cell by the lattice translation, so that, e.g. $\frac{4}{3}+ \equiv \frac{1}{3}+$.

Normal to paper   Parallel to paper

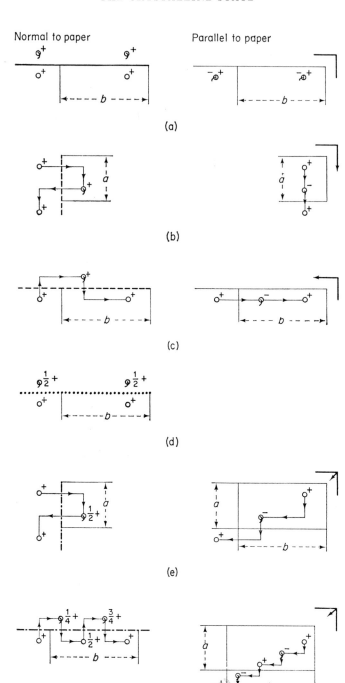

(a)

(b)

(c)

(d)

(e)

(f)

when the translational movement is in the x-axis direction it is called an *a*-glide plane, in the y-axis direction it is a *b*-glide plane, and in the z-axis direction it is a *c*-glide plane; when, after reflection, translations occur simultaneously parallel to two cell edges the element is known as a diagonal *n*-glide plane. The orientation of a particular plane (e.g. normal to the x-axis) also needs to be specified in some symmetry groupings; this will be taken up at a relevant point in the discussion of space groups in the next section. All these glide elements may replace and supplement the mirror symmetry planes of point groups in repetitive pattern types. There is, however, a further glide plane also illustrated as Fig. 7.11(f) which is restricted to some multiply primitive lattices; this results from the effects of diagonal gliding with reflection across a centred lattice plane. In the diamond *d*-glide plane, the two simultaneous translations are only $\frac{1}{4}$ of the cell edges, whereas an *n*-glide plane would cause movements through distances of $\frac{1}{2}$ of these edges; in space groups *d*-glide planes need only be considered in orientations parallel to the faces of an *F*-cell or the $\{110\}$ planes of an *I*-cell.

In Chapter 2.1 it was shown that combinations of rotation axes with lattice translations in planes normal to the direction of the axes did not give any new kinds of symmetry elements in planar patterns; but in atomic structures combinations can also occur with translations in the axial directions. These lead to a range of new symmetry operators which are best considered separately for $n = 2, 3, 4$ and 6. Fig. 7.12 shows the effect for diad symmetry; a lattice point after rotation through 180° is translated through $\frac{1}{2}$ of the cell edge in the direction of the symmetry axis. This is a new symmetry element known as a *screw diad* (2_1); conventional symbols used in space group diagrams are also shown in the figure. Extending this to triad symmetry, there are now two possible distributions of lattice points around the axial direction depending on the sense of rotation (Fig. 7.13), the two arrangements are enantiomorphously related with opposing senses of screw. Conventionally we choose to

---

Fig. 7.11. Symmetry planes in space groups. In all diagrams, a right-handed axial set is taken with the xy plane in the plane of the paper; a height z above this plane is indicated conventionally by the symbol $+$. (a) Mirror (m) plane. (b) Glide plane (a) gliding in x-axis direction. (c) Glide plane (b) gliding in y-axis direction. (d) Glide plane (c) gliding in z-axis direction; this cannot occur parallel to the xy plane. (e) Diagonal glide plane (n) gliding in two axial directions. (f) Diamond glide plane (d) gliding in two axial directions with translations half those of *n*-glide plane. Note the conventional symbols used for these elements in space group diagrams. When planes are parallel to the paper, heights other than zero are indicated by writing the z-co-ordinate by the symbol (e.g. $\daleth \frac{1}{4}$). In the left-hand diagram of (f) the arrow gives the directions of positive glide in the z-direction; in the right-hand diagram the arrow gives the actual direction of glide (no confusion with the n symbol can occur for this d-plane must be accompanied by another parallel plane with a height difference of $\frac{1}{4}$ with the symbol $\daleth^\searrow \frac{1}{4}$).

describe both sets of points in the same way by anti-clockwise rotations and upward translations; in these terms one axis translates through $\frac{1}{3}$ of the cell edge after each 120° rotation, whilst the other has a translation of $\frac{2}{3}$ of the cell edge. They are then respectively described as the 3_1 and 3_2 *screw triads* (by analogy with the 2_1 screw diad which has a motion through $\frac{1}{2}$ of the cell edge

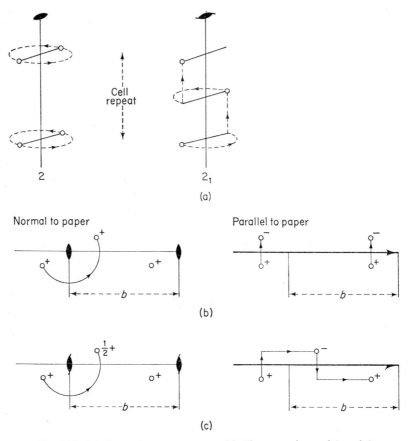

Fig. 7.12. Diad axes in space groups. (a) The operations of 2 and 2_1 axes. (b) Diad axes (2) in projections. (c) Screw diad axes (2_1) in projections. As in Fig. 7.11 the symbol + indicates a height z above the xy plane. Note the conventional symbols for these elements in space group diagrams. When axes are parallel to the paper, heights other than zero are indicated by writing the z-co-ordinate by the symbol (e.g. $\rightarrow\frac{1}{4}$).

after each rotation). With tetrad symmetry (Fig. 7.14(a)), the screw axes 4_1 and 4_3 have the same enantiomorphous character as the two screw triads; these have upward translations of $\frac{1}{4}$ and $\frac{3}{4}$ of the cell edge respectively after each 90° anticlockwise rotation. But there can also be a 4_2 screw axis with a translation of two-fourths of the cell edge after each rotation; like the 2_1 axis, this has no definite sense of screw. Finally, for hexad symmetry the implications are clear (Fig. 7.14(b)); the five screw axes form two distinct enantiomorphous sets, 6_1, 6_5 and 6_2, 6_4, with the 6_3 axis having no sense of screw. Screw axes of the appropriate degree can replace and supplement the rotation axes of point groups in crystal structures.

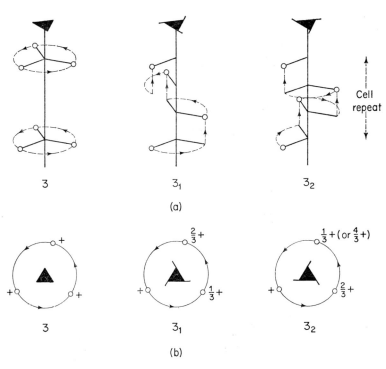

Fig. 7.13. Triad axes in space groups. (a) The operations of 3, 3_1 and 3_2 axes; note the opposite senses of rotation for 3_1 and 3_2. (b) The axes normal to the plane of projection; 3_1 and 3_2 are both drawn with anti-clockwise rotation and upward translation. The symbol + indicates a height z above the xy plane. Note the conventional symbols for these elements in projection.

   In some atomic patterns glide planes and screw axes are the only symmetry elements relating the units of the structure, but such materials still develop crystalline forms for which the point group symmetry has such elements replaced by mirror planes and rotation axes. The angular relations between the faces of a crystalline form are unaffected by the nature of the axes or planes; screw axes and glide planes must cause a slight displacement of faces due to their translational components, but any displacement intervals are so

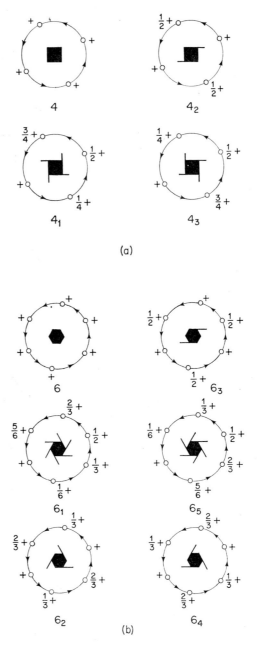

Fig. 7.14. Tetrad and hexad axes in space groups. (a) Tetrad axes normal to the plane of projection. (b) Hexad axes normal to the plane of projection. The symbol + indicates a height z above the xy plane. Note the conventional symbols for these elements.

small as to be undetectable for a macroscopic crystal. This emphasises again the relation between point groups and space groups, and leads us on to the derivation of space groups from isomorphous point groups outlined in the next section.

7.4. Space groups

In Chapter 2.4 we described how the seventeen planar space groups could be derived by combining the translational repetitions of the lattice type with those of the symmetry operations of the point group. Crystallographic space groups may be determined in much the same way; every permutation of permissible lattice type and related symmetry combinations (expressed as rotation, inversion or screw axes, mirror or glide planes, and centres) must be considered to see if it leads to a distinctive pattern type. This is a formidable task, and, although there are some ways to ease the labour, a full treatment is quite inappropriate in an introductory text. In the event, many possible permutations prove to be identical, and the total number of space groups is limited to 230. Even so, it is impossible to give attention to every one of these here; we shall only attempt any kind of detailed description within the triclinic and monoclinic systems, and confine discussion of the space groups of more symmetrical systems to brief comments on the symbolism and any general explanations that are demanded.

The common symbolism is similar to that used for planar groups; each space group is designated by a lattice type and a point group symbol, but modified as necessary to embrace translational symmetry elements and descriptions of orientation. Triclinic space groups are very simple; there is only one lattice type with a P-cell, and the two point groups 1 and $\bar{1}$ have no replaceable elements. There are only two space groups, symbolised by $P1$ and $P\bar{1}$. In the derivation of monoclinic space groups there are many different alternatives to be examined; each of the two lattice types, P and C, must be combined with all possible symmetry operations within the point groups 2, m and $2/m$. These combinations may be set out thus:

Class 2		Class m*		Class $2/m$*			
$P2$	$C2$	Pm	Cm	$P2/m$	$P2_1/m$	$C2/m$	$C2_1/m$
$P2_1$	$C2_1$	Pa	Ca	$P2/a$	$P2_1/a$	$C2/a$	$C2_1/a$
		Pc	Cc	$P2/c$	$P2_1/c$	$C2/c$	$C2_1/c$
		Pn	Cn	$P2/n$	$P2_1/n$	$C2/n$	$C2_1/n$

The equivalence of some of these symbols is immediately obvious; for example, those which only differ in the kind of glide plane which is combined

---

* With the conventional choice of axes in monoclinic crystals, symmetry planes in point groups m and $2/m$ are normal to the y-axis; in associated space groups the only permissible glide planes are therefore a, c and n.

with a primitive lattice must give equivalent arrangements, for any choice and labelling of the x- and z-axes is arbitrary (i.e. $Pc \equiv Pa \equiv Pn$, etc.). In fact, there are only thirteen different monoclinic space groups listed by their conventional symbols in Table 7.3; equivalences with some of the other symbols

Table 7.3. Conventional symbols for monoclinic space groups

Class 2		Class m		Class 2/m		
$P2$	$C2$	Pm	Cm	$P2/m$	$P2_1/m$	$C2/m$
$P2_1$		Pc	Cc	$P2/c$	$P2_1/c$	$C2/c$

listed above are demonstrated by the diagrams of the conventional groups in Figs. 7.15 and 7.16. As with planar examples, it is usual to draw two diagrams for each space group; that on the right shows all symmetry elements in the cell, whilst that on the left contains the distribution of general equivalent positions (G.E.P.'s) produced by the operations of these elements on the points of the lattice. The diagrams of these figures illustrate some general points applicable to space groups of all systems which can be conveniently mentioned here. Firstly, whenever possible, the cell origin is chosen at a centre of symmetry; this simplifies co-ordinates of equivalent points, and is helpful in the calculation of diffracted intensities for any structure based on the space group (see Chapter 8.3(b)).

If the group has no centres of symmetry, the origin is usually chosen to lie on symmetry elements appearing in the symbol; in more symmetrical systems convenient positions are often at the intersections of elements (though in some cases an origin is better placed symmetrically situated between the elements). In every space group additional symmetry to that specified by the symbol is developed; this can simply be extra elements of the same kind (e.g. the diad through the mid-point of the cell in $P2$) or new elements not mentioned in the symbol (e.g. the screw diads in $C2/m$). In all cases these extra elements are in accord with point group symmetry both in nature and orientation (e.g. in space groups developed from 2/m all axes must be diads parallel to the y-axis, all planes must be perpendicular to this direction and there must always be centres of symmetry). We notice that identical symmetry elements are always developed at separations of $\frac{1}{2}$ of the appropriate cell edge, and in any space group such repetitions may be inserted immediately; when elements are parallel to the paper, heights other than 0 (and $\frac{1}{2}$), are written beside the symbol (e.g. the diads at heights $\frac{1}{4}$ (and by implication $\frac{3}{4}$) in $C2/c$). In groups with non-primitive lattices, further additional elements often not mentioned in the symbol are found in parallel interleaving positions (e.g. the screw diad axes and glide planes in $C2/m$); particular attention must be given to such possibilities in multiply primitive cells.

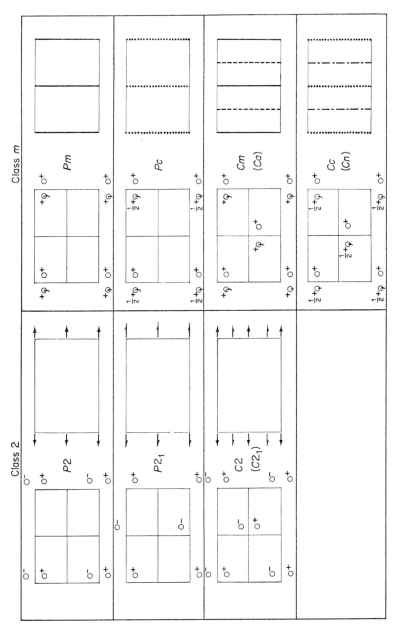

Fig. 7.15. Monoclinic space groups derived from classes 2 and *m* projected on (001) plane. Alternative non-standard symbols are given in brackets; diagrams of general equivalent positions on the left have cells quartered for convenience. In this orientation the *z*-axis is not normal to the paper; displacements of points due to the inclination of this translation is omitted to avoid confusion.

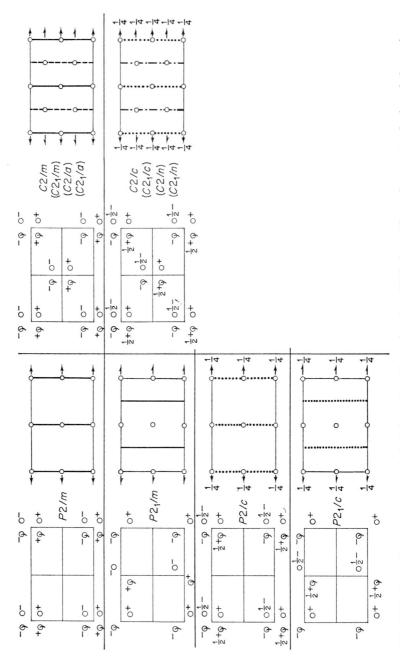

Fig. 7.16. Monoclinic space groups derived from class 2/m projected on (001) plane. Alternative non-standard symbols are given in brackets; diagrams of general equivalent positions on the left have cells quartered for convenience; in diagrams of symmetry elements on the right note the centres of symmetry indicated symbolically by small open circles. In this orientation the z-axis is not normal to the paper; displacement of points due to the inclination of this translation is omitted to avoid confusion.

Diagrams of G.E.P's show how the pattern motif (an atom or atom group) would be reproduced by the particular scaffolding of symmetry elements in the space group; in every case the number of G.E.P's per cell is the product of the multiplicity of the lattice cell and the number of symmetry repetitions in the point group (e.g. *P2* has $1 \times 2 = 2$ G.E.P's per cell whereas *C2/c* has $2 \times 4 = 8$ G.E.P's per cell). For most space groups the number of equivalent positions per cell can be reduced if the lattice points are in special positions with respect to certain symmetry elements (rotation axes, mirror planes and centres of symmetry). In *C2/c*, for example, there are only four equivalent positions per cell when lattice points are on diad (2) axes due to the coalescence of G.E.P's in pairs; notice that this reduction cannot take place when the lattice points are on screw axes or glide planes due to the translational components of these elements. There may be several different sets of special equivalent positions (S.E.P's), and a complete study of a space group requires the evaluation of all different possibilities. Conventionally equivalent positions, general and special, are listed for each space group in terms of fractional cell co-ordinates; Table 7.4 shows the list for *C2/m*. The presenta-

Table 7.4. List of general and special equivalent positions for space group *C2/m*

Number of positions and point symmetry		Co-ordinates of equivalent positions $(0, 0, 0; \frac{1}{2}, \frac{1}{2}, 0)+$
8	1	$x, y, z;\ \ x, \bar{y}, z;\ \ \bar{x}, y, \bar{z};\ \ \bar{x}, \bar{y}, \bar{z}.$
4	m	$x, 0, z;\ \ \bar{x}, 0, \bar{z}.$
4	2	$0, y, \frac{1}{2};\ \ 0, \bar{y}, \frac{1}{2}.$
4	2	$0, y, 0;\ \ 0, \bar{y}, 0.$
4	$\bar{1}$	$\frac{1}{4}, \frac{1}{4}, \frac{1}{2};\ \ \frac{1}{4}, \frac{3}{4}, \frac{1}{2}.$
4	$\bar{1}$	$\frac{1}{4}, \frac{1}{4}, 0;\ \ \frac{1}{4}, \frac{3}{4}, 0.$
2	$2/m$	$0, \frac{1}{2}, \frac{1}{2}.$
2	$2/m$	$0, 0, \frac{1}{2}.$
2	$2/m$	$0, \frac{1}{2}, 0.$
2	$2/m$	$0, 0, 0.$

tion, a simplified version of the format of the standard reference text *International tables for X-ray crystallography*, needs a little clarification. Listing of equivalent positions for multiply primitive cells is abbreviated by quoting the translations of the lattice type separately. In this case they are $(0, 0, 0; \frac{1}{2}, \frac{1}{2}, 0)+$ which implies that these translations must be added to those co-ordinates listed in each row to obtain the total set; thus in this example the co-ordinates of the eight G.E.P's are $x, y, z; x, \bar{y}, z; \bar{x}, y, \bar{z}; \bar{x}, \bar{y}, \bar{z}$ (quoted in the table) and $\frac{1}{2} + x, \frac{1}{2} + y, z; \frac{1}{2} + x, \frac{1}{2} - y, z; \frac{1}{2} - x, \frac{1}{2} + y, \bar{z}; \frac{1}{2}-x, \frac{1}{2} - y, \bar{z}$ (obtained by the addition of the lattice centring $\frac{1}{2}, \frac{1}{2}, 0$ to the listed co-ordinates). Each set of co-ordinates is accompanied by the number of equivalent positions and the point symmetry; often, as in this group, there are several sets of S.E.P's with the same point symmetry. A distinct set will arise for each position with the same point symmetry which is not repeated by the lattice translations and symmetry; for example, the interleaving diads (2) in the xy plane are equivalent but are not related to those at height $\frac{1}{2}$. Similar tabulations of equivalent positions clearly grow more complex for space groups of higher symmetry systems.

Even space groups of the orthorhombic system cannot be treated here in comparable detail. They are formed by the combination of the space lattices P, C, I and F and the symmetry operators of the point groups $mm2$, 222 and mmm; the standard symbols for the fifty-nine different groups are given in Table 7.5, and we shall only make general comments on their interpretation.

Table 7.5. Conventional symbols for orthorhombic space groups

Class 222			Class $mm2$					Class mmm			
$P222$	$C222$	$F222$	$Pmm2$	$Cmm2$	$Amm2$	$Fmm2$	$Imm2$	$Pmmm$	$Cmmm$	$Fmmm$	$Immm$
$P222_1$	$C222_1$		$Pcc2$	$Ccc2$	$Ama2$	$Fdd2$	$Ima2$	$Pmma$	$Cmma$	$Fddd$	$Imma$
$P2_12_12$			$Pma2$	$Cmc2_1$	$Aba2$		$Iba2$	$Pmmn$	$Cccm$		$Ibam$
$P2_12_12_1$			$Pnc2$		$Abm2$			$Pccm$	$Ccca$		$Ibca$
		$I222$	$Pba2$					$Pmna$	$Cmcm$		
		$I2_12_12_1$	$Pnn2$					$Pbam$	$Cmca$		
			$Pmc2_1$					$Pbcm$			
			$Pca2_1$					$Pnnm$			
			$Pmn2_1$					$Pnma$			
			$Pna2_1$					$Pnnn$			
								$Pban$			
								$Pnna$			
								$Pcca$			
								$Pccn$			
								$Pbcn$			
								$Pbca$			

It is conventional in this system that in listing the symmetry elements, the first place in the symbol refers to the x-axis, the second to the y-axis and the third to the z-axis; thus $P222_1$ has rotation diads parallel to the x- and y-axes and screw diads parallel to the z-axis, whilst $Pban$ has a b-glide plane normal to the x-axis, an a-glide plane normal to the y-axis and an n-glide plane normal to the z-axis. In the list of groups derived from $mm2$ it will be noticed that a distinction has been made between one-face centred lattices C and A. With this type of point group symmetry, the symmetry planes are normal to the centered face for the C-cell, and parallel to the centred face for $A(\equiv B)$-cells; new distributions of G.E.P's can therefore be formed which cannot be described using a conventionally constructed symbol and a C-lattice. The standard symbols of the table depend on giving certain precedences in groups where there are different kinds of parallel elements (e.g. $F222 \equiv F2_12_12_1$) but also more importantly they depend on the orientations of particular elements with respect to the choice of x-, y- and z-axes. We have already mentioned the difficulties of labelling crystallographic axes in orthorhombic crystals in Chapter 5.4(c), and we must expect to find non-standard forms of space group symbols arising in some descriptions. Alternatives are simply worked out in any particular case; suppose that we wish to discover the symbol for *Pban* when the axes are re-selected so that the old $x \rightarrow$ new z, the old $y \rightarrow$ new x and the old $z \rightarrow$ new y (Fig. 7.17). The diagram shows that the plane normal to the new x-axis glides in the new z-axis direction, that normal to the new y-axis is a diagonal glide plane, and that normal to the new z-axis glides in the direction of the new x-axis. With this change of axes *Pban* $\rightarrow$ *Pcna*;

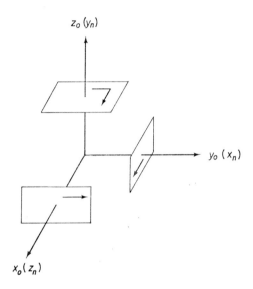

Fig. 7.17. Axial transformation for *Pban* → *Pcna*. Directions of glide are shown by arrows on the planes normal to the axes.

similarly *Cmca* → *Bbcm*, *Iba2* → *Ic2a*, *C222*$_1$ → *B22*$_1$*2*, etc. Possible permutations of symbol for each orthorhombic group are listed in the International Tables; the standard form is said to be in the *abc* orientation, and the new choice of axes in Fig. 7.17 would be described as the *bca* orientation.

Combinations of the hexagonal *P*-lattice and the rhombohedral *R*-lattice with the symmetry of the point groups 3, $\bar{3}$, 3*m*, $\bar{3}$*m* and 32 give the twenty-five trigonal space groups listed in Table 7.6 by their conventional symbols. In the

Table 7.6. Conventional symbols for trigonal space groups

Class 3		Class $\bar{3}$		Class 32			Class 3*m*			Class $\bar{3}$*m*		
*P*3	*R*3	*P*$\bar{3}$	*R*$\bar{3}$	*P*312	*P*321	*R*32	*P*3*m*1	*P*31*m*	*R*3*m*	*P*$\bar{3}$1*m*	*P*$\bar{3}$*m*1	*R*$\bar{3}$*m*
*P*3$_1$				*P*3$_1$12	*P*3$_1$21		*P*3*c*1	*P*31*c*	*R*3*c*	*P*$\bar{3}$1*c*	*P*$\bar{3}$*c*1	*R*$\bar{3}$*c*
*P*3$_2$				*P*3$_2$12	*P*3$_2$21							

Table 7.7. Conventional symbols for tetragonal space groups

Class 4	Class $\bar{4}$	Class $4/m$	Class $4mm$	Class $\bar{4}2m$	Class 422	Class $4/mmm$
$P4$ $I4$	$P\bar{4}$ $I\bar{4}$	$P4/m$ $I4/m$	$P4mm$ $I4mm$	$P\bar{4}2m$ $I\bar{4}2m$	$P422$ $I422$	$P4/mmm$ $I4/mmm$
$P4_1$ $I4_1$		$P4_2/m$ $I4_1/a$	$P4bm$ $I4cm$	$P\bar{4}2c$ $I\bar{4}2d$	$P42_12$ $I4_122$	$P4/mcc$ $I4/mcm$
$P4_2$		$P4/n$	$P4cc$ $I4_1md$	$P\bar{4}2_1m$ $I\bar{4}m2$	$P4_122$	$P4/nbm$ $I4_1/amd$
$P4_3$		$P4_2/n$	$P4nc$ $I4_1cd$	$P\bar{4}2_1c$ $I\bar{4}c2$	$P4_12_12$	$P4/nnc$ $I4_1/acd$
			$P4_2cm$	$P\bar{4}m2$	$P4_222$	$P4/mbm$
			$P4_2nm$	$P\bar{4}c2$	$P4_22_12$	$P4/mnc$
			$P4_2mc$	$P\bar{4}b2$	$P4_322$	$P4/nmm$
			$P4_2bc$	$P\bar{4}n2$	$P4_32_12$	$P4/ncc$
						$P4_2/mmc$
						$P4_2/mcm$
						$P4_2/nbc$
						$P4_2/nnm$
						$P4_2/mbc$
						$P4_2/mnm$
						$P4_2/nmc$
						$P4_2/ncm$

two lowest symmetry classes the symbolism is self-explanatory, but it is necessary to comment on those space groups with a P-lattice type in classes 32, $3m$ and $\bar{3}m$. With this hexagonal cell, the symmetry can be oriented in two ways depending on whether the planes (or diads) are normal or parallel to the prism faces of the cell; the two orientations give different space groups, and the symbolism must be adapted to describe them. The device used is similar to that employed in the tetragonal class $\bar{4}2m$ (or $\bar{4}m2$) already discussed in point group symbolism (Chapter 4.3); as $\bar{4}2m$ the symmetry is oriented so that the diads are parallel to the x- and y-axes, but $\bar{4}m2$ indicates symmetry planes normal to these axes. Following this, $P3m1$ has symmetry planes normal to hexagonal x- and y-axes inclined at $120°$, i.e. the planes are rotated through $30°$ from the xz, yz, uz planes; in $P31m$ the symmetry planes are parallel to these axial planes. Similar relationships exist between $P\bar{3}m1$ and $P\bar{3}1m$ (N.B. Although there are horizontal diad axes, their use in the symbol would only cause confusion; if $P\bar{3}2m$ were written for $P\bar{3}1m$, it could be taken to mean that the diads are parallel to the x- and y-axes; which they are not.) In other groups $P321$ places the diads parallel to the x- and y-axes, whereas $P312$ has them normal to these directions.

There are sixty-eight tetragonal space groups, the largest number associated with any one system; their conventional symbols, given in Table 7.7, are constructed from the two lattice types P and I and the operations of the seven point groups 4, $\bar{4}$, $4/m$, $4mm$, $\bar{4}2m$, 422 and $4/mmm$. In the three classes of lowest symmetry the space group symbolism is self-evident; in the next three classes the second symbol (after the tetrad) refers to elements parallel to the x- and y-axes and the final symbol refers to directions bisecting these axes. Thus for class $\bar{4}2m$, $P\bar{4}2m$ has the diads along the axes to give a different space group from $P\bar{4}m2$ in which the diads are at $45°$ to the x- and y-axes; $P4bm$ has b-glide planes normal (and parallel) to these axes which are bisected by mirror planes. In the holosymmetric class such directions are in the last two places after a symbol for planes normal to the tetrad axis direction; thus $P4/nbm$ has an n-glide normal to the tetrad, vertical b-glide planes normal (and parallel) to the x- and y-axes, with vertical mirror planes at $45°$ to these axes. It will be noticed that d-glide planes can occur at $45°$ to the x- and y-axes in some space groups with an I-lattice.

The twenty-seven space groups of the hexagonal system all have the P-lattice cell in combination with the symmetry operations of 6, $\bar{6}$, $6/m$, 622, $6mm$, $\bar{6}2m$ and $6/mmm$ (Table 7.8). As in the tetragonal system, interpretation of the standard symbols for the lowest symmetry classes 6, $\bar{6}$ and $6/m$ is straightforward, and the conventional order of symbols in the other four classes adapts the method just discussed for tetragonal space groups. For point group $\bar{6}2m$ (or $\bar{6}m2$), the symmetry planes contain the diad axes, and such planes can be set in two orientations with respect to the cell edges, permitting the occurrence of different space groups; by analogy with earlier

Table 7.8. Conventional symbols for hexagonal space groups

Class 6	Class $\bar{6}$	Class $6/m$	Class 622	Class $6mm$	Class $\bar{6}2m$	Class $6/mmm$
$P6$	$P\bar{6}$	$P6/m$	$P622$	$P6mm$	$P\bar{6}2m$	$P6/mmm$
$P6_1$		$P6_3/m$	$P6_122$	$P6cc$	$P\bar{6}2c$	$P6/mcc$
$P6_2$			$P6_222$	$P6_3cm$	$P\bar{6}m2$	$P6_3/mcm$
$P6_3$			$P6_322$	$P6_3mc$	$P\bar{6}c2$	$P6_3/mmc$
$P6_4$			$P6_422$			
$P6_5$			$P6_522$			

conventions, $P\bar{6}2m$ has the diads (and the planes) along the x- and y-axes, whilst $P\bar{6}m2$ has the symmetry planes (and the diads) normal to these axes.

Table 7.9 lists the conventional symbols for the 36 cubic space groups formed from the lattices P, I and F and the point groups 23, $m3$, $\bar{4}3m$, 43 and $m3m$; interpretation of these symbols requires no special elaboration, though the resultant groups are very complex. This is due to the multiplicity of equivalent points (up to 192 per cell) produced by a large number of symmetry elements, some of which are inclined to any plane of projection; the normal diagrammatic representation is not usually attempted for cubic crystals, though they can be treated as if they were derived from certain orthorhombic groups (of the 222 and mmm types) and certain tetragonal groups (of the $\bar{4}2m$, 422 and $4/mmm$ types) with the imposition of the dimensional regularity of the cubic cell. In general, however, a list of the co-ordinates of equivalent positions is sufficient to characterise each group.

This brief discussion of crystallographic space groups has only allowed an introductory account of their nature and symbolic representation; the pattern types which they represent are fundamental to any analysis of crystal structure and fuller accounts both of their derivation and properties will be found in the bibliography. In particular we must make a special reference to Vol. 1 of *International tables for X-ray crystallography*, a definitive text on this subject; its nomenclature (based on the Hermann–Mauguin notation) has been followed here, with no reference to the older Schoenflies notation, though this is still occasionally used.

The practical determination of space groups is briefly outlined in Chapter 8.5 and later chapters; for a particular substance, this is not just an arid cataloguing exercise, for some knowledge of the space group is an essential part of the preparation for the determination of atomic structure, in which the detailed arrangement of cell contents is worked out. Uses of space group information in such work are manifold, and we cannot here penetrate far into the vast subject of crystal structure determination; nevertheless, as an elementary illustration, we will consider two hypothetical examples concerning the use of S.E.P's. Let us choose the orthorhombic space group *Pccn*, which can be recognised with certainty experimentally; the usual diagrams are shown

Table. 7.9. Conventional symbols for cubic space groups

Class 23			Class m3			Class 432			Class $\bar{4}3m$			Class m3m		
P23	F23	I23	Pm3	Fm3	Im3	P432	I432	F432	P$\bar{4}$3m	I$\bar{4}$3m	F$\bar{4}$3m	Pm3m	Im3m	Fm3m
P2$_1$3		I2$_1$3	Pn3	Fd3	Ia3	P4$_1$32	I4$_1$32	F4$_1$32	P$\bar{4}$3n	I$\bar{4}$3d	F$\bar{4}$3c	Pm3n	Ia3d	Fm3c
			Pa3			P4$_2$32						Pn3m		Fd3m
						P4$_3$32						Pn3n		Fd3c

in Fig. 7.18 and the co-ordinates of equivalent positions are given in Table 7.10. If the material is an element A, and Z (the number of formula units per cell) calculated from the density and cell dimensions, as in Chapter 1, is 4, all atoms A must be situated on one of the sets of S.E.P's. With a single atom as pattern motif, positions with any point symmetry can be occupied; but examination of the co-ordinates of those S.E.P's with point symmetry 2 as in

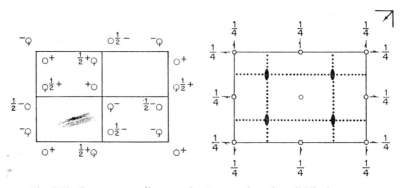

Fig. 7.18. Space group diagrams for *Pccn* projected on (001) plane.

Table 7.10. List of general and special equivalent positions for space group *Pccn*

Number of positions and point symmetry		Co-ordinates of equivalent positions
8	1	$x, y, z;$ $\frac{1}{2} - x, \frac{1}{2} - y, z;$ $\frac{1}{2} + x, \bar{y}, \frac{1}{2} - z;$ $\bar{x}, \frac{1}{2} + y, \frac{1}{2} - z.$ $\bar{x}, \bar{y}, \bar{z};$ $\frac{1}{2} + x, \frac{1}{2} + y, \bar{z};$ $\frac{1}{2} - x, y, \frac{1}{2} + z;$ $x, \frac{1}{2} - y, \frac{1}{2} + z.$
4	2	$\frac{1}{4}, \frac{3}{4}, z;$ $\frac{3}{4}, \frac{1}{4}, \bar{z};$ $\frac{1}{4}, \frac{3}{4}, \frac{1}{2} + z;$ $\frac{3}{4}, \frac{1}{4}, \frac{1}{2} - z.$
4	2	$\frac{1}{4}, \frac{1}{4}, z;$ $\frac{3}{4}, \frac{3}{4}, \bar{z};$ $\frac{1}{4}, \frac{1}{4}, \frac{1}{2} + z;$ $\frac{3}{4}, \frac{3}{4}, \frac{1}{2} - z.$
4	$\bar{1}$	$0, 0, \frac{1}{2};$ $\frac{1}{2}, \frac{1}{2}, \frac{1}{2};$ $0, \frac{1}{2}, 0;$ $\frac{1}{2}, 0, 0.$
4	$\bar{1}$	$0, 0, 0;$ $\frac{1}{2}, \frac{1}{2}, 0;$ $0, \frac{1}{2}, \frac{1}{2};$ $\frac{1}{2}, 0, \frac{1}{2}.$

Fig. 7.19(a), shows that sole occupation of either set would lead to a halving of the c-dimension of the cell. We must therefore reject these possibilities and site the atoms on S.E.P's with point symmetry $\bar{1}$ as in Fig. 7.19(b); with sole occupation these two sets are equivalent with only a change of origin, and lead to the conclusion that atoms A are arranged on an *F*-lattice. Although this would cause certain systematic absences in diffraction maxima (see Chapter 8.), the cell dimensions would be unchanged. A complete solution of an atomic structure in this manner is rare, and it is more likely that this use of space groups will only lead to some partial structural data, as in our second illustration. In this we will keep to the same space group and value of Z, but now assume that the specimen is a compound ABX_4; within each cell there are 4 A atoms, 4 B atoms and 16 X atoms, so that, whilst the A and B atoms

are in S.E.P's, the X atoms will occupy two sets of G.E.P's. Once again, individual atoms of A and B can be placed in positions of any point symmetry, and with two types of atom, arguments about the halving of cell dimensions are no longer valid. In this situation one is sometimes helped by a known form of atomic grouping expected to occur in the structure; for example, it could be that our compound is likely to have atomic complexes

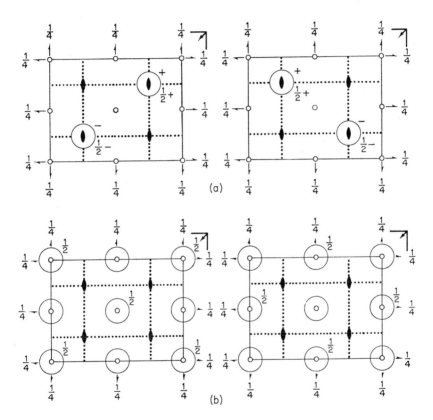

Fig. 7.19. Sole occupation of S.E.P.'s in space group *Pccn*. (a) Point symmetry 2. (b) Point symmetry $\bar{1}$.

BX_4 with a tetrahedral arrangement of the 4X atoms around a central B atom, as in an SO_4^{2-} radical. With this grouping, the central B atoms cannot be placed at S.E.P's with point symmetry $\bar{1}$, for the tetrahedra of which they are a part do not have this kind of symmetry; however, they could be located on the S.E.P's with point symmetry 2, and this would allow the X atoms in G.E.P's to form tetrahedra around them. We deduce that the B atoms must occupy one of this type of S.E.P., and that the A atoms are either on the other

S.E.P. with point symmetry 2 or on one of the sets with point symmetry $\bar{1}$. These conclusions, though more restricted than in the previous example, are still helpful in beginning to decipher the atomic structure of this compound; they could lead, for example, to the hypothetical structure of the compound ABX_4 shown in Fig. 7.20, which shows atomic co-ordinates for the three types of atoms consistent with the space group considerations. Whether such a configuration is a realistic model of the most probable atomic pattern can be assessed by checking its consistency with other data on the packing sizes of

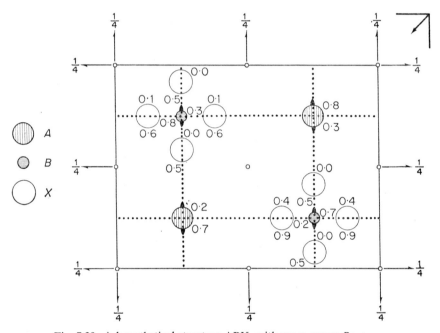

Fig. 7.20. A hypothetical structure ABX_4 with space group *Pccn*.

the particular atoms to be contained in the cell of known dimensions, the co-ordination and bond distances that are implied, the physical properties of the compound, experience of known related structures, etc. When the structure which accords best with all available data (including space group deductions that we have outlined) is found, it can then be tested by comparison of the observed and predicted intensities of diffraction maxima as is discussed in later chapters.

7.5. Exercises and problems

1. What are the lattice types for the structures of gold, SnO_2 and $CO(NH_2)_2$ described in Chapter 1.3? Determine the values of S_{110}, S_{111}, S_{112}, d_{110},

d_{111} and d_{222} for (a) gold ($a = 4\cdot08$ Å), and (b) SnO_2 ($c = 3\cdot19$, $a = 4\cdot74$ (in Å)).

2. $CdCl_2$ is trigonal, and the co-ordinates of the atoms in terms of a triply primitive hexagonal cell are:

$$(0, 0, 0; \tfrac{2}{3}, \tfrac{1}{3}, \tfrac{1}{3}; \tfrac{1}{3}, \tfrac{2}{3}, \tfrac{2}{3})+$$
$$Cd: 0, 0, 0; \quad Cl: \pm (\tfrac{2}{3}, \tfrac{1}{3}, \tfrac{1}{12})$$

Draw a plan of several unit cells on (0001), and identify the minimum volume rhombohedral cell; write down the atomic co-ordinates for Cd and Cl atoms within this cell.

3. Show that two perpendicular intersecting screw diad axes imply the existence of a rotation diad normal to their plane. If the space group $P2_12_12_1$ contains only screw diad axes how must these be arranged? If the three screw diad axes of this symbol are intersecting, what distribution of G.E.P's is produced? To which of the space groups listed in Table 7.5 does this distribution belong?

4. In a centric space group, the co-ordinates of G.E.P's could be listed as:

$$x, y, z; \quad x, \bar{y}, z; \quad \tfrac{1}{2} - x, y, \bar{z}; \quad \tfrac{1}{2} - x, \bar{y}, \bar{z};$$
$$\tfrac{1}{2} + x, \tfrac{1}{2} + y, z; \quad \tfrac{1}{2} + x, \tfrac{1}{2} - y, z; \quad \bar{x}, \tfrac{1}{2} + y, \bar{z}; \quad \bar{x}, \tfrac{1}{2} - y, z.$$

Identify the space group, and re-draw it with an origin on a centre of symmetry; list the co-ordinates of G.E.P's with respect to the new origin.

5. $Cnnb$ is an unconventional form of an orthorhombic space group listed in Table 7.5. Determine the conventional symbol, and write down the G.E.P's. Give equivalent symbols for other choices of axes.

6. Draw conventional space group diagrams for $R3c$ and $P6/mcc$. In each of these groups how many G.E.P's are there? Show that in both cases the S.E.P's can be reduced to 2, and list their co-ordinates.

7. Repeat (6) for the space group $P4/nmm$.

8. In an idealised representation of the structure of aragonite ($CaCO_3$), the appropriate fractional co-ordinates of atoms in an orthorhombic cell ($a = 4\cdot94$, $b = 7\cdot94$, $c = 5\cdot72$ (in Å)) are quoted as follows:

$$Ca: 0, \tfrac{1}{12}, \tfrac{1}{4}; \quad \tfrac{1}{2}, \tfrac{7}{12}, \tfrac{1}{4}; \quad 0, \tfrac{5}{12}, 0; \quad \tfrac{1}{2}, \tfrac{11}{12}, 0.$$
$$C: \tfrac{1}{2}, \tfrac{1}{4}, \tfrac{1}{6}; \quad 0, \tfrac{3}{4}, \tfrac{1}{3}; \quad \tfrac{1}{2}, \tfrac{1}{4}, \tfrac{2}{3}; \quad 0, \tfrac{3}{4}, \tfrac{5}{6}.$$
$$O: \tfrac{1}{4}, \tfrac{1}{6}, \tfrac{1}{6}; \quad \tfrac{3}{4}, \tfrac{1}{6}, \tfrac{1}{6}; \quad \tfrac{1}{2}, \tfrac{5}{12}, \tfrac{1}{6};$$
$$\tfrac{1}{4}, \tfrac{5}{6}, \tfrac{1}{3}; \quad \tfrac{3}{4}, \tfrac{5}{6}, \tfrac{1}{3}; \quad 0, \tfrac{7}{12}, \tfrac{1}{3};$$
$$\tfrac{1}{4}, \tfrac{1}{3}, \tfrac{2}{3}; \quad \tfrac{3}{4}, \tfrac{1}{3}, \tfrac{2}{3}; \quad \tfrac{1}{2}, \tfrac{1}{12}, \tfrac{2}{3};$$
$$\tfrac{1}{4}, \tfrac{2}{3}, \tfrac{5}{6}; \quad \tfrac{3}{4}, \tfrac{2}{3}, \tfrac{5}{6}; \quad 0, \tfrac{11}{12}, \tfrac{5}{6}.$$

Draw a plan of the structure on (001). Identify the space group, and specify the different equivalent positions occupied by the Ca, C and O atoms. What is the conventional symbol for this space group?

SELECTED BIBLIOGRAPHY

Derivation of lattice types and space groups
BUERGER, M. J. 1956. *Elementary crystallography*. Wiley.
HILTON, H. 1963. *Mathematical crystallography*. Dover Publications.
KOSTER, G. F. 1957. *Space groups and their representations*. Academic Press.
TERPSTRA, P. 1955. *Introduction to the space groups*. Groningen.

Representation of all 230 space groups in standard form
INTERNATIONAL UNION OF CRYSTALLOGRAPHY. 1965. *International tables for X-ray crystallography*, vol. 1. Kynoch Press, Birmingham.

Use of space groups in structure determination
LIPSON, H. and COCHRAN, W. 1966. *The crystalline state:* Vol. III, *The determination of crystal structure*. Bell.

8

X-RAYS AND THEIR INTERACTION WITH CRYSTALLINE MATTER

8.1. X-rays

(a) The Nature of the Radiation

X-rays were discovered in 1895 by Röntgen during the operation of a cathode ray tube. Early experiments on their properties showed that they travelled in straight lines from which they were not deflected by electric or magnetic fields, and that they had considerable powers of penetration and could ionise gases and blacken photographic plates. Later work confirmed that X-rays are part of the spectrum of electromagnetic radiation with transverse waves similar to light; they occupy a band of wavelengths from about 10^{-6} to 10^{-9} cm whereas visible light has wavelengths between 10^{-4} and 10^{-5} cm. Despite their formal similarities, visible light and X-rays have different interactions with matter. One of the most striking is in the power of penetration, where the term 'X-ray' has become synonymous with an examination of features underlying the surface, like bone structure, etc. But over the range of X-ray wavelengths the penetrating power decreases as the wavelength increases; this variation provides a crude division of the X-ray region into 'hard' X-rays of shorter wavelengths and 'soft' X-rays of longer wavelengths. Medical X-rays are hard, and many soft X-rays are so strongly absorbed by matter as to severely limit their use in experimental work; nevertheless, there is a small region of the spectrum with wavelengths between about 5×10^{-7} and 3×10^{-8} cm which are neither too heavily nor too lightly absorbed and whose scattering by the atoms of an absorber is detectable without great experimental difficulties. It is in this region that the X-rays are suitable for crystallographic work, and it should be understood that henceforward whenever we talk of X-rays we refer only to this limited sector of the spectrum.

(b) Production of X-rays

Like other X-rays, the narrow spectral range of crystallographic radiation is produced by bombarding a target of a suitable material with fast moving electrons; the construction of X-ray sources (or tubes) will be described shortly, and for the moment we will confine discussion to the nature of the

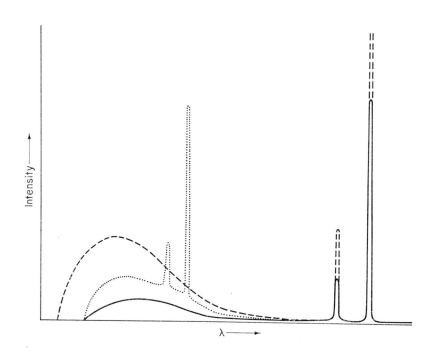

Fig. 8.1. Emission spectra for X-rays. (a) Target A, applied voltage V (full line). (b) Target A, applied voltage $2V$ (broken line). (c) Target B, applied voltage V (dotted line).

radiation emitted under these conditions. All emitted X-ray spectra are similar in general but detailed differences depend on two experimental factors: (i) the target material, and (ii) the conditions of excitation, in particular the accelerating voltage applied to the electrons. Fig. 8.1 shows diagrammatically the effects of these two factors; the full line (a) is the spectrum for target material A with an applied voltage V, the broken line (b) is from the same target but with a voltage of $2V$, whilst the dotted line (c) represents the spectrum for a different target material B with the original voltage V. We see from this that an emission spectrum has two separable components, with sharp peaks at wavelengths dependent on the target material but also a continuous weaker background at all wavelengths greater than a minimum value related to the applied voltage. This background containing a continuous range of wavelengths is said to form *white radiation*. The intensity of white radiation rises sharply from the minimum wavelength but decreases again at longer wavelengths, with a spectral distribution that depends more on the conditions of excitation than the nature of the target material; the precise form of the curve is affected by the target, but the minimum wavelength is constant and there is always a peak at the lower wavelengths. A change in experimental conditions leads to a different minimum wavelength and the peak is modified both in height and wavelength even for the same target material. Superimposed on the white radiation spectrum are relatively sharp spectral lines at wavelengths determined by the target material; this other part of the emission spectrum is known as the *characteristic radiation*. Although their intensities depend on the conditions of excitation, the positions of characteristic lines are always unchanged for a given material. The number of characteristic lines also depends on this material, but for most targets used in crystallographic work there are usually two peaks, the stronger of which at higher wavelengths can be resolved into a doublet.

White and characteristic radiations are formed in different physical processes within the bombarded target. White X-rays are emitted when bombarding electrons lose energy in deceleration caused by collisions, by the internal fields of target atoms, etc.; a decelerating electron loses energy as radiation and so provides a continuous range of wavelengths, for the electrons may suffer many differing energy losses before they are finally arrested. It is also possible for an electron of maximum energy to lose it all as radiation in one packet; when this occurs it will fix a lower limit to the wavelengths emitted under given operating conditions. If the accelerating potential is V, this energy loss can be represented as

$$eV = h\nu_{max}$$

where e is the electronic charge, h is Planck's constant and ν_{max} is the maximum frequency of white radiation emitted. Substituting numerical values and writing the *minimum wavelength* λ_{min} as c/ν_{max}, this is expressed in convenient terms as

$$\lambda_{min} = 12 \cdot 4/V$$

where λ_{min} is measured in Ångstrom units (1 Å = 10^{-8} cm, a unit commonly employed in crystallographic work) and V is measured in kilovolts (KV); thus with an applied potential of 50 kV, the cut-off limit of the white radiation spectrum is about $\frac{1}{4}$ Å.

The separate sharp lines of characteristic radiation have many parallels in other forms of atomic spectra in different wavelength regions; all originate in the discontinuous energy changes that must occur as extranuclear electrons of an atom move from one orbital of fixed energy to another. Within the range of X-ray wavelengths we are concerned mainly with changes affecting the innermost shells of electrons of the atoms of the target material; normally these are fully occupied, and the bombarding electrons must be sufficiently energetic to dislodge electrons bound in these orbitals in order to create vacancies. Excitation of characteristic spectra from a particular target cannot occur until the accelerating voltage has been increased to a value at which bombarding electrons can acquire this minimum energy; this is known as the *minimum excitation voltage* for the target. Once vacancies in the innermost shells have been created, they will be filled by other extranuclear electrons of higher energies in the same atom; in such changes the excess energy is released as radiation to give characteristic X-rays at wavelengths determined by the energy levels of the target atoms. The common notation for these discrete spectral lines is descriptive of the physical changes which cause emission. Suppose that an electron has been removed from the innermost K shell of a target atom by bombardment; this vacancy can be filled by a variety of extranuclear rearrangements. The most probable process is one in which it is filled by an electron from the adjacent L shell of the atom; this requires a loss of energy $E_L - E_K$, a fixed value for the atoms of a particular target, say Cu. Characteristic radiation is emitted of wavelength λ ($= E_L - E_K/hc$), and is described as Cu$K\alpha$, denoting the fixed energy levels of the target material by Cu, the vacancy by K, and the filling of this vacancy from an adjacent shell by α. In the same way Cu$K\beta$ denotes radiation emitted by a copper target when a vacancy in the K shell is filled by an electron from the next but one (M) shell; this process is less likely than that which produces Cu$K\alpha$ radiation and so the corresponding spectral line is not as intense. Of course vacancies created by the bombarding electrons are not confined to the K shell, and simultaneously one might expect CuL, etc. lines; but for most common target materials (except tungsten) the energy levels are such that wavelengths corresponding to such changes are so long as to be outside the normal range of the X-ray region and are completely absorbed within the target. In general, two characteristic peaks are observed due to $K\alpha$ and $K\beta$ radiations, though there is a fine structure which can be important in crystallographic studies. Whilst energy levels within the K shell are the same for all electrons, they are slightly different for various electrons in the L shell; this gives minor variations in the energy transitions of $K\alpha$ radiation to form a doublet (e.g. Cu$K\alpha_1$ and Cu$K\alpha_2$ have a wavelength difference of about 0·004 Å). It might also be expected that $K\beta$ radiation should be resolved into multiple peaks, but for most common target materials such resolution as occurs can be neglected in practice and only a mean value is quoted. Table 8.1 gives a list of common targets with their characteristic wavelengths and other data.

Table 8.1. Common X-ray target materials

Element	Line	λ (in Å)	Minimum excitation voltage (in KV)	Absorption edge of target (in Å)	β-filter
Mo	$K\alpha_1$	0·709			
	$K\alpha_2$	0·714	20·0	0·620	Zr
	$K\beta$	0·632			
Cu	$K\alpha_1$	1·541			
	$K\alpha_2$	1·544	9·0	1·380	Ni
	$K\beta$	1·392			
Ni	$K\alpha_1$	1·658			
	$K\alpha_2$	1·662	8·3	1·487	Co
	$K\beta$	1·500			
Co	$K\alpha_1$	1·789			
	$K\alpha_2$	1·793	7·7	1·607	Fe
	$K\beta$	1·621			
Fe	$K\alpha_1$	1·936			
	$K\alpha_2$	1·940	7·1	1·743	Mn
	$K\beta$	1·757			
Cr	$K\alpha_1$	2·290			
	$K\alpha_2$	2·294	6·0	2·070	V
	$K\beta$	2·085			

(c) X-ray Tubes

Technical problems involved in producing satisfactory X-ray sources have been gradually overcome until most modern X-ray tubes can be compared to bulbs for electric light and are inserted into a standard electrical system as a particular radiation is desired. The insert tube and its predecessors, gas and demountable tubes, all produce the kind of X-ray spectrum we have described by electron bombardment of a target. In the early cold cathode gas tube a controlled air leak maintained a pressure of about 10^{-3} cm of mercury between the cathode and anode (target). Electrons and positive ions, from ionised gas particles, move towards target and cathode respectively when an accelerating voltage is applied; further electrons can be produced by positive ion bombardment of a suitable cathode (e.g. aluminium). Such a simple system has many disadvantages, not least its low output of X-rays; but it does achieve a purity of radiation often lost in modern hot-cathode designs. In these tubes there is no discharge in the residual gas (at pressures of less than 10^{-5} cm of mercury) and the bombarding electrons are emitted from a hot tungsten filament acting as cathode. Many of the earlier hot-cathode tubes were of a demountable variety, so-called because they were continuously evacuated by pumping; repairs such as the replacement of filaments and targets could be undertaken by allowing the system to come to atmospheric pressures. Despite their flexibility and indefinite life, demountable tubes have gradually declined in popularity because of the maintenance that is essential for smooth running over long periods; many that now remain are used for particular specialised purposes. A modern insert tube has the filament and target permanently sealed off at low pressure; Fig. 8.2 shows schematically the essentials of its design. Electrons are emitted from a heated filament and focused on to the target material, usually into a rectangular area about 10 mm × 1 mm in dimensions; this source is viewed through windows of a lightly absorbing material (often Be or a glass of light elements). There are commonly four windows at 90° arranged so that foreshortening of the focal area by viewing at a slight inclination gives either a line or square focus; each

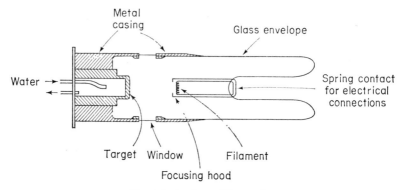

Fig. 8.2. An insert X-ray tube.

of these two shapes of source is advantageous for certain experimental arrangements. As in all X-ray tubes, heat is generated in the target, which is cooled by the circulation of water through its hollow interior. In all designs the efficiency of the cooling system is an important factor in limiting the flux density of electrons which can be directed into the focal area on the target; this restricts the maximum output of X-rays from the tube, for the target must not be overheated and punctured; in insert tubes the maximum loadings are of the order of 0.1 kW/mm^2. There have been a number of successful attempts to improve the efficiency of X-ray sources in demountable tubes by the use of very small focal areas, sometimes located on the periphery of a rotating target disc; such equipment has nearly always been developed by the demands of specialised experiments, and the normal output of an insert tube is quite adequate for routine work, particularly as the detection of scattered X-rays has become much more efficient. The practical advantages of sealed insert tubes in ease of operation greatly outweigh their disadvantages in capital cost, limitation of life and lack of interchangeability of targets. Like all hot filament tubes, they suffer from the gradual deposition of tungsten evaporating from the filament; on the target this can cause unwanted W radiation, whilst on the windows it will reduce the transmission of X-rays, both of which limit the useful life of an expensive insert tube.

(d) Properties of X-rays

Like other electromagnetic radiations, X-rays are absorbed in their passage through a medium and refracted in passing from one medium to another. Refractive indices are less than 1, but in most cases differ from unity in only the fifth or sixth decimal place; the effects of refraction are only important in experiments of the highest accuracy, and refraction corrections can be neglected in routine studies.

Absorption, however, can be of practical importance in any kind of work, and it requires rather fuller discussion. In their passage through a material the X-rays lose energy (i) to the secondary X-rays scattered in all directions, and (ii) by transformation of the radiation. In (i), the scattered X-rays can either be identical with the incident radiation (i.e. they can be regarded as diffraction of the incident X-rays by atoms of the material), or they can be changed in wavelength by a collision with an electron (the Compton effect); in (ii) the energy of the X-rays is transformed into β-rays, fluorescent radiation, heat, etc. The sum total of all processes leads to an expression of the effects of absorption as

$$I = I_o \exp(-\mu t)$$

where I, I_o are the transmitted and incident intensities, t, the path length of X-rays of wavelength λ, and μ is a constant for a particular material, its *linear absorption coefficient*. In practice, it is rather more convenient to express the path within the absorber as the number of grams per square cm; this eliminates any variation in the absorption coefficient for a given element due to the state of aggregation, a complication which affects the values of linear absorption coefficients. To do this we define a *mass absorption coefficient* (μ_M) independent of physical state such that

$$\mu_M = \mu/\rho$$

where ρ is the density. The mass absorption coefficients of elements can then be used to calculate linear absorption coefficients of compounds in which the proportions of the component elements are known, for

$$\mu = \rho \Sigma p \mu_M$$

where p is the fraction by weight of an element of mass absorption coefficient μ_M.

Values of μ_M for all the elements are tabulated in reference books in the bibliography; for a given element, μ_M increases with wavelength, often said to be as a λ^3 variation though a smaller exponent between 2·5 and 3 would be more appropriate in many cases. Within the range of X-ray wavelengths many elements show sharp discontinuities in this variation (Fig. 8.3), so that there is a sharp rise in μ_M with decreasing λ. These are known as *absorption edges*, and occur at wavelengths for which the energy of an incident quantum (hc/λ) is sufficient to dislodge an electron in, say, the K shells of atoms of the absorber element; this causes a sharp drop in the intensity of the transmitted X-ray beam, so that μ_M increases rapidly at these wavelengths. This process must also be accompanied by the simultaneous emission of a characteristic radiation by the absorber atoms (the fluorescent radiation in (ii) above) as the vacancy in the K shell is filled and the scattering power of the absorber will change. Values of the K absorption edges for the elements of common target materials are quoted in Table 8.1. The existence of absorption edges in the X-ray range can be both troublesome and advantageous in practical application. When studying diffraction, we must avoid the emission of fluorescent

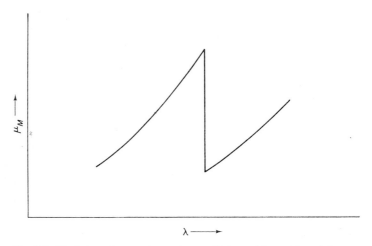

Fig. 8.3. Variation of mass absorption coefficient with wavelength for an element. Note the sharp discontinuity at the wavelength of an absorption edge.

radiation by the crystal for this can mask diffraction of the primary beam; this requires a careful selection of the wavelength of primary radiation used in any diffraction experiments. By the nature of the electronic changes an absorption edge wavelength is necessarily shorter than that of the excited fluorescent radiation; to avoid fluorescence we must choose a primary radiation whose wavelength is longer than any absorption edge of the atoms of the crystalline specimen. For example the data of Table 8.1 shows that Fe atoms (absorption edge, 1·743 Å) will fluoresce with Cu$K\alpha$ radiations ($\lambda \sim 1·54$ Å) but not with Fe$K\alpha$ radiations ($\lambda \sim 1·94$ Å); any compound containing significant numbers of Fe atoms should not therefore be studied with Cu$K\alpha$ radiation, but its diffraction pattern can be recorded with Fe$K\alpha$ radiation. In practice it is not possible to select a primary radiation which will eliminate all fluorescent effects for many compounds; fortunately for many elements fluorescent radiation is of very long wavelength and is absorbed by the specimen or the air before it reaches a photographic plate.

Absorption edges can, however, be turned to advantage in other circumstances; in subsequent chapters we shall see that many diffraction techniques require incident radiation with a narrow wavelength band, such as that provided by a characteristic peak. Strictly monochromatic radiation can only be obtained when the spectrum emitted by an X-ray tube is reflected from a suitably oriented crystal monochromator; but the intensity of crystal reflected radiation is rather weak, and as incident radiation it will considerably lengthen exposure times necessary to detect even weaker scattered radiation. For most practical purposes it is sufficient to place a *filter* in the path of the X-rays from the tube to preferentially reduce the intensity of, say, unwanted $K\beta$ radiation (and also white radiation to some extent). Suitable filters can be made from materials with absorption edges between the wavelengths of the $K\alpha$ and $K\beta$ radiations; for example from Table 8.1, Ni (absorption edge $1\cdot487$ Å) will preferentially reduce the intensities of $CuK\beta$ ($\lambda = 1\cdot392$ Å) compared to $CuK\alpha$ radiations ($\lambda \sim 1\cdot54$ Å). β filters, listed for common target materials in this table, are simply foils inserted into the beam from the X-ray tube; the thickness of the foil determines the reduction in the ratio of intensities $K\beta/K\alpha$, so that with a suitable material the ratio of transmitted intensity to incident intensity is, say, $\frac{1}{2}$ for $K\alpha$ radiation but only $1/1000$ for $K\beta$ radiation. Filtered radiation will also contain some regions of the white X-ray spectrum, but this can usually be discounted in interpreting diffraction patterns. There are also some occasions when an experiment requires the use of white radiation. In practice, this is normally provided by using unfiltered radiation from the tube, for superposed effects due to characteristic peaks are again easily identified and discounted; if white radiation alone is essential, this can be obtained from a tungsten target under running conditions that do not excite characteristic lines.

Finally, it is well known that all X-rays can be harmful to the tissues of the human body; even the hard X-rays of medical radiography must be used with strictly controlled dosages. For the longer wavelengths of the diffraction range, there are much greater dangers; burns and the cumulative effects of relatively small doses can be very damaging to health, and all X-ray tubes and apparatus are shielded (usually with lead) to prevent direct or scattered radiation reaching the body. Nevertheless, regular medical checks must always be undertaken on any person in danger of frequent exposure.

8.2. Optical principles of diffraction

In Chapter 8.1(d) we mentioned that part of the interaction of X-rays with matter consists of the scattering of unmodified radiation; this occurs when extranuclear electrons of the atoms are set into forced vibration by the incident X-rays and subsequently re-radiate X-rays of the same wavelength in all directions. Such behaviour is common to many types of electromagnetic radiation, and the study of scattered radiation can tell us about the nature of the scattering atoms and the way in which they are arranged. Crystalline matter, with its regular and repetitive atomic arrangements, might reasonably be expected to scatter into a series of relatively sharp discrete maxima analogous to those formed by the passage of light through a diffraction grating. In fact the optical principles of the diffraction of light by a regular array of scatterers are paralleled by those which control the more complex diffraction of X-rays by crystals. The purpose of this section is to re-state and clarify optical principles in such a way that they are relevant and instructive in the discussion of X-ray diffraction in the next section.

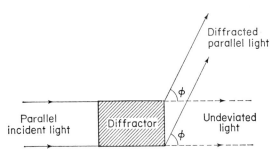

Fig. 8.4. Geometry of Fraunhofer diffraction.

The optical patterns that are of value are those produced by Fraunhofer diffraction, using parallel light with planar wavefronts; Fig. 8.4 shows the essential geometry of the arrangement, and full experimental details for the observation of these patterns will be found in textbooks of physical optics. In analysing the nature of the diffracted beams formed from a repetitive diffractor like a grating, it is advantageous in the present context to consider the scattering by a single diffractor and then see how this is modified by regular repetition. The simplest unit is a single slit of finite width a; with normal parallel illumination, we can regard each element dx of the slit as scattering light in the direction ϕ with a particular amplitude and phase (Fig. 8.5(a)). If the elements are of the same length, they will scatter with identical amplitudes $k\,dx$; but the relative phases of the scattered radiations will depend on the positions of elements within the slit. Let us assign a phase angle $0°$ to the element at the bottom of the slit and phase angle 2α to that at the top; the phase angle for any other element at a distance x from the bottom of the slit must be $\dfrac{x}{a} \cdot 2\alpha$. We can then sum all the contributions from the elements on a

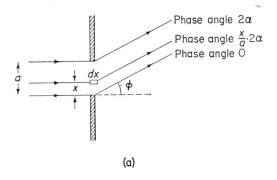

(a)

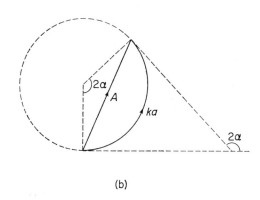

(b)

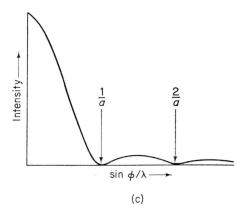

(c)

Fig. 8.5. Fraunhofer diffraction by a single slit of width a. (a) The passage of light through the slit. (b) The phase-amplitude diagram. (c) The variation of intensity from the central maximum.

phase-amplitude diagram to give an arc of a circle of radius $ka/2\alpha$ (Fig. 8.5(b)), so that the resultant amplitude of scattering at angle ϕ is

$$A = \frac{ka}{\alpha} \sin \alpha$$

But the total phase difference can be expressed in terms of the path difference between the top and bottom rays as

$$2\alpha = \frac{2\pi}{\lambda} \cdot a \sin \phi$$

so that the resultant intensity becomes

$$I = A^2 = \frac{k^2}{\pi^2} \cdot \frac{\sin^2 (\pi/\lambda \cdot a \sin \phi)}{\sin^2 \phi/\lambda^2}$$

Fig. 8.5(c) shows the variation of this intensity function plotted against the scattering angle expressed as $\sin \phi/\lambda$; the central maximum is flanked by sub-sidiary maxima with minima of zero intensity whenever $\sin \phi/\lambda = m/a$ (m is an integer). This single slit pattern with its broad maxima is well known; the narrower the slit, the wider the central maxima and so on.

The first step towards grating repetition involves the introduction of a second identical slit at a distance c from the first (Fig. 8.6(a)). Each separate slit will scatter as above at angle ϕ, but their contributions must be combined in accordance with the phase difference introduced by the separation c; in other words between corresponding points on the two slits there is a phase difference $2\beta \left(= \frac{2\pi}{\lambda} \cdot c \sin \phi \right)$. Fig. 8.6(b) shows the phase-amplitude diagram for the two slit system, from which the intensity function is

$$I = \bar{A}^2 = 4A^2 \cos^2 \left(\frac{\pi}{\lambda} \cdot c \sin \phi \right)$$

i.e. it has the single slit distribution (A^2) modulated by a squared cosine function dependent on c; Fig. 8.6(c) shows the intensity function again plotted against $\sin \phi/\lambda$. In effect, the single slit pattern determines the relative intensities of peaks which occur at positions fixed by the dependence of squared cosine fringes on c.

In a similar way we should find that with three, four, five, etc. slit systems, the intensity function for a single slit forms an envelope which limits the height of maxima whose positions and widths are respectively determined by the slit separation c and the number of slits in the system. Moving on to a full grating of N slits of width a separated by c, the final phase-amplitude diagram is shown in Fig. 8.7(a); the contribution of each element of the grating is an

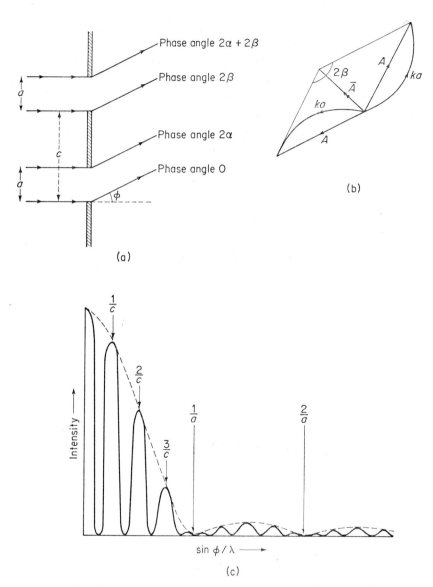

Fig. 8.6. Fraunhofer diffraction by a pair of slits of width a separated by a distance c. (a) The passage of light through the system. (b) The phase-amplitude diagram. (c) The variation of intensity from the central maximum; the pattern is drawn for the conditions $4a = c$. The single slit pattern is indicated by the dashed line.

infinitesimal chord of a circle, and the resultant amplitude $\bar{A}$ is also a chord of the same circle, so that

$$I = \bar{A}^2 = A^2 \frac{\sin^2 N\beta}{\sin^2 \beta}$$

i.e. the intensity distribution has the single slit function modulated by the grating term $\sin^2 N\beta / \sin^2 \beta$ dependent on the slit separation c. The grating term has maxima of two different kinds.

    (i) When $\beta = 0, \pi, 2\pi$, etc. there are maxima of height N^2, often called the principal maxima; they will occur when $\sin \phi / \lambda = 0, 1/c, 2/c$, etc.

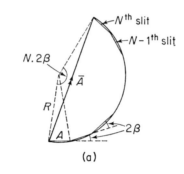

(a)

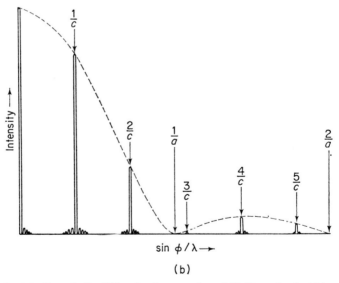

(b)

Fig. 8.7. Fraunhofer diffraction by a grating of N slits each of width a, separated by distance c. (a) The phase-amplitude diagram; if N is large the polygon may be approximated to the arc of a circle. (b) The variation of intensity from the central maximum. The single slit pattern is indicated by the dashed line.

(ii) When $N\beta = 0$, π, 2π, etc. there will be minima with zero intensity; these will occur when $\sin \phi/\lambda = 1/Nc$, $2/Nc$, $3/Nc$, etc. to give $N - 1$ minima between two principal maxima, equally spaced except for those on either side of a principal maximum. Between these minima, the intensity function rises to give secondary maxima of very weak intensity; these secondary maxima are not of equal intensities but fall off away from principal maxima.

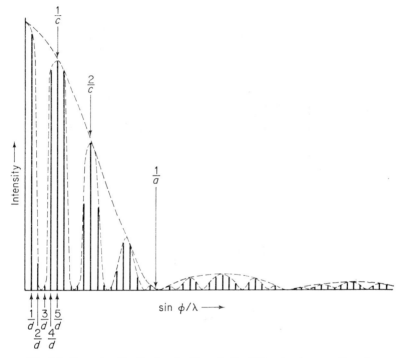

Fig. 8.8. Intensity distribution for Fraunhofer diffraction by a grating of N units separated by distance d. Each grating unit is a pair of slits of width a separated by a distance c; the dashed line is the pattern for the double slit system (cf. Fig. 8.6(c)). The pattern is drawn for $20a = 5c = d$.

Such a grating pattern is schematically illustrated in Fig. 8.7(b); although some secondary maxima are shown in this diagram, for the large values of N of an actual grating they are either too weak or too close to principal maxima to be observed in practice. Compare the sharpness of the principal maxima with those in the same positions for a simple two-slit system.

This analysis of diffraction by a simple grating shows that the pattern produced by any repetitive diffractor can be described as the sampling of a scattering curve due to a single diffractor (the *slit factor*) at points determined

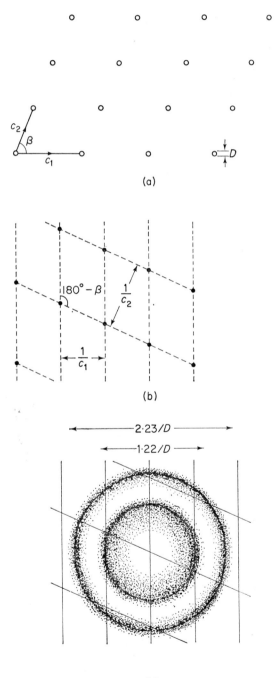

(a)

(b)

(c)

by the repetition interval (the *grating factor*). The unit diffractor can be more complex than a single slit, as for example in Fig. 8.8 which shows the intensity distribution in the pattern from a grating formed from units of two slits of width a at a distance c apart; each pair is repeated at intervals of d to build up the grating. The 'slit factor' now has an intensity distribution similar to that of a double slit system in Fig. 8.6; the 'grating factor' determines the points at which this distribution is sampled at unit intervals corresponding to the principal maxima. We can extend these arguments to planar gratings with a regular two-dimensional arrangement of circular scatterers; Fig. 8.9(a) shows part of a simple oblique pattern of holes each of diameter D. The 'slit factor' is the diffraction pattern of a single aperture, which is a central maximum with a number of fainter concentric rings of weaker intensity; the spacing of these circular fringes depends on D. The 'grating factor' is determined by the geometry of the oblique pattern; rows of holes with spacing c_1 will give perpendicular line fringes spaced at $1/c_1$, whilst the rows with spacing c_2 give perpendicular line fringes spaced at $1/c_2$. Diffraction by a grating with rows of this kind will combine the two sets of line fringes to give a 'grating factor' that will only be significant at points where they intersect (Fig. 8.9(b)); the spacings c_1 and c_2 define a grid of sampling points at which diffraction maxima can be observed, but the intensity of any maximum is determined by the underlying pattern of the single hole on which this grid is superposed (Fig. 8.9(c)). Clearly the 'slit factor' can be made more complex (two holes, a hexagonal pattern of holes, holes of different diameter, etc.) but in every case the diffraction pattern from a planar grating will consist of a grid of maxima whose positions are determined by the repeat intervals of the unit diffractors and whose intensities are determined by the relation of this grid to the pattern from a single unit of the grating.

We are now in a position to relate optical diffraction by planar gratings with a two-dimensional periodicity and X-ray diffraction by crystals with a three-dimensional periodicity. In the latter the various atoms and their arrangement within the cell may be compared to the 'slit factor', whilst the periodic repetitions by the lattice translations correspond to the 'grating factor'. In any X-ray diffraction pattern we should therefore expect the positions of discrete maxima to be fixed by the shape and dimensions of the unit cell but the intensities of individual maxima will be determined by the details of the structural arrangements within the cell. In practice, this aspect of the optical analogy can be usefully applied to help with the solutions of problems in

---

Fig. 8.9. Fraunhofer diffraction by a planar grating with an oblique pattern of holes. (a) A part of the grating. (b) The two sets of interference fringes from rows of spacing c_1 and c_2; maxima are observed only at the points of intersection. (c) The grid of (b) superposed on the diffraction pattern of a single hole to determine the intensities of maxima at the points of the grid.

X-ray diffraction, particularly in the field of crystal structure determination; any discussion of these topics is outside the scope of this present book but they are pursued in some of the texts cited in the bibliography.

8.3. Diffraction of X-rays by crystals

(a) *Positions of Maxima*

Attempts to diffract X-rays began almost from the time of their discovery in efforts to demonstrate the wave nature of the radiation; in 1899 a pattern was obtained with a very narrow aperture but the diffraction effects were so slight that they were not generally accepted as conclusive. Some years later von Laue suggested that the spacings of atoms or molecules in crystalline matter were of the same order as some X-ray wavelengths and that diffraction could be achieved by using crystals as diffraction gratings. In 1912 Friedrich and Knipping passed a narrow beam of white radiation through a thin crystal of zincblende (ZnS); a photographic plate beyond the crystal showed a symmetrical pattern of sharp maxima of different intensities in addition to blackening due to the undeviated beam. From this simple experiment the sophisticated techniques of modern X-ray crystallography have developed; they have been supplemented in recent times by the diffraction of electrons and neutrons to give a most powerful range of investigative tools for many aspects of solid state studies. From the discussion of optical diffraction in the previous section it is clear that any analysis of diffraction patterns from crystals can be considered in two parts, those factors which influence the positions of maxima and those which determine the intensities of individual maxima. We shall start by considering the positions of maxima which, from the optical analogy, will be fixed by the lattice translations, i.e. the dimensions and shape of the unit cell. For simplicity we may consider the scattering matter within a cell as associated with lattice points which act as units of the diffraction grating; it is only the separations of these units which are important in deducing the directions of maxima. In an elementary treatment the lattice may be regarded either as a collection of parallel lattice rows or as a stack of equidistant parallel planar nets of lattice points. We shall demonstrate that both methods inevitably lead to the same results, though it is the latter conception which can be more readily developed and which dominates the descriptions of diffraction today.

In the first approach, usually associated with von Laue, the grating of the crystal is regarded as a three-dimensional array of lattice rows so that diffraction is an extension of the optical problem for a planar grid discussed in the preceding section. The optical pattern had maxima at the intersections of fringe systems whose positions were determined by the lattice repeats of two rows; in the same way the maxima in the scattering of X-rays by a crystal will be found at the intersections of fringe systems due to the lattice repeats of three non-co-planar rows. For simplicity, we choose these rows to be the lattice translations defining the cell edges; Fig. 8.10(a) shows parallel monochromatic X-rays of wavelength λ incident at angle i_a upon a row of point scatterers separated by the lattice translation a from which they are scattered at angle d_a. The path difference for X-rays diffracted from successive points is

$$a(\cos d_a - \cos i_a)$$

so that the condition for reinforcement to give a maximum requires

$$a(\cos d_a - \cos i_a) = h\lambda$$

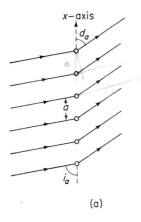

(a)

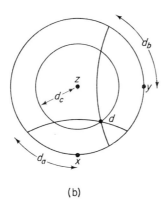

(b)

Fig. 8.10. The Laue conditions. (a) Diffraction by a row parallel to the x-axis. (b) Stereographic representation of the Laue conditions. Cones of diffracted radiation are small circles about the x, y and z axes which are shown as orthogonal for simplicity; maxima will occur only at points such as d, where small circles about the three axes intersect.

where h is an integer. Such lattice rows will produce fringes whose intensity maxima lie on a series of cones about the directions of a; each cone corresponds to a particular value of h. For the two other lattice translations there will be comparable equations

$$b(\cos d_b - \cos i_b) = k\lambda$$
$$c(\cos d_c - \cos i_c) = l\lambda$$

where k and l are integers; as before, the intensity maxima will lie on cones about the directions of b and c determined by the values of k and l. For a crystal lattice only when all three conditions are simultaneously satisfied in a direction common to cones about all three cell edges will a diffracted maximum be observed; the three equations which must be satisfied are often known as the *Laue equations*. The geometry of a solution to these equations is represented on the stereogram of Fig. 8.10(b). It is clear that the equations impose such specific conditions that for a given set of cell constants and a single X-ray wavelength it is probable that few (if any) maxima will occur; in the next chapter we will discuss how experimental techniques overcome this problem either by using a range of wavelengths or by varying the orientation of the crystal.

Whilst the Laue treatment is helpful in interpreting certain features of diffraction patterns, it is distinctly more cumbersome in other respects than another approach due to Bragg in which the crystal lattice is regarded as a collection of equi-distant parallel nets. It is instructive to develop the Bragg method from the Laue equations and this may be done geometrically using the stereogram of Fig. 8.11(a). On this the x-axis, with its rows of lattice points separated by a, is plotted at the centre of the projection, and the poles of the incident (i) and diffracted (d) beams are also shown; a great circle through i and d is constructed to represent the plane containing these beams, and a point n on this circle is found so that $\widehat{in} = \widehat{nd}$. Now this point n must represent the normal to a family of lattice planes, as we can show in the following way:

$$\cos(\pi - i_a) = \cos \widehat{xn} \cos \widehat{in} + \sin \widehat{xn} \sin \widehat{in} \cos \widehat{n}$$

$$\cos d_a = \cos \widehat{xn} \cos \widehat{dn} + \sin \widehat{xn} \sin \widehat{dn} \cos(\pi - \widehat{n})$$

from the spherical triangles xin and xdn respectively (see Appendix B). By addition

$$\cos d_a - \cos i_a = 2 \cos \widehat{xn} \cos \widehat{in}$$

since $\widehat{in} = \widehat{nd}$ by construction. One of the Laue equations will therefore be satisfied when

$$h\lambda/a = 2 \cos \widehat{xn} \cos \widehat{in}.$$

By a similar procedure we can get equations

$$k\lambda/b = 2 \cos \widehat{yn} \cos \widehat{in}$$

$$l\lambda/c = 2 \cos \widehat{zn} \cos \widehat{in}$$

for the other two Laue conditions relating the positions of i and d. In these equations, the angles $\widehat{xn}$, $\widehat{yn}$ and $\widehat{zn}$ define the direction of n relative to the crystallographic axes (Fig. 8.11(b)); a plane which is normal to n must have

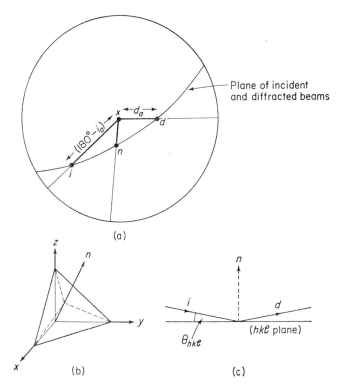

Fig. 8.11. Development of the Bragg law from the Laue equations. (a) Stereogram showing diffraction by a row parallel to the x-axis and the point n in the plane of the incident and diffracted beams. (b) The orientation of the plane defined by $\widehat{xn}$, $\widehat{yn}$ and $\widehat{zn}$. (c) The geometry of reflection.

intercepts which are in the ratios of $1/\cos \widehat{xn}:1/\cos \widehat{yn}:1/\cos \widehat{zn}$ on the x, y and z axes respectively. From the equations above, the values of these ratios must be $a/h:b/k:c/l$, which are just those defined by any member of the (hkl) set of lattice planes; thus, at conditions which satisfy the three Laue equations, the pole n is in the direction of the normal to the (hkl) planes of a lattice in which the cell is defined by a, b and c. The geometry is shown in Fig. 8.11(c); the directions of the incident and the diffracted beams are symmetrically disposed with respect to the normal to a plane of the (hkl) set, and all three directions are co-planar. This is analogous to simple reflection as from a mirror, and it is for this reason that the term '*X-ray reflection*' is commonly used to describe a diffraction maximum. It cannot be too strongly stressed that this terminology implies no physical reflection of the incident X-rays in the sense that they bounce off planes of lattice points (or atoms) in a structure; it is merely that the concept is convenient in locating and describing positions of maxima produced by the physical process of diffraction. Moreover, an analogy with simple mirror reflection is incomplete, for the geometry of Fig. 8.11(c) is critical; 'reflection' will only occur when $\widehat{in}$ has a particular value determined by the crystal constants and the X-ray wavelength. The value of the angle which will allow us to observe a diffraction maximum in some direction as if incident X-rays were reflected from a plane parallel to (hkl) lattice planes is called the *Bragg angle*, θ_{hkl}; it is defined as the glancing angle of incidence so that

$$\theta_{hkl} = 90° - \widehat{in}.$$

The factors which determine θ_{hkl} can be identified by choosing the plane of Fig. 8.11(c) to be the first plane ($p = 1$) of the set (hkl); this is legitimate, since i, n and d in this figure only represent directions. For such a plane the interplanar spacing $d_{hkl} = \dfrac{a}{h} \cos \widehat{nx}$, etc. (see Chapter 5.1), so that the equations expressing the Laue conditions derived from Fig. 8.11(a) become

$$\lambda = 2d_{hkl} \sin \theta_{hkl}$$

which is known as the *Bragg law*, an equation basic to all X-ray diffraction phenomena treated in this way.

The Bragg equation suggests a simple method of identifying the various diffraction maxima from the values of the integers h, k and l; these locate the planes from which 'reflection' takes place when the conditions of incidence are satisfied. In solving this equation interplanar spacings are needed; as we described in Chapter 7.2, d_{hkl} values for a particular substance depend on the size and shape of the unit cell, but a lower limit (with the largest h, k and l values) is imposed by the X-ray wavelength on those that can be brought into a reflecting position. In older usage, the Bragg law was often written $n\lambda = 2d_{hkl} \sin \theta_{hkl}$ so that various orders of reflection (with $n = 1, 2, 3...$) were said to occur from a given family of planes, rather as is done in some branches of spectroscopy. Nowadays it is more usual to regard these different orders of reflection as arising from new families, parallel to the parent planes but with spacings which are fractional parts of their spacing. Thus in the old terminology a maximum might be described as the 2nd order reflection from (100) planes, i.e. $2\lambda = 2d_{100} \sin \theta''_{100}$; nowadays it is considered as a reflection from (200) planes, which are parallel to the (100) set but with only half their interplanar spacing, i.e. $\lambda = 2d_{200} \sin \theta_{200}$. Possible positions of maxima are computed from the Bragg law for all d_{hkl} values greater than the limit $(\lambda/2)$, without the removal of common factors in (hkl); whether an individual maximum is observed depends on its intensity and the factors which control this are discussed in the next section.

The concepts of the Bragg law and its reflecting planes are universally employed in X-ray diffraction work, where maxima are described as hkl reflections (without brackets); another view of their significance can be obtained from Fig. 8.12, which shows the conditions at reflection from a given set of

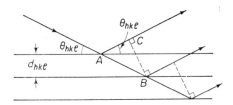

Fig. 8.12. Path difference between radiation reflected from successive planes at Bragg position.

planes. Incident X-rays lose energy as they penetrate the crystal, but meet any member of the family at the correct angle for reflection demanded by the Bragg law. For a particular plane there is no phase difference between X-rays scattered from any point on its surface, but clearly phase differences must occur between radiation scattered from one plane and any other plane of the set. For example, the path difference for the two uppermost planes of the figure is $AB - AC$; under the geometrical conditions of the Bragg law

$$AB - AC = AB(1 - \cos 2\theta_{hkl}) = 2AB \sin^2 \theta_{hkl} = 2d_{hkl} \sin \theta_{hkl} = \lambda,$$

so that the path difference between successive planes of a set is one wavelength. We can therefore regard the reflecting planes of the Bragg treatment as representing planes of constant phase for X-rays scattered in the direction of this maximum; a phase change of 2π is introduced when radiation is scattered from successive planes of a set. This interpretation of the significance of reflecting planes is of value when we consider the intensities of maxima later.

The techniques of interpreting X-ray diffraction patterns are greatly assisted by transforming the Bragg conditions into a convenient geometrical form in a construction first proposed by Ewald. In this the reflecting planes of the direct Bravais lattice are represented by points of a *reciprocal* (or *polar*) *lattice*; reciprocal lattice points are clearly analogous to the points of the grid on which maxima can be observed in optical diffraction by a planar oblique pattern of holes (Fig. 8.9). Possible solutions of the Bragg conditions are determined by considering the intersection of points of the reciprocal lattice with a spherical surface constructed according to the conditions of the experiment. In this way a simpler treatment of the geometrical aspects of complex diffraction problems can be achieved, and the concept of reciprocal space is widely used in X-ray work; unfortunately, its full development and application must be regarded as beyond the scope of an introductory account of X-ray methods but excellent descriptions will be found in books listed in the bibliography.

The geometry of diffraction has been discussed here for the scattering of X-rays, but diffraction patterns for the scattering of electrons and neutrons by crystalline matter have similar general features. These patterns are interpreted in terms of the wave nature of such particles; the scale may be quite different (e.g. 50 kV electrons have a wavelength of about 0·05 Å) and the physical process of diffraction may not be the same (e.g. it is not extranuclear electrons which are important in neutron scattering), but the positions of maxima can be located by solution of the Bragg law as above. Electron and neutron diffraction can give information which is unobtainable by X-ray methods; for example the scattering power of atoms may be quite different for neutrons and X-rays, so giving quite different relative intensities in patterns from the same crystal; sometimes this permits a much readier resolution of some lighter elements which scatter neutrons more effectively than X-rays. The technical problems of obtaining satisfactory electron and neutron diffraction patterns are much greater than for X-rays, which remain the commonest means of structural investigation in the solid state.

(b) Intensities of Maxima

The intensity of a particular diffracted beam as recorded experimentally depends on a number of factors; apart from its relation to the incident intensity an experimental value can be affected by the method of measurement, and also by the micro-texture of the crystal under examination. We shall discuss these and other factors briefly in the next section, but when they have been taken into account, the intensity of any maximum must be fundamentally related to the various scattering atoms and the way in which these form the crystal structure. In terms of the optical analogy, the intensities are determined by the scattering pattern of a unit of the grating; in a crystal structure it is the content and arrangement of atoms in a unit cell which determine this scattering pattern, and in this section we shall set up a method of calculating the expected intensities for structures in which atomic co-ordinates within a cell are known. To do this, the first stage is to devise a means of assigning suitable scattered amplitudes to each atom in the unit cell; these must then be summed with reference to the different locations of the various atoms, bearing in mind that we have just seen that this scattering pattern is only observable in directions determined by the cell dimensions.

Like other diffraction phenomena, the atomic scattering of X-rays implies that the atoms become sources of secondary wavelets of the incident radiation. The nucleus takes no part in this physical process both because of its small radius ($\sim 10^{-14}$ cm) and its large mass; any scattering must be due to forced oscillations of the extranuclear electron clouds whose radii are of the order of Ångstroms. It is therefore reasonable to choose to measure amplitudes scattered by atoms in terms of the scattering by an isolated electron. Any given atom has, of course, a characteristic number of extranuclear electrons, but we cannot just use this number (Z) as a simple factor to obtain its scattering power under all circumstances; particular arrangements of electrons in various orbitals of different radii distinguish one kind of atom from another so that scattered amplitudes depend on the angle of scattering with a form of variation that is different from one atom to the next. Fig. 8.13 shows

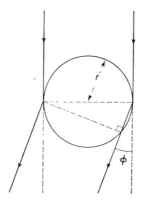

Fig. 8.13. Scattering of X-rays by an electron cloud of radius r in a direction given by ϕ.

scattering by a hypothetical atom in which all electron clouds are concentrated in one particular orbital of radius r; for two limiting rays there must be a finite path difference of $2r \sin \phi$ for any non-zero value of the scattering angle ϕ. When $\phi = 0$ (or 180°) all parts of the cloud scatter X-rays in phase; but in any other direction phase lags are introduced between radiation scattered from different parts of the cloud and these will cause interference; the amplitude scattered by this simple atom will consequently be reduced in a way that depends on the angle of scattering. In any general direction we can express this scattered amplitude as the product of a numerical factor and the scattering of an isolated electron under the same conditions; this numerical factor is known as the *atomic scattering factor* (f) and its variation with scattering angle will depend on the nature of the extranuclear electron clouds for any given atom. When $\phi = 0, f = Z$; but for all other values of $\phi, f < Z$; the actual value of f denotes the number of electrons which would give an equivalent scattered amplitude. Atomic scattering factor curves showing the variations for different atoms can be calculated, though the methods are too complex for discussion here; values of f for different atoms are tabulated in standard reference texts (see *International tables for X-ray crystallography* in particular) usually as a function of $\sin \theta/\lambda$ (= $1/2d$), which provides a convenient measure of scattering angle in diffraction work. Fig. 8.14 shows parts of the atomic f curves for three atoms, light, medium and heavy; there are significant differences in the fall-off of amplitudes as the scattering angle increases.

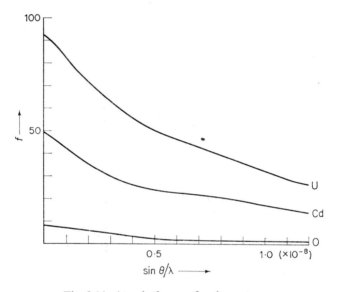

Fig. 8.14. Atomic f curves for three atoms.

Before proceeding to the second stage of intensity calculation, there are two general observations to be made on f values. Firstly, the amplitude scattered by an electron is a very small fraction of any incident amplitude; the scattering by atoms, even of the heaviest elements, is still quite small and consequently the total amplitude for all atoms in a unit cell must remain rather small even under the most favourable conditions; thus the fractional part of the incident intensity diverted into any diffraction maximum is small even for the strongest reflections. Secondly, the shape of atomic f curves will be dependent on any thermal motion of the atoms at finite temperatures. Standard tabulations of f values are quoted for atoms at rest; and, in general, thermal vibrations will cause the electron distribution to spread out and decrease the scattering power. The form of the relationship between the scattering factor

(f) for a vibrating atom and that (f_0) for the same atom at rest is usually expressed as

$$f = f_0 \exp(- B \sin^2 \theta / \lambda^2)$$

in which B is taken to be a constant at a given temperature. This is not strictly valid, for B varies from atom to atom and with the direction of scattering; but the relation is sufficiently good for the accuracy of most work, and a value of B can be found from a comparison of observed and calculated intensities in any particular case. Corrections for thermal motion are only required in accurate work on structure determination.

From atomic f curves we can obtain the amplitude of scattering (in terms of electron equivalence) for each kind of atom in the cell at a given angle of scattering; the next step in the calculation of the expected intensities involves the summation of these amplitudes for all atoms. Let us suppose that there are n atoms of various kinds with fractional co-ordinates $x_1 y_1 z_1$; $x_2 y_2 z_2$; ... $x_n y_n z_n$ in the cell; for a particular scattering angle defined by a set of reflecting planes (hkl) these will scatter with amplitudes $f_1, f_2, \ldots, f_n$, but, owing to their different locations, the radiation diffracted by each atom will have a different phase; we will specify the phase angles as $\phi_1, \phi_2, \ldots, \phi_n$ relative to the cell origin. Our first problem is to express these phase angles in terms of the atomic co-ordinates and the angle of scattering; this may be done by remembering that planes of the set (hkl) can be regarded as planes of constant phase spaced at phase intervals of 2π under conditions that satisfy the Bragg law. The equations of these planes are

$$x \Big/ a/h + y \Big/ b/k + z \Big/ c/l = 0, 1, 2, 3, \text{etc.}$$

An atom with fractional co-ordinates $x_n y_n z_n$ will lie on a parallel plane, so that

$$x \Big/ a/h + y \Big/ b/k + z \Big/ c/l = ax_n \Big/ a/h + by_n \Big/ b/k + cz_n \Big/ c/l = \frac{hx_n + ky_n + lz_n}{} = m$$

where m can be irrational, so that the phase difference between radiation scattered from this atom and the origin is

$$\phi_n = 2\pi m = 2\pi(hx_n + ky_n + lz_n)$$

Now we know both the phase and amplitude of the radiation scattered by individual atoms in the cell; the summation can be made on a phase-amplitude diagram (Fig. 8.15), from which we see that

$$OR^2 = OS^2 + RS^2$$

$$= \left(\sum_n f_n \cos \phi_n\right)^2 + \left(\sum_n f_n \sin \phi_n\right)^2$$

$$= \left[\sum_n f_n \cos 2\pi(hx_n + ky_n + lz_n)\right]^2 + \left[\sum_n f_n \sin 2\pi(hx_n + ky_n + lz_n)\right]^2.$$

$\overrightarrow{OR}$, the total amplitude scattered in the particular direction defined by *hkl* by all atoms in the unit cell is usually called the *structure amplitude* (less desirably, the 'structure factor') and written *F(hkl)*. For the various diffraction maxima, the scattered radiation has structure amplitudes which can differ both in magnitudes (*OR*) and phases (ϕ); experimentally relative phases for different reflections cannot be observed and we can only measure intensities. These can be related to the magnitudes of *F(hkl)*'s, though due account must be taken of other factors which influence experimental values, as we shall see shortly; the absence of experimental data on relative phases is particularly important in structure determination where many methods are concerned with overcoming the problems of this handicap.

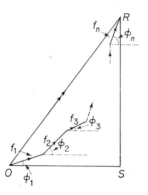

Fig. 8.15. Phase-amplitude diagram summing contributions to the total scattering by each of the *n* atoms in the unit cell.

The use of the structure amplitude expression to calculate $|F(hkl)|^2$ (and hence intensities) for a known atomic assemblage is a formidable task if the arrangement is not very simple, but modern computational methods relieve the calculations of much of their labour. They are simplified when the space group is known, for each space group requires the repetition of atoms or atom groups at equivalent positions within the cell at co-ordinates determined by the symmetry and lattice operators. Calculations may proceed through *geometrical structure factors* formed by independently summing the trigonometrical terms for the set of related co-ordinates in the particular space group; the product of the geometrical structure factor and the *f* value for atoms occupying these positions in the cell gives the contribution of this set of atoms to the structure amplitude. For example, the geometrical structure factor for the G.E.P's in *C2/m* listed in Table 7.4 is

$$A = 8 \cos^2 2\pi \frac{h + k}{4} \cos 2\pi(hx + lz) \cos 2\pi ky; \quad B = 0,$$

where A is the summation of the cosine terms in the structure amplitude expression, and B^*, the summation of the sine terms; contributions of all the atoms at different co-ordinates may then be included in the expression

$$|F(hkl)|^2 = \left(\sum_n f_n A_n\right)^2 + \left(\sum_n f_n B_n\right)^2$$

where A_n, B_n are the values calculated from the geometrical structure factor for the nth set of atoms scattering with amplitude f_n. Geometrical structure factors for all space groups are listed in Vol. 1 of *International tables for X-ray crystallography*.

8.4. The recording of diffraction patterns

A complete study of diffraction by a particular crystal requires the location of all possible reflections and measurement of their intensities; to do this we must employ an experimental technique which allows Bragg conditions to be fulfilled for every family of planes, and have some means of detecting the X-rays scattered in various directions. Some of the simpler experimental methods are described in subsequent chapters; in this section we are concerned only with the manner of recording patterns without reference to the technique by which they are obtained. Historically the first diffracted X-rays were detected by the blackening of a photographic plate, and its successor—the photographic film—is still used for much modern work. The response to incident quanta depends upon the speed of the film, but most X-ray reflections are so weak that exposure times of hours (and sometimes days) are normally necessary to obtain detectable blackening on the developed film. (Recently polaroid films have been introduced, but these are not yet suitable for accurate intensity work.) To measure X-ray intensities from films it is essential to relate the number of X-ray quanta to a particular *blackening*, defined as a function of the ratio of incident and transmitted light; experimentally it is found for most films that blackening is linearly related to the number of quanta per unit area of film, provided that a limiting blackening (depending on the film and its processing) is not exceeded. If all recorded reflections on the film are spots with blackening within this linear range, relative intensities of the various maxima can be compared from their blackenings by visual estimation or by photometry, as accuracy and circumstances demand. Much of the experimental data on intensities needed for structure determination has been (and still is) collected in this rather simple manner; but almost from the earliest days of X-ray crystallography attempts were made to develop quantitative *direct methods* to detect the scattered radiation so as to eliminate the

---

* One of the advantages of listing co-ordinates of equivalent positions with respect to an origin sited on a centre of symmetry is that they must occur in pairs x_n, y_n, z_n and $\bar{x}_n$, $\bar{y}_n$, $\bar{z}_n$; when these pairs are substituted in the summation of the sine terms (B) they vanish, since $\sin(-\theta) = -\sin\theta$, so that $B = 0$.

uncertainties inherent in photographic films and their processing. Initially attention was concentrated on ionisation chambers in which currents due to gas ionisation by the X-rays were measured. These were cumbersome and often difficult in use, but the development of stable modern counters (Geiger, proportional and scintillation) with their associated electronic counting circuitry has led to a more reliable and accurate method of measuring X-ray intensities; these are now increasingly employed when the highest accuracies are required. Naturally, this complex equipment poses many technical problems, and we can do no more here than refer to the many books which describe in detail these techniques of X-ray detection.

We emphasised earlier that the positions of reflections are determined by the cell size and shape; their intensities depend on the arrangement of the cell contents as expressed by the structure amplitude, but there are many other factors which can influence values observed experimentally. Some of these involve the particular experimental arrangement and depend critically on the geometry of the camera; other geometrical factors are related to the partial polarisation of the reflected beams and the speed at which reflecting planes pass through the Bragg positions. Moreover, we must expect the properties of the specimen itself to be of some importance. For example, diffracted beams will be absorbed by the crystalline specimen, and the extent of this absorption for a particular reflection will depend in a complex way on the camera geometry and the size and shape of the specimen; often the best policy is to employ crystals whose dimensions are so small that any absorption for the chosen radiation can be regarded as negligible (the upper limit of size is usually assumed to be $2/\mu$). In the single crystal and powder methods described in subsequent chapters, any intensity formulae for powder records include a *multiplicity factor* (p) due to the superposition of reflections from different planes of the same form; in single crystal patterns all such reflections are separated. All these factors can be corrected, or minimised, so as to allow a comparison between observed intensities and those expected from calculations of structure amplitudes; but there is a further factor, in some ways of rather a more fundamental nature, whose estimation is much more uncertain.

When X-rays are re-radiated by the atoms of a crystal, the scattered radiation for any planes (hkl) is 90° out of phase with the incident beam; this is due to the difference between the frequency of the X-rays and the lower natural frequencies of the atoms. The geometry of the Bragg law allows multiple reflection of X-rays at adjacent parallel planes of a set, so that some scattered radiation travels in the same direction as the incident beam but is exactly out of phase with it; this causes destructive interference, which can lead to a rapid decrease in the amplitude of the incident beam as it penetrates into the crystal. The effect is most marked in a perfect crystal set in the reflecting position for a family of planes, and it will be most noticeable for those planes which scatter most strongly, i.e. those with the largest values of $F(hkl)$; this progressive reduction in the intensity of the incident beam as it passes through successive reflecting planes is known as *primary extinction*, and it can have a profound effect on observed intensities. The *total integrated intensity* (i.e. the total energy scattered at all angles as a crystal passes through the reflecting position) can be calculated for a very small perfect crystal; relative to unit incident intensity, this can be written as QdV, where dV is the volume of the crystal and Q is a complex factor including the geometrical terms, etc. that we have mentioned as well as $|F(hkl)|^2$. In this very small crystal, the effect of primary extinction is negligible, but for a large perfect experimental crystal of volume V containing many hundreds of thousands of reflecting planes, we would not be justified in writing the integrated intensity as QV. Due to the shielding effects of primary extinction, the incident beam does not penetrate to the deeper regions of the crystal and the effective volume V' contributing to the integrated intensity is less than V. The value of V' will depend on the particular reflecting planes, and to a reasonable approximation we can write that V' is inversely related to $|F(hkl)|$, with primary extinction having more effect the stronger the reflection. The total integrated intensity is then QV', so that for a large perfect crystal it depends on $|F(hkl)|$. Early attempts to verify this dependence showed that it was reasonably valid for only very few crystals, and there were considerable difficulties in reconciling experimental intensity measurements with those expected from calculations of this kind. Explanation of these discrepancies lies in the realisation that real crystals rarely achieve the perfection of atomic arrangement that has been assumed (see Chapter 12); at the extreme, we may imagine that actual crystals have a mosaic texture in which small perfect blocks of structure are slightly misorientated from one another by structural imperfections. If individual blocks in the mosaic are small enough for primary extinction to be negligible, the total integrated intensity is QV, so that for a large imperfect crystal it depends on $|F(hkl)|^2$. A rigorous treatment of extinction is more complex, e.g. in ideally imperfect crystals there can be some shielding of deeper mosaic blocks by those nearer to the surface in parallel orientations; fuller discussion of these difficult problems is found in books listed in the bibliography, particularly that by James. Nevertheless, the present simplified exposition illustrates how profoundly experimental intensities can depend on the state of perfection of the crystal specimen. Schematically drawn in Fig. 8.16 are the reflection curves from the same family of planes for the same crystal structure (a) for an ideally perfect crystal, and (b) for an ideally imperfect crystal; in (a) the peak height is considerable, with a limited range of reflecting angle; but the total integrated intensity under the curve is very much smaller than for (b), which has a lower peak height but a much broader width of maximum since the slightly misorientated blocks reflect over a greater range of angles.

In any practical context, a value of $|F(hkl)|$ calculated from atomic parameters can be compared with an experimental intensity (after correction for geometrical factors) only if some assumption is made about the state of perfection of the crystal specimen. Most real crystals are neither ideally perfect nor ideally imperfect in their behaviour, but experience shows that they are more likely to approximate to the imperfect condition; indeed, we may attempt to destroy any suspected perfection before experimental measurement

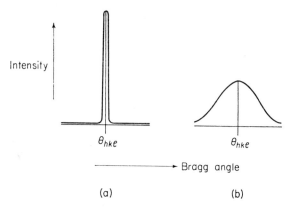

Fig. 8.16. Idealised reflection curves. Both are for the same structure and the same reflection. (a) Shows the narrow peak of a perfect texture and (b) the broad low peak of an imperfect texture.

by a simple treatment (e.g. by dipping the crystal in liquid air). In practice, the crystal is then assumed to be ideally imperfect for all reflections, so that integrated intensities are related to $|F(hkl)|^2$; even so there will remain discrepancies between observed and calculated intensities; but, provided these disagreements are confined to the strongest maxima, they may be reasonably written off as due to the uncorrected effects of primary extinction for such reflections.

8.5. The determination of space groups

In discussing computation of the structure amplitude expression earlier we mentioned that lattice and symmetry operators demand relationships between the co-ordinates of the atoms of the structure in which they occur; the geometrical structure factors formed from these co-ordinates imply that intensities of different X-ray reflections are not always independent. Indeed, the form of the amplitude expression itself ensures that $|F(hkl)| \equiv |F(\bar{h}\bar{k}\bar{l})|$, so that the intensities of X-rays reflected from the opposite sides of families of planes are identical for any set of atomic co-ordinates.* This is known as

* There are certain exceptional experimental conditions when this is not valid, but these are sufficiently unusual to justify omission here.

Friedel's law, and means that it is not possible to deduce the absence of a centre of symmetry by comparing the intensities of the hkl and $\bar{h}\bar{k}\bar{l}$ reflections; consequently, it may be said that X-rays insert a centre into a diffraction pattern when one is not present in the crystal. Clearly this limits any deductions of symmetry from diffraction patterns, but more positively other symmetry operators in point groups can be identified by intensity comparisons; the presence of axes and planes necessitates atomic distributions giving identical intensities to maxima whose indices are related by the symmetry element. For example, the intensities hkl and $h\bar{k}\bar{l}$ must be identical when the crystal has a diad parallel to the x-axis but would be unrelated in triclinic crystals where this element is not present. Broader categories of point group symmetries can be distinguished by a study of the relationships between the intensities of the various maxima and convenient experimental techniques are described in the next chapter.

Once the point group has been established (or at least the broader Laue group as in Chapter 9.2), we can discover non-primitive lattices, screw axes, and glide planes by the *systematic absences* which they cause in certain classes of reflections; in this way we can assign the structure to a particular space group or, at the worst, deduce that it can belong only to a very limited number of groups. Systematic absences are observed because certain lattice and symmetry operators relate atomic co-ordinates so as to reduce to zero the structure amplitudes (and intensities) of certain classes of reflections; they should not be confused with accidental absences of individual reflections whose structure amplitudes are so small that they are not detectable. As an example of a set of systematic absences, let us assume that our structure has an I-lattice, in which identical atoms must occur in pairs with co-ordinates related by the translation $(\frac{1}{2}, \frac{1}{2}, \frac{1}{2})+$; if one atom is at x_n, y_n, z_n, the other member of the pair must be at $\frac{1}{2} + x_n, \frac{1}{2} + y_n, \frac{1}{2} + z_n$. The geometrical structure factor for the space group must contain terms in pairs as

$$\cos 2\pi(hx_n + ky_n + lz_n) + \cos 2\pi\left(hx_n + ky_n + lz_n + \frac{h + k + l}{2}\right)$$

in the A summation, and

$$\sin 2\pi(hx_n + ky_n + lz_n) + \sin 2\pi\left(hx_n + ky_n + lz_n + \frac{h + k + l}{2}\right)$$

in the B summation.

On simple manipulation, such pairs become

$$2 \cos 2\pi\left(\frac{h + k + l}{4}\right) \cos 2\pi\left(hx_n + ky_n + lz_n + \frac{h + k + l}{4}\right)$$

and

$$2 \cos 2\pi\left(\frac{h + k + l}{4}\right) \sin 2\pi\left(hx_n + ky_n + lz_n + \frac{h + k + l}{4}\right)$$

respectively; so that the geometrical structure factor has a common trigono-metrical factor $\cos 2\pi \left(\dfrac{h + k + l}{4} \right)$, which is independent of any particular atomic co-ordinates. Since h, k and l are integers, this cosine term can only have values of 0 or ± 1, depending on whether $h + k + l$ is odd $(2n + 1)$ or even $(2n)$ respectively; when it is zero, the structure amplitude $F(hkl)$ will be zero. Thus all reflections for which $h + k + l = 2n + 1$ are absent from the diffraction patterns of crystals with body-centred cells; from these systematic absences we can deduce that a specimen has this lattice type. Other multiply primitive cells lead to different sets of systematic absences as can be demonstrated by similar arguments. Table 8.2 summarises the effects

Table 8.2. Conditions imposed upon possible X-ray reflections by lattice type

Lattice symbol	Conditions for reflections of class hkl
P	None
I	$h + k + l = 2n$
F	h, k and l all even or all odd
C	$h + k = 2n$
B	$h + l = 2n$
A	$k + l = 2n$
R[a]	$-h + k + l = 3n$
	or $h - k + l = 3n$

[a] The reflections referred to are $hk \cdot l$ obtained on the Miller–Bravais system of indexing using a triply primitive hexagonal cell; with rhombohedral indices pqr of the Miller system of indexing there are no restrictions. The upper condition is that for the obverse orientation, the lower is for the reverse orientation.

of lattice operators on diffraction patterns by specifying the conditions for possible reflections; owing to accidental absences not all reflections satisfying these conditions may actually be observed. Systematic absences due to lattice type may be accompanied and supplemented by those due to glide planes; these occur in restricted reflection classes in a manner which depends both on the translational element and orientation of the plane. For example, if the structure contains a c-glide plane through the origin normal to the x-axis, this requires pairs of identical atoms with co-ordinates x_n, y_n, z_n and $\bar{x}_n, y_n, \frac{1}{2} + z_n$. Following the same arguments as for an I-lattice, the pairs in the geometrical structure factor become

$$2 \cos 2\pi \left(hx_n - \frac{l}{4} \right) \cos 2\pi \left(ky_n + lz_n + \frac{l}{4} \right)$$

in the A summation, and

$$2 \cos 2\pi \left(hx_n - \frac{l}{4} \right) \sin 2\pi \left(ky_n + lz_n + \frac{l}{4} \right)$$

in the B summation; there is no common trigonometrical factor in the terms needed to calculate $|F(hkl)|^2$, for they will contain the different x co-ordinates of the various atoms. However, if we confine our attention to $0kl$ reflections, we eliminate these co-ordinates from the expressions and have the factor $\cos 2\pi \dfrac{l}{4}$ common to all maxima. Since l is integral, this term will be zero if l is odd $(2n + 1)$ and ± 1 if l is even $(2n)$; as before, a zero term means that the structure amplitude $F(0kl)$ is zero, so that reflections of the type $0kl$ do not occur when this glide plane is present in this orientation unless $l = 2n$. By similar arguments we can discover all the sets of systematic absences caused by glide planes in various orientations; Table 8.3 lists the various conditions for

Table 8.3. Conditions imposed upon possible X-ray reflections by planes of symmetry

Plane symbol	Orientation (as ( ) planes)	Reflection class	Conditions for reflection
m	Any	Any	None
a	(010) planes	$h0l$	$h = 2n$
	(001) planes	$hk0$	$h = 2n$
b	(100) planes	$0kl$	$k = 2n$
	(001) planes	$hk0$	$k = 2n$
c	(100) planes	$0kl$	$l = 2n$
	(010) planes	$h0l$	$l = 2n$
	(110) planes	hhl	$l = 2n$
	($1\bar{1}00$) planes	$hh \cdot l$	$l = 2n$
	($11\bar{2}0$) planes	$h\bar{h} \cdot l$	$l = 2n$
n	(100) planes	$0kl$	$k + l = 2n$
	(010) planes	$h0l$	$h + l = 2n$
	(001) planes	$hk0$	$h + k = 2n$
d	(100) planes	$0kl$	$k + l = 4n \ (k, l = 2n)$
	(010) planes	$h0l$	$h + l = 4n \ (h, l = 2n)$
	(001) planes	$hk0$	$h + k = 4n \ (h, k = 2n)$
	(110) planes	hhl	$2h + l = 4n$

possible reflections under the common circumstances of normal space group orientations, with the omission of rhombohedral indexing. Further systematic absences are due to screw axes, and these too depend on the orientation of the axis and its nature. Taking a simple example of a screw diad through the origin parallel to the x-axis, the pairs of related co-ordinates are x_n, y_n, z_n and $\frac{1}{2} + x_n, \bar{y}_n, \bar{z}_n$ which lead to terms in the geometrical structure factor of

$$2 \cos 2\pi \left(ky_n + lz_n - \frac{h}{4} \right) \cos 2\pi \left(hx_n + \frac{h}{4} \right)$$

in the A summation and

$$2 \cos 2\pi \left(ky_n + lz_n - \frac{h}{4} \right) \sin 2\pi \left(hx_n + \frac{h}{4} \right)$$

in the B summation. If we restrict attention to the $h00$ reflections, the factor $\cos 2\pi \dfrac{h}{4}$ is common to all maxima, so that $h00$ reflections will occur only if $h = 2n$; other conditions dependent on screw axes to be found in the normal space group orientations are listed in Table 8.4; again rhombohedral indexing is omitted.

Table 8.4. Conditions imposed upon possible X-ray reflections by axes of symmetry

Axis symbol	Orientation (in [ ] direction)	Reflection class	Conditions for reflections
$\left.\begin{array}{l}2, 3, 4, 6 \\ \bar{2}, \bar{3}, \bar{4}, \bar{6}\end{array}\right\}$	Any	Any	None
2_1	[100]	$h00$	$h = 2n$
	[010]	$0k0$	$k = 2n$
	[001]	$00l$	$l = 2n$
$3_1, 3_2$	[00 . 1]	$00 . l$	$l = 3n$
$4_1, 4_3$	[100]	$h00$	$h = 4n$
	[010]	$0k0$	$k = 4n$
	[001]	$00l$	$l = 4n$
4_2	[100]	$h00$	$h = 2n$
	[010]	$0k0$	$k = 2n$
	[001]	$00l$	$l = 2n$
$6_1, 6_5$	[00 . 1]	$00 . l$	$l = 6n$
$6_2, 6_4$	[00 . 1]	$00 . l$	$l = 3n$
6_3	[00 . 1]	$00 . l$	$l = 2n$

The conditions for reflection imposed by the various lattice types and symmetry elements listed in these tables can be used to identify these operators, and so a survey of diffraction maxima can lead to the deduction of the possible space group or groups. In practice, this should proceed from the more general to the more particular reflection classes, i.e. from lattice absences to glide plane absences to screw axis absences, remembering that absences which are sub-conditions of a more general condition cannot be regarded as significant. Thus $hk0$ and $h00$ reflections must have $h + k = 2n$ and $h = 2n$ respectively for a C-face centred lattice which imposes the general condition for all reflections hkl that $h + k = 2n$; they cannot be taken as indicating n-glide planes parallel to (001) (Table 8.3) and 2_1 axes parallel to [100] (Table 8.4), elements which may or may not be present in the space group. Some space groups are uniquely characterised by their systematic absences; $Pccn$ with conditions $0kl$ with $l = 2n$, $h0l$ with $l = 2n$, and $hk0$ with $h + k = 2n$ is one such group. Some have the same kind of reflection patterns as several other groups so that an individual space group cannot be recognised by surveys of reflections alone; this is due to a lack of positive evidence for rotation or inversion axes, and mirror planes, so that for example Pm, $P2$ and $P2/m$ all have no restrictions on any reflection class. Nevertheless, we should at least recognise that our structure belonged to one of these three monoclinic space groups; any attempt to differentiate among them would require some means other than systematic absences. The limit of information that can be obtained in such diffraction studies is formally expressed by the *diffraction symbol*; this has three parts denoting the point group symmetry as determined by diffraction (the Laue group of Chapter 9.2), the lattice type (from Table 8.2), and any symmetry elements which can be deduced from systematic absences (Tables 8.3 and 8.4). Hence the diffraction symbol for a crystal of space group *Pccn* is *mmmPccn*, the completion of all places in the symbol implying an unambiguous interpretation of the diffraction data; the symbol for the three monoclinic space groups above is *2/mP–/–*, the incomplete places in the symbol embracing the symmetry elements whose possible presence to give either *P2* or *Pm* or *P2/m* is unresolved. All diffraction symbols are listed in Vol. 1 of *International tables for X-ray crystallography*, together with their associated alternative space groups; diffraction symbols are a convenient formality in a subject which generates formal symbolism; but they are not much used in practice, perhaps because the information they convey is stated more directly and usefully by quoting any alternative space groups permitted by the survey of diffraction maxima.

8.6. Exercises and problems

1. (i) Ag is sometimes used as the target material in an X-ray tube. If its K absorption edge is at 0.485 Å, what is the minimum excitation voltage for AgK radiations? At this voltage what is the cut-off limit of the white radiation spectrum?

   (ii) In Table 8.1, Fe is suggested as a suitable β-filter for Co radiation. Find what thickness of Fe foil is needed to reduce the $K\beta/K\alpha$ intensity ratio for Co from 1/5 to 1/500. What intensity reduction would this filter produce for Co$K\alpha$ radiation?

   (μ_M for Fe is 59.5 for Co$K\alpha$ radiation, and 375 for Co$K\beta$ radiation; density of Fe is 7.9 gm/cm^3.)

   (iii) Satisfactory diffraction patterns from alloys rich in Co cannot be obtained using Cu$K\alpha$ radiation. Why not? What radiation would be suitable?

   (iv) Calculate the approximate thickness of lead needed to absorb at least 99.95% of the most penetrating radiation emitted by an X-ray tube running at 50 kV.

   (μ_M for Pb for Mo$K\alpha$ radiation is 141; the density of Pb is 11.4 gm/cm^3.)

2. (i) Sketch the intensity distribution for Fraunhofer diffraction by a complex line grating, like that of Fig. 8.8, in which $50a = 10c = d$.

   (ii) Repeat (i) for a planar grating, like that of Fig. 8.9, in which a pair of equal holes separated by a distance a ($\parallel c_1$) are repeated with spacings c_1 and c_2 inclined at $90°$.

(iii) In a miniature venetian blind, slats 2 mm wide just meet when the blind is closed. If the blind is used as a diffraction grating with a slit source, describe and sketch the variations in the observed Fraunhofer patterns as the slats are turned through 90°. At what angle are all the even orders of diffraction missing?

3. In examining diffraction by a cubic crystal ($a = 4\cdot08$ Å, $\lambda = 1\cdot54$ Å), one maximum was recorded when $i_a = 150\frac{1}{2}°$, $i_b = 90°$, $i_c = 60\frac{1}{2}°$ and $d_a = 119\frac{1}{2}°$, $d_b = 90°$ and $d_c = 29\frac{1}{2}°$. Determine the indices of the family of planes which would give this maximum by reflection, and the value of the Bragg angle. Calculate the values of the Bragg angles for all other families of parallel planes which could be set into reflecting positions.

4. Evaluate the geometrical structure factors for the space groups (a) *Pm*, (b) *Pnnn*, (c) *Ccca*. For each, list the conditions limiting possible reflections.

5. (i) Diamond has a cubic structure in which the C atoms occupy the following positions:

$$0,0,0; \quad 0,\tfrac{1}{2},\tfrac{1}{2}; \quad \tfrac{1}{2},0,\tfrac{1}{2}; \quad \tfrac{1}{2},\tfrac{1}{2},0; \quad \tfrac{1}{4},\tfrac{1}{4},\tfrac{1}{4}; \quad \tfrac{3}{4},\tfrac{3}{4},\tfrac{1}{4}; \quad \tfrac{1}{4},\tfrac{3}{4},\tfrac{3}{4}; \quad \tfrac{3}{4},\tfrac{1}{4},\tfrac{3}{4}.$$

Draw a plan of several unit cells projected on (001), and insert traces of the planes (100), (200), (300), (400), (110), (220), (330) and (440). By considering the positions of the atoms relative to these families of planes, decide which of these reflections are systematically absent for this structure.

(ii) Repeat (i) for NaCl, another cubic structure in which the atomic coordinates are:

$$\text{Na: } \tfrac{1}{2},\tfrac{1}{2},\tfrac{1}{2}; \quad \tfrac{1}{2},0,0; \quad 0,\tfrac{1}{2},0; \quad 0,0,\tfrac{1}{2};$$
$$\text{Cl: } 0,0,0; \quad 0,\tfrac{1}{2},\tfrac{1}{2}; \quad \tfrac{1}{2},0,\tfrac{1}{2}; \quad \tfrac{1}{2},\tfrac{1}{2},0.$$

(iii) By calculating the structure amplitude expression for each of these structures, confirm your deductions in (i) and (ii); use these expressions to generalise conclusions concerning systematically absent reflections for both diamond and NaCl.

6. What conclusions can be drawn about possible space groups from the following observations on the classes of absent reflections?

(i) Monoclinic cell:

00*l* reflections observed only if $l = 2n$.

(ii) Orthorhombic cell:

hkl reflections observed only if			$h + k, k + l, l + h = 2n$
0*kl* „	„	„ „	$k + l = 4n$
*h*0*l* „	„	„ „	$l + h = 4n$
*hk*0 „	„	„ „	$h, k = 2n$
*h*00 „	„	„ „	$h = 4n$
0*k*0 „	„	„ „	$k = 4n$
00*l* „	„	„ „	$l = 4n.$

(iii) Hexagonal cell:

$hk \cdot l$ reflections observed only if $-h + k + l = 3n$
$hh \cdot l$ „   „   „ „ $l = 3n$
$h\bar{h} \cdot l$ „   „   „ „ $h + l = 3n, l = 2n$
$00 \cdot l$ „   „   „ „ $l = 3n$.

7. Stokesite ($CaSnSi_3O_9 \cdot 2H_2O$) is orthorhombic with four formula units per cell. The diffraction patterns show that

$0kl$ reflections observed only if $k + l = 2n$
$h0l$ „   „   „ „ $h + l = 2n$
$hk0$ „   „   „ „ $h = 2n$
$h00$ „   „   „ „ $h = 2n$
$0k0$ „   „   „ „ $k = 2n$
$00l$ „   „   „ „ $l = 2n$.

It was also observed that reflections with $h = 2n$, $k + l = 2n$ tend to be very strong, particularly at high Bragg angles. Deduce the probable space group, and draw what conclusions you can about the atomic positions.

8. Gahnite ($ZnAl_2O_4$) belongs to class $m3m$, and has a cell dimension $a = 8\cdot1$ Å. The diffraction patterns show that

hkl reflections observed only if $h + k, k + l, l + h = 2n$
hhl „   „   „ „ $l + h = 2n$
$0kl$ „   „   „ „ $k + l = 4n(k, l = 2n)$.

Determine the probable space group, and draw what conclusions you can about the atomic positions.

(Density of gahnite, $4\cdot62$ gm/cm^3; atomic weights Zn, $65\cdot4$, Al, $27\cdot0$, O, $16\cdot0$.)

SELECTED BIBLIOGRAPHY

General reading
AZAROFF, L. V. 1968. *Elements of X-ray crystallography*. McGraw-Hill.
BUNN, C. W. 1961. *Chemical crystallography*. Clarendon Press.
JAMES, R. W. 1953. *X-ray crystallography*. Methuen.
LIPSON, H. S. 1970 *Crystals and X-rays*. Wykeham.

Advanced text on diffraction theory
JAMES, R. W. 1962. *The crystalline state*: Vol. II, *The optical principles of the diffraction of X-rays*. Bell.

Optical diffraction
DITCHBURN, R. W. 1963. *Light*. Blackie.
JENKINS, F. A. and WHITE, H. E. 1957. *Fundamentals of optics*. McGraw-Hill.
LIPSON, H. and TAYLOR, C. A. 1958. *Fourier transforms and X-ray diffraction*. Bell.

Space group and structure determination
BUERGER, M. J. 1942. *X-ray crystallography*. Wiley.
BUERGER, M. J. 1960. *Crystal structure analysis*. Wiley.

LIPSON, H. and COCHRAN, W. 1966. *The crystalline state:* Vol. III, *The determination of crystal structures.* Bell.

STOUT, G. H. and JENSEN, L. H. 1968. *X-ray structure determination.* Macmillan.

Related topics

BACON, G. E. 1966. *X-ray and neutron diffraction.* Pergamon.

BEECHING, R. 1946. *Electron diffraction.* Methuen.

PINSKER, Z. G. 1953. *Electron diffraction* (in translation). Butterworths.

Standard text for G.S.F's, diffraction symbols, etc.

INTERNATIONAL UNION OF CRYSTALLOGRAPHY. 1965. *International tables for X-ray crystallography*, Vol. I. Kynoch Press, Birmingham.

Source book for physical and chemical constants relevant to X-ray diffraction

INTERNATIONAL UNION OF CRYSTALLOGRAPHY. 1968. *International tables for X-ray crystallography*, Vol. III. Kynoch Press, Birmingham.

9

DIFFRACTION BY SINGLE CRYSTALS

9.1. General considerations

The experimental techniques of X-ray diffraction can be broadly divided into two groups according to the state of the specimen; within each of these groups there are many different methods of examination, and in the next two chapters we shall consider some of the simpler experimental arrangements and discuss the nature of the diffraction patterns which they produce. Our objective will be to locate the positions of maxima to see how these can be used to give basic information (symmetry, cell constants, etc.) about the specimen; we shall not pursue in any detail the study of intensity relations among the various indexed maxima which is essential to structure determination. The present chapter is concerned with one group of techniques, often known as *single crystal methods* because the specimen is a small fragment of a homogeneous crystalline material.

The specific conditions for the occurrence of any diffraction maximum can be expressed by the Bragg law derived in the previous chapter; the geometry of a reflection in this treatment is represented simply and conveniently on the stereogram of Fig. 9.1(a). The pole i shows the direction of the incident X-rays; for any planes (hkl) a reflection will only occur when the Bragg condition $\sin \theta_{hkl} = \lambda/2d_{hkl}$ is satisfied, i.e. for the normals to these planes to be in a position to reflect they must lie on the small circle of radius $90° - \theta_{hkl}$ drawn about the pole i. If the crystal is so oriented that an (hkl) normal does lie on this small circle, a reflection will occur in accordance with the simple geometry of the Bragg law, i.e. in the direction d in a symmetrical position on the great circle through i and the normal. When a stationary crystal is irradiated by X-rays of wavelength λ, this stringent Bragg geometry will be fulfilled for very few reflecting planes; for most planes the situation is likely to be that of $(h'k'l')$ in the figure where the orientation of the crystal places the normal off the appropriate small circle so that no reflection can take place. Under these conditions any diffraction pattern shows few (if any) maxima,

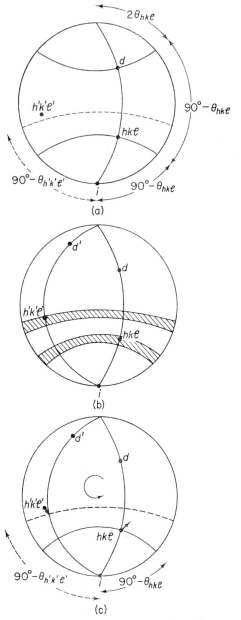

Fig. 9.1. The Bragg conditions for reflection. (a) Stationary crystal and X-ray wavelength λ; *hkl* reflects but *h'k'l'* does not. (b) Stationary crystal and white radiation; the shaded areas represent the limiting positions of small circles set by the wavelengths of the white radiation spectrum. *hkl* and *h'k'l'* both reflect but for different wavelengths. (c) Crystal rotated about axis normal to X-ray beam of wavelength λ; the movements of *hkl* and *h'k'l'* are indicated by arrows. *hkl* and *h'k'l'* both reflect but at different positions of rotation.

and their sporadic angular distribution makes interpretation difficult; it is far easier if many more maxima are permitted to occur by changing the conditions of the experiment. More planes will have the opportunity to reflect if we use an incident beam with a range of wavelengths and if we change the orientation of the crystal relative to the incident beam; in practice, interpretation is simplified if one or other (but not both) of these methods are employed, and with suitable experimental arrangements either technique yields well defined reflection patterns which allow fundamental crystallographic data to be obtained. In view of this single crystal methods may be subdivided into those in which there is a stationary crystal and those in which the crystal moves.

Laue methods, in which white radiation is diffracted by a stationary crystal, are similar to the original experiments of Friedrich and Knipping; we shall discuss the experimental techniques and the interpretation of the patterns in the next section but the essential principles are illustrated by the stereogram of Fig. 9.1(b). In this, for each set of planes a range of small circles is drawn about the pole of the incident beam between limits of $90° - \theta_{hkl}$ set by the maximum and minimum wavelengths of the white radiation spectrum; the poles of both (hkl) and $(h'k'l')$ now lie within the shaded areas which show fulfilment of Bragg conditions and so both reflections hkl and $h'k'l'$ are observed. In this kind of experiment the two families of planes are likely to be reflecting X-rays of different wavelengths within the continuous range; moreover some observed maxima will have the superposition of reflections due to different wavelengths. The value of $\sin \theta \, (= \lambda/2d)$ is the same for λ and the planes (hkl) as it is for $\lambda/2$ and the planes $(2h, 2k, 2l)$, so both reflections are in the same direction; superposed reflections are due to those wavelengths in the continuous range which are harmonics of the longest wavelength contributing to the maximum. Such problems are avoided in *moving crystal methods*, where the incident radiation is restricted to a single wavelength but the orientation of the crystal specimen is varied so as to permit more and more planes to pass through a reflecting position; reflections from the various planes are not simultaneous (as in the Laue method) but occur consecutively at different instants of time as the crystal is in motion. Moving crystal techniques can be very sophisticated with complex motions of the crystal; the nature of the pattern and its detailed interpretation depend on the kind of movement made by the crystal and other experimental factors and we shall give some account of these in later sections. In the simplest method the crystal is rotated about an axis perpendicular to the incident beam; such a movement is illustrated in the stereogram of Fig. 9.1(c). Rotation causes the (hkl) planes to move out of a reflecting position, but a little later the pole $(h'k'l')$ intersects the appropriate small circle to give a reflection; the two maxima are due to the reflection of the single wavelength radiation but they occur at different positions of rotation.

Specimens for single crystal work are usually chips taken from a larger crystal or fragments obtained in a separation of one component in a crushed mixture; a lower limit of size is set by the need to manipulate the specimen crystal with reasonable ease under a binocular microscope; if the observation of intensities is important it is best to be below the optimum size set by absorption considerations (see Chapter 8.4). An experienced experimenter can work with fragments as small as 0·1 mm diameter, though it is preferable for the specimen to be rather larger than this (up to $\frac{1}{2}$–1 mm diameter) if possible. In practice, it is desirable that diffraction patterns are recorded when the incident beam is oriented with respect to the lattice geometry of the crystal; random patterns are much more difficult to interpret than those obtained with some alignment of the crystal. This implies some preliminary examination of the fragment (and the material from which it comes). Whilst the fragment may have little external morphology, the occurrence of recognised cleavage planes (or traces) can give some help in orientation; often physical properties (particularly optical observations) such as those outlined in Chapter 11 can help to establish the kind of symmetry (if unknown) and orientation of the specimen. We should emphasise that all this is done only to smooth the path of the subsequent X-ray investigation, for the Laue methods described in the next section are most valuable in the determination of symmetry (and orientation) even when there is no other data; but, like all X-ray methods, they can be most expeditiously applied in conjunction with any results from other work on the specimen.

9.2. Laue methods

The first X-ray diffraction record was a Laue photograph, and, as this suggests, Laue methods need only the simplest apparatus; however, there are some practical advantages if extra refinements (particularly for adjusting and orienting the specimen crystal) are included in the camera. In fact, most Laue photographs are taken on moving crystal cameras with the motion arrested and filters removed from the incident X-ray beam. The small crystal is mounted by a suitable gum at the tip of a glass fibre embedded in a small piece of plasticine; the plasticine is fixed to the rotatable central spindle of the camera so that the crystal can be brought into the X-ray beam from the collimator by lateral movements on two perpendicular translation slides and a height adjustment which raises or lowers the top of the spindle. Laue photographs with a symmetrical distribution of maxima require the crystal to be correctly oriented with respect to the X-ray beam; it is roughly set into an appropriate position (as suggested by the preliminary examination), and final adjustment to achieve a satisfactory pattern is made by tilts on two perpendicular arcs carried on the central spindle. The essential components in the mounting of the crystal on the camera are shown later in Fig. 9.3. Diffraction of white radiation by a stationary crystal occurs in many different directions simultaneously and is invariably recorded on a photographic film; this film is either cylindrical with axis coincident with the central spindle, or a flat plate normal to the incident beam before (the 'back reflection' method) or after (the 'transmission' method) it strikes the crystal. Cylindrical films record more of the diffraction pattern than flat plates, though the simpler geometry of the latter makes detailed interpretation a little easier; whatever the film mounting, holes (or beam traps of some kind) must be provided to avoid fogging of the central areas of the film by the primary beam.

Laue photographs taken with the X-ray beam travelling in a general direction in a crystal lattice show a random distribution of spots on exposed films, but in certain directions patterns can usually be obtained in which the arrangement of diffracted maxima is symmetrical to a greater or lesser extent (Plate 1). *Symmetrical Laue patterns* occur when the X-ray beam is incident along a direction of symmetry in the crystal lattice; such patterns are commonly used for the determination of crystal symmetry. We can best illustrate this by outlining a practical example; we will assume that photographs are taken by transmission and recorded on a flat plate film. On any Laue photograph of this kind the diffraction spots lie on conic sections all of which pass through the intersection of the undeviated primary beam with the film; each curve marks the locus of reflections from planes in a particular zone. Let us suppose that after a little trial and error the first symmetrical Laue pattern from our crystal shows a distribution of spots consistent with horizontal mirror symmetry both in position and intensity; the appearance of a few zones of spots is sketched in Fig. 9.2(a). The symmetry shown by diffraction maxima was discussed in Chapter 8.5, and this photograph allows us to begin an assessment of the point group of the specimen which apparently contains a mirror plane (*m*). We can continue by investigating the degree of any axis normal to this plane; if photographs are taken after the crystal has been rotated through 60°, 90°, 120° and 180° about the vertical axis, patterns identical in spot distribution with Fig. 9.2(a) will be obtained at any of these settings only if $n = 6$, 4, 3 and 2 respectively. This will establish any symmetry grouping X/m and begin to clarify the crystal system; with possible implications of this grouping in mind we can extend our investigation by searching for any symmetry axes contained in the horizontal *m*-plane. When the X-ray beam is incident along the direction of a symmetry axis, the patterns of Laue spots display the appropriate symmetry; for example Fig. 9.2(b) shows the appearance of a few zones consistent with the symmetry *mm2* when the X-rays are incident along the direction of the diad axis. In this way we can gradually build up a picture of the spatial arrangement of the point group symmetry elements; it is sometimes necessary to obtain further symmetrical patterns after the crystal is re-mounted about another axis before the symmetry can be confirmed.

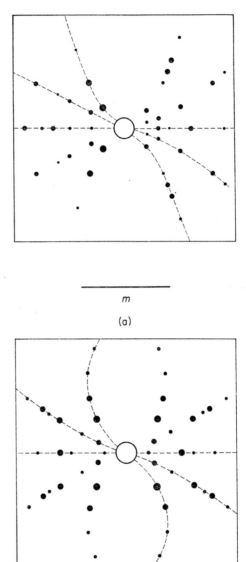

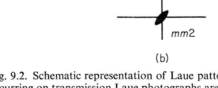

Fig. 9.2. Schematic representation of Laue patterns. Some of the spots occurring on transmission Laue photographs are shown; differing intensities are indicated by the size of circles; the broken lines (not present on photographs) indicate zones of reflecting planes. (a) Shows a horizontal *m* in the direction of travel of X-rays. (b) Shows the symmetry *mm2* in the direction of travel of X-rays.

It must be stressed, however, that such procedures cannot provide an un-ambiguous determination of crystal class, for Friedel's law implies that every crystal diffracts radiation as if it were centrosymmetric; with the addition of a centre of symmetry, each of the twenty-one acentric crystal classes becomes identical with one or other of the eleven centric classes, and so it is only the symmetry elements of these eleven classes which can be detected on Laue photographs. We can only divide crystalline matter into the eleven *Laue groups* listed in Table 9.1 by these methods; if the photographs of Fig. 9.2 had

Table 9.1. The Laue groups

System	Laue group	Other indistinguishable acentric classes
Triclinic	1	$\bar{1}$
Monoclinic	2/m	2, m
Orthorhombic	mmm	222, mm2
Trigonal	$\bar{3}$	3
	$\bar{3}m$	3m, 32
Tetragonal	4/m	4, $\bar{4}$
	4/mmm	4mm, $\bar{4}m2$, 42
Hexagonal	6/m	6, $\bar{6}$
	6/mmm	6mm, $\bar{6}m2$, 62
Cubic	m3	23
	m3m	$\bar{4}3m$, 432

finally led to the conclusion of Laue group *mmm* for the specimen, the crystal class of the material could be *mmm* or 222 or *mm*2. Occasionally, further study of diffraction patterns and the systematic absences which they display will allow the particular class within the Laue group to be resolved, e.g. the absences for space group *Pccn* are unambiguous, so that if our specimen has a pattern with these characteristic absences it must belong to class *mmm*; but more often we must resort to some means other than diffraction (morphology, other physical tests, etc.) if it is essential to distinguish among the possible classes within a Laue group. Laue methods however can identify the crystal system in all cases and allow the recognition of those symmetry elements which can define a suitable choice of crystallographic axes.

Other applications of Laue photographs include the determination of the orientation of a single crystal or a grain in an aggregate when the essential crystallography of the specimen material is known; it is particularly difficult to orient crystals of cubic symmetry in which most other physical tests cannot recognise any differences between the various lattice directions, and adap-tions of the Laue method are much used in the problems of metal single crystal orientation. Since any Laue photograph is characteristic of the lattice direction along which the X-ray beam is travelling, specimen orientation is established either by the use of standard photographs of the particular material or by an analysis of the indices of zones of Laue spots. It is only in these rather unusual orientational problems that indices are assigned to the various reflections on a Laue photograph, although it is not particularly difficult to index them on symmetrical patterns (see Henry, Lipson and Wooster, Chapter 6). The lattice constants which determine the positions of maxima can usually be obtained more easily and reliably by other methods, as can any systematic absences; moreover, intensities of individual indexed spots are complicated by the possibility of contributions from several planes and the variation in the intensities of different wavelengths within the white radiation spectrum which have been selected for reflection. Indeed a routine determination of Laue symmetry from Laue photographs needs some care if it is thought that the specimen may have some pseudo-symmetry. Suppose, for example, that a monoclinic crystal has a dimensionally orthorhombic cell (i.e. $\beta = 90°$), as can happen with a reasonable degree of approximation; the diffraction maxima will be found in positions corresponding to the ortho-rhombic Laue symmetry mmm. Any recognition of the true Laue group as $2/m$ would come from a study of the relative intensities of symmetry related spots, e.g. on photographs with the X-ray beam parallel to the x and z axes, the intensity distribution must have the symmetry m (rather than $mm2$). Intensity differences which distinguish these distributions may be quite small; they should, of course, be apparent on symmetrical Laue photographs, but the slightest mis-setting of the crystal can have a marked effect on the intensities of individual spots and lead to erroneous conclusions. When complications like these are suspected, it is more reliable to make intensity comparisons (and Laue symmetry determinations) by single crystal methods which have diffraction patterns formed by radiation of a single wavelength.

9.3. Rotation and oscillation methods

These are the simplest of the many moving crystal methods, all of which employ either monochromatic or, more usually, filtered radiation, and satisfy the Bragg conditions for a large number of planes by some motion of the crystal. The crystal is rotated (or oscillated over a small angular range) about an axis normal to the incident X-ray beam; the crystal is oriented so that a prominent zone axis (often a cell edge) is parallel to the rotation direction. The mounting of a crystal in a suitable orientation has been described in the previous section and the essentials of the camera are displayed schematically in Fig. 9.3. A motor drive can be set to rotate the central spindle carrying the crystal or to oscillate it through a fixed angle (usually 5°, 10° or 15°) about a predetermined position. A final setting of the crystal axis parallel to the spindle and any orientation with respect to the X-ray beam is achieved by preliminary photographs (often obtained in a preceding Laue examination) after the crystal has been roughly aligned by visual observation through the telescope.

In some modern instruments used to obtain the accurate intensity values required in certain kinds of structural work, diffracted beams are detected by counters, often with the fullest automation of the whole apparatus; more usually reflections are recorded on photographic films either as flat plates in the back-reflection or transmission positions, or preferably on cylindrical films concentric with the axis of rotation.

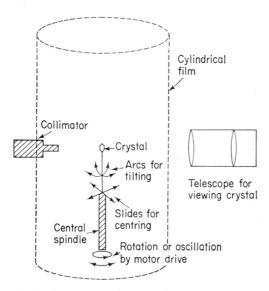

Fig. 9.3. The essential features of a rotation camera.

The nature of the diffraction pattern from this experimental arrangement is clear from consideration of the Laue equations (Chapter 8.3(a)). Let us suppose that the specimen has been oriented so that the crystallographic x-axis is parallel to the rotation axis; in this case $i_a = 90°$ at all positions of rotation, so that the Laue equation for the rows parallel to this direction becomes

$$a \cos d_a = h\lambda$$

For lattice rows along the directions of the other two axes we have

$$b(\cos d_b - \cos i_b) = k\lambda$$
$$c(\cos d_c - \cos i_c) = l\lambda$$

where i_b and i_c vary continuously as the crystal rotates about the x-axis. From the first equation, the cones of reinforcement for the various integral values of h are of fixed semi-vertical angles about the x-axis direction for all positions of the rotating crystal; for the other two equations with given values of k and l the cones of reinforcement will have variable semi-vertical angles as i_b and

i_c change. All three equations must be simultaneously satisfied for a reflection to occur; for particular values of k and l the direction common to the two cones about the y- and z-axes will change as the specimen rotates. As this common direction changes, at some positions it will intersect the fixed cones about the x-axis; when this happens, all three Laue conditions are met and we shall get reflections. Any reflections, therefore, must be in directions along the surfaces of the fixed cones drawn about the x-axis around which the

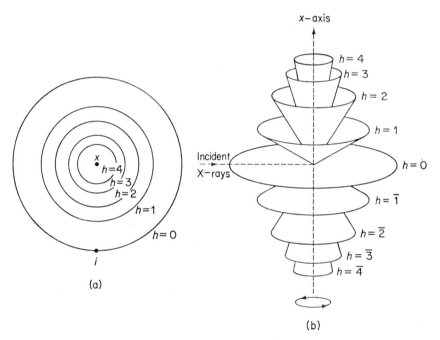

Fig. 9.4. Cones corresponding to fulfilment of Laue condition for lattice row parallel to rotation axis. (a) Stereographic projection with pole of incident beam (i) normal to x-axis as rotation direction at centre of projection. (b) Spatial arrangement of cones on which all reflections must lie.

crystal is rotating; the diffraction pattern consists of separate reflections occurring at different positions of rotation whose directions must diverge from the crystal along the surfaces of cones whose semi-vertical angles are determined by a, λ and h (Fig. 9.4). In a complete rotation reflection conditions for many different maxima are satisfied to give a distribution of diffracted beams which is symmetrical in space; the number of maxima on a particular cone depends on the cell dimensions b and c, and is greater the larger these repeats. There can be superposition and overlapping of different maxima, which is undesirable when we wish to identify individual reflections and examine their intensities; this can be avoided by studying the diffraction pattern in stages in each of which the crystal is oscillated through small angles (usually up to 15°) until all reflections possible in a complete rotation have been recorded. Naturally, oscillation patterns are generally similar to the rotation patterns of

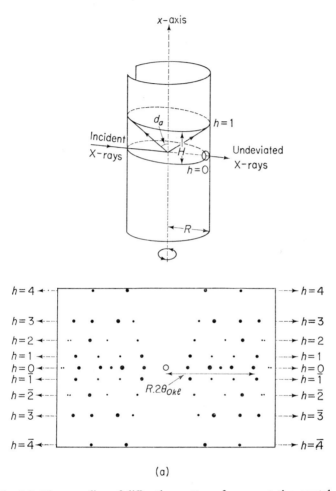

(a)

Fig. 9.5. The recording of diffraction patterns from a rotating crystal. (a) Cylindrical film. (b) Flat plate film in transmission position (overleaf). In each case the upper diagram shows the film mounting and the intersection of the cones of Fig. 9.4; the lower diagram shows the appearance of the processed film.

which they are a part, though the spatial distribution of maxima in any one range need not be so regular; any symmetry depends on the choice of oscillation range, and symmetrical oscillation photographs are often used for Laue symmetry determinations in which intensity comparisons are important.

Apart from direct recording by counter diffractometers, most rotation or oscillation patterns are examined by cameras which have a cylindrical film of small radius (usually 3 cm) whose axis is coincident with the rotation axis; the incident beam enters through a gap in the film and undeviated radiation passes through a hole punched in the centre. A cylindrical film enables a large angular range of diffracted beams to be recorded on one exposure, and has

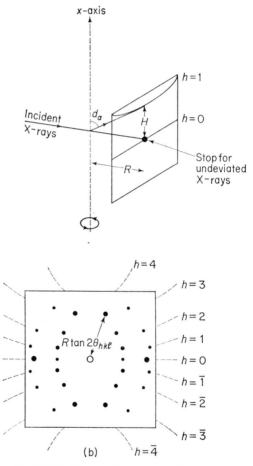

Fig. 9.5(b). (For caption see preceding page.)

the advantage that it is intersected in horizontal lines by the cones on which maxima must lie (Fig. 9.5(a)); we can readily identify which reflections are on the different cones from these lines, usually known as *layer lines*. With flat-plate films a much more restricted portion of the pattern is recorded on a single exposure, and layer lines, although recognisable, are curved unless $h = 0$ (Fig. 9.5(b)). Usually photographs are taken with filtered radiation

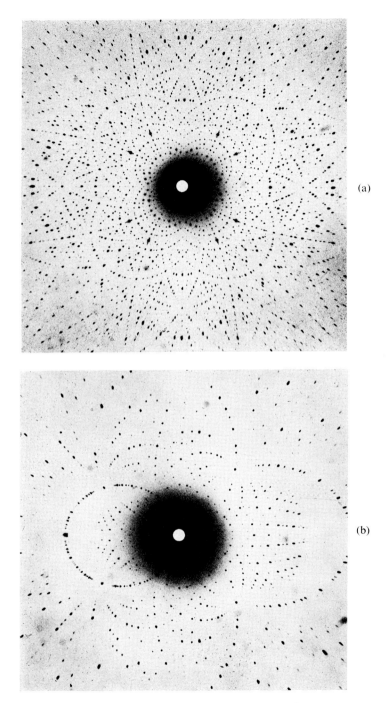

(a)

(b)

Plate 1. Laue photographs of vesuvianite (a complex alumino-
silicate); (a) X-rays parallel to [001]: symmetry 4mm;
(b) X-rays parallel to [UV0]: symmetry m.

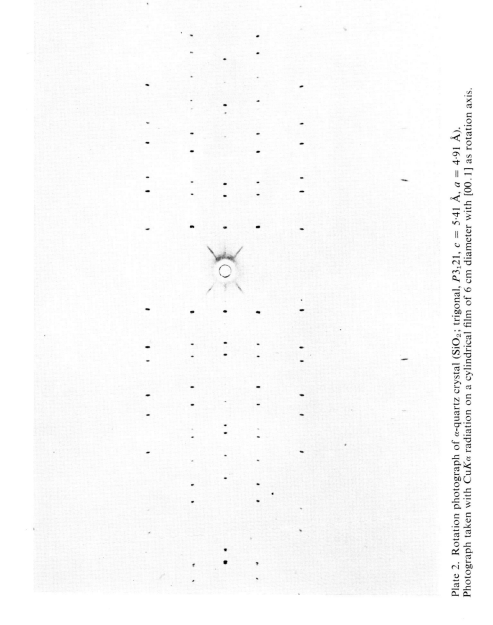

Plate 2. Rotation photograph of α-quartz crystal (SiO_2; trigonal, $P3_121$, $c = 5\cdot41$ Å, $a = 4\cdot91$ Å). Photograph taken with Cu$K\alpha$ radiation on a cylindrical film of 6 cm diameter with [00.1] as rotation axis.

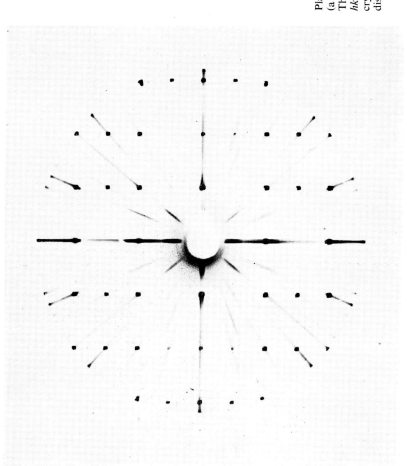

Plate 3. Precession photograph of stokesite (a complex silicate) crystal. The camera has been set to record only *hk0* reflections from this orthorhombic crystal; note the undistorted orthogonal distribution of the characteristic maxima.

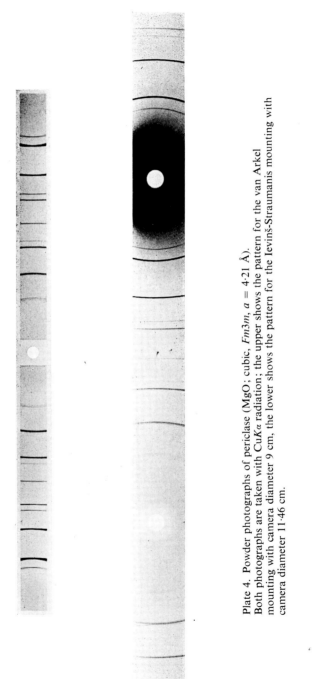

Plate 4. Powder photographs of periclase (MgO; cubic, *Fm3m*, $a = 4·21$ Å).
Both photographs are taken with Cu$K\alpha$ radiation; the upper shows the pattern for the van Arkel
mounting with camera diameter 9 cm, the lower shows the pattern for the Ievinš-Straumanis mounting with
camera diameter 11·46 cm.

(Plate 2), and require exposures of the orders of hours rather than minutes; on heavily exposed films the effects of residual white radiation (and occasionally β radiation) are seen as streaks around stronger low angle reflections. Resolution of the two doublet wavelengths of $K\alpha$ radiation is found at high angles (see Chapter 10.4) often causing the appearance of two closely spaced spots for some planes towards the outer edge of the film.

No matter how the pattern is recorded it is possible to identify individual spots as reflections from particular planes and use their positions to obtain dimensional data about the crystal lattice. In fact, the repeat distance S_{UVW} (Chapter 7.2) between points on a lattice row $[UVW]$ parallel to the rotation axis can be found from simple measurements on the film to determine the semi-vertical angles of cones corresponding to the layer lines. For example, when rotation is about the x-axis (or [100]), the height (H) of the hth layer line on a cylindrical film is $R \cot d_a$, where R is the radius of the film (Fig. 9.5(a)): $S_{100} = a$, so that from the Laue equation the semi-vertical angle of the corresponding cone is

$$d_a = \cos^{-1}(h\lambda/a),$$

to give $$a = h\lambda/\cos(\cot^{-1} H/R).$$

On flat-plate films a measurement of H must be made on the vertical median line to avoid distortions due to the curvatures of layer lines. Other cell dimensions, b and c, can be obtained by similar measurements on photographs taken with the y- and z-axes along the rotation direction. For more general zone axes, photographs will yield S_{UVW} from which any non-orthogonal interaxial angles could be determined as well as the lattice type; for example in a monoclinic crystal, $S_{101} = (a^2 + c^2 - 2ac \cos(180° - \beta))$ so that measurements on patterns with [101] as the rotation axis can be used to find the angle β; also $S_{110} = (a^2 + b^2)^{1/2}$ for a P-lattice but $S_{110} = \frac{1}{2}(a^2 + b^2)^{1/2}$ for a C-lattice, so that layer line positions on [110] photographs can differentiate between the two cell types. In more sophisticated single-crystal techniques this kind of information can be more easily obtained without the necessity of re-mounting a crystal about each of the zone axes; nevertheless an understanding of simple rotation methods is essential to the efficient use of these more complex cameras and much preliminary and survey work is still carried out on rotation cameras.

Individual spots on the films can be identified with reflections from particular families of planes when this is necessary. When indexing, the Laue condition controlling the cones of reinforcement which define layer lines, ensures that, for rotation about $[UVW]$, spots with indices hkl are segregated on the layer lines in such a way that the condition

$$Uh + Vk + Wl = n$$

relates reflections on the nth layer line. On rotation about [100] (the x-axis) indices on the zero layer line must be $0kl$, on the first layer above the zero layer $1kl$, on the first below $\bar{1}kl$ and so on. Indexing spots on particular layer lines requires us to find the k and l values for each and this is best carried out by using the Bragg law in some way. In principle, if θ_{hkl} can be determined from the spot position, we can calculate the interplanar spacing d_{hkl} ($= \lambda/2 \sin \theta_{hkl}$); since this is a function of the known cell constants and h, k and l (Table 7.2) it should be possible to assign the appropriate values of k and l. In practice, a direct measurement of θ_{hkl} values from spot positions is simple enough on flat plate films (Fig. 9.5(b)), but is rather more difficult on cylindrical films, for which the general relation between spot position and Bragg angle is rather complex, though on the zero layer reflections are situated at $R \cdot 2\theta_{0kl}$ (radians) from the centre of the film (Fig. 9.5(a)). Even if we could determine all θ_{hkl} values with sufficient accuracy this method of indexing would prove to be very tedious and often ambiguous, and simpler systematic procedures employing a graphical solution of Bragg conditions in terms of the reciprocal lattice and Ewald sphere (Chapter 8.3(a)) are preferable. Full accounts of these techniques of indexing rotation and oscillation patterns will be found in references mentioned in the bibliography.

9.4. Other methods

Rotation of the specimen about a zone axis to give a diffraction pattern recorded upon a stationary cylindrical or flat film is the simplest and earliest moving crystal method; as we have seen, it is particularly suitable for a preliminary determination of cell constants and is still much used for rapid general surveys of diffraction patterns. Inevitably the recording of a three-dimensional pattern on the two dimensions of a film presents difficulties of interpretation and we have already mentioned that relatively cumbersome graphical methods are necessary to identify and index individual maxima. Subsequent developments in camera design have had the objective of simplifying the interpretation of photographs to the point at which indexing is clear even on a simple visual inspection; this involves complex instrumentation to satisfy certain geometrical requirements, and we do not propose to discuss the detailed design of any modern single crystal camera here. However, it is relevant to give some consideration to the principles which control camera construction and to mention briefly the techniques which are in commonest use.

We can regard the rotation pattern from a particular zone axis as consisting of layers of reflections; maxima on a particular layer line constitute one of these layers and the new cameras are designed to analyse the complete pattern layer by layer in separate exposures. The nature of each layer of reflections is apparent from the optical analogy. In the discussion of Chapter 8.3(a) optical diffraction by a planar two-dimensional grating has maxima at the intersection of two linear fringe systems when the pattern is plotted against sin ϕ/λ; similarly the maxima of a pattern from a crystal are to be found at the mutual intersections of the three linear fringe systems formed by the cell repeats when plotted in a corresponding fashion (the reciprocal space of the Ewald treatment of the Bragg law). Each layer of reflections constitutes a layer of the reciprocal points formed by these mutual intersections, and the cameras are designed to record one of these layers so that we can recognise the arrangement of the maxima and index them easily; in an undistorted form the patterns of spots on the film from a given layer is comparable to the grid shown for a planar optical grating in Fig. 8.9. In all these cameras both the film and the crystal move so that reflections occurring at different angular positions are spread out on to the two-dimensional surface of the film at positions which depend on its movement and the way in which this is coupled to the motion of the crystal; for this reason they are often said to be *moving-film cameras*. In addition, all cameras must have some kind of absorbing screen placed between the crystal and the film to ensure that only reflections from a selected layer reach the film on each exposure.

The simplest moving film method is that of a Weissenberg camera; essentially this is a simple rotation camera in which the cylindrical film travels backwards and forwards parallel to the rotation axis as the crystal oscillates, the angular motion of the crystal and the translation of the film being coupled by a fixed gearing. A cylindrical screen with a circumferential slot is set to select the layer line for each exposure. The geometry of reflection in a Weissenberg camera is such that the pattern of spots on the film is a distorted projection of the grid of reciprocal points forming the layer; nevertheless, interpretation is relatively simple when aided by specially constructed charts, and an experienced user can often index by inspection alone. Later moving film cameras are designed to give undistorted projections of this grid, and the two most commonly used direct projection methods are those of the rotation-retigraph and precession-retigraph cameras*; in both these, undistorted projection is achieved by using a flat film which is always kept in the same relation to the selected layer of reflections while Bragg conditions are being satisfied. Both have a flat screen with an annular aperture to eliminate all unwanted reflections; this is stationary in the rotation-retigraph, but follows the motion of the selected zone axis in the precession method. Indeed an important difference between these two methods, as their names imply, is that in a rotation-retigraph the crystal movement is simple rotation about the selected zone axis, whereas in a precession-retigraph the zone axis precesses about the incident beam, rather as the motion of a top. Naturally, the mechanisms of these cameras are rather complicated; full details will be found in the references in the bibliography, together with discussions of the ways in which

---

* The term 'retigraph' is an adaptation of the Greek for 'a drawer of nets'.

geometrical conditions for undistorted direct projections of layers are ful-filled. For our part, we will merely note that modern instrumentation has re-duced the problem of identifying and indexing maxima in single crystal diffraction patterns to one which can be undertaken with little or no inter-pretative skill (see Plate 3). In these latter two methods, the use of flat-plate films means that only limited regions of each layer of the rotation pattern can be observed, at least without further mechanical adjustment; with a crystal set to oscillate over a 180° range, the Weissenberg camera will record all possible reflections for a layer on one film. This can be embarrassing, how-ever, when spots are too closely spaced (due to large cell repeats), and little can be done about this with a Weissenberg camera of fixed diameter; on both direct projection methods, the scale of the projection can be increased, where necessary, by simple adjustments to the camera. All of these common moving-film techniques have their advantages and drawbacks, and a selection of the most suitable method must be made according to experimental demands.

9.5. Exercises and problems

1. What symmetry could be observed on Laue photographs taken with the X-ray beam incident along [100], [011] and [111] directions for a crystal with (a) an orthorhombic cell, (b) a tetragonal cell and (c) a cubic cell?

2. Gallium belongs to the Laue group mmm, and single-crystal rotation photographs taken with CuKα radiation ($\lambda = 1.542$ Å) on a cylindrical camera of radius 3.0 cm provided the following data:

Axis of rotation	[100)	[010]	[001]	[011]	[101]	[110]
Order of layer line	2nd	2nd	3rd	2nd	3rd	3rd
Distance from zero layer line (cm.)	2·81	2·81	2·27	2·89	1·83	3·16

Determine the cell dimensions and lattice type for gallium.

3. A rotation photograph of NaCl (cubic, $a = 5.64$ Å) is taken with [001] as the axis of rotation; unfiltered radiation from a Cu target is incident on the crystal, and the diffraction pattern is recorded on a flat-plate film in the transmission position 3 cm from the crystal. On the zero layer, a strong spot due to characteristic radiation is found at 1·855 cm from the centre of the film; a streak due to white radiation passes through this spot and extends to within 0·33 cm of the centre of the film. Estimate the voltage applied to the X-ray tube during the exposure.

4. A certain crystal belongs to the Laue group $2/m$, and single-crystal rota-tion photographs taken with CuKα radiation ($\lambda = 1.542$ Å) on a cylindri-cal camera of radius 3.0 cm provided the following data:

Axis of rotation	[100]	[010]	[001]	[101]	[011]	[110]	[111]
Order of layer line	2nd	4th	6th	3rd	7th	5th	8th
Distance from zero layer line (cm)	2·25	4·00	3·40	3·36	3·32	4·50	4·75

Determine the cell dimensions and the lattice type for this crystal. It is possible to choose a smaller conventional cell. Find the dimensions and lattice type for such a cell, and re-index the rotation axes above in terms of your choice.

5. The following data were obtained during the examination of single crystals of anatase (TiO_2):

Laue photographs	Direction of incidence	[001]	[100]	
	Symmetry	4mm	mm2	
Rotation photographs ($\lambda = 1\cdot542$ Å camera radius 3 cm)	Axis of rotation	[001]	[100]	[111]
	Order of layer line	3rd	1st	2nd
	Distance from zero layer line (cm)	1·70	1·36	2·10

Determine the Laue symmetry, cell dimensions, lattice type and the number of formula units per cell for anatase.

(Atomic weights Ti, 47·9; O, 16·0; density of anatase, 3·9 gm/cm³.)

6. A crystal of class $m3m$ (lattice type, F) showing faces of the form {110} was set up on an oscillation camera of radius 3 cm with [110] parallel to the axis of rotation; the height of the first layer line was measured as 1·90 cm ($\lambda = 1\cdot542$ Å). Determine the inclination of the X-ray beam to [1$\bar{1}$0] at which the 1$\bar{1}$1 reflection occurs.

7. The morphological description of a crystal is quoted as: Class $2/m$, $a:b:c = 0\cdot7143:1:1\cdot714$, $\beta = 109° 28'$, showing the forms {010}, {$\bar{1}$01} and {12$\bar{1}$}. For this crystal rotation photographs on a cylindrical camera of radius 3 cm, $\lambda = 1\cdot542$ Å, give the following data:

Axis of rotation	[010]	[101]
Order of layer line	3rd	3rd
Distance from zero layer line (cm)	2·64	4·21

Reconcile the X-ray and morphological observations, and give a description of the crystal in terms of the cell suggested by the X-ray examination.

SELECTED BIBLIOGRAPHY

Practical methods and the interpretation of diffraction patterns

AZAROFF, L. V. 1968. *Elements of X-ray crystallography*. McGraw-Hill.

BUERGER, M. J. 1942. *X-ray crystallography*. Wiley.

HENRY, N. F. M., LIPSON, H. and WOOSTER, W. A. 1960. *The interpretation of X-ray diffraction photographs*. Macmillan.

NUFFIELD, E. W. 1966. *X-ray diffraction methods*. Wiley.

WOOLFSON, M. M. 1970. *An introduction to X-ray crystallography*. Cambridge University Press.

Specialised topic

BUERGER, M. J. 1964. *The precession method in X-ray crystallography*. Wiley.

Source book for crystallographic data

DONNAY, J. D. H., DONNAY, G., COX, E. G., KENNARD, O. and KING, M. V. 1963. *Crystal data (determinative tables)*. American Crystallographic Association.

10

DIFFRACTION BY POWDERS

10.1 General considerations

In this chapter we deal with the second main group of diffraction techniques, usually known as the *powder methods*, in which the specimen consists of a randomly oriented aggregate of very small single crystals. Powder specimens are often prepared by crushing larger single crystals to a fine powder and then cementing these particles together to form the random aggregate; sometimes the growth process itself produces a polycrystalline aggregate (as in the bulk crystallisation of metals or metallic alloys) which can serve as a suitable specimen. We shall only consider some aspects of simpler experimental arrangements and be mainly concerned with the use of positions of diffraction maxima to determine lattice geometry; powder methods are extensively employed in a variety of technological problems and a selection of some of the more important applications is discussed in the last section.

The character of the diffraction pattern from a powder specimen irradiated by X-rays of a single wavelength can be simply deduced by examining the Bragg condition for reflection represented by the stereogram of Fig. 10.1. For any family of planes (*hkl*) and a specimen with a completely random array of crystallites, we must regard the whole stereogram as occupied by possible positions of normals to these planes. When the small circle of radius $90° - \theta_{hkl}$ is drawn about the pole of the incident beam *i*, every point on it must correspond to a small crystallite in a reflecting position; diffracted radiation is therefore transmitted in all directions represented by the upper small circle of the figure. Physically this diffracted radiation forms a cone of semi-vertical angle $2\theta_{hkl}$ about the direction of the undeviated X-rays; this cone is the locus of all beams reflected from *hkl* planes for the many different crystallites which happen to be in an orientation suitable for such a reflection. In any crystalline material there are many different discrete values of d_{hkl} (related to the values of *h*, *k* and *l* and the cell constants as in Table 7.2); for each of these there is a distinct Bragg angle to satisfy the conditions for reflection. The diffraction pattern from the powder specimen will therefore consist of a series of cones about the direction of the undeviated beam, each with its characteristic semi-vertical angle related to the d_{hkl} value, and limited only by the condition that $\sin \theta_{hkl} = \lambda/2d_{hkl} \leqslant 1$; Fig. 10.2 shows the general form of a powder pattern.

This ideal powder pattern requires that crystallites in every possible orientation are to be found within the finite irradiated volume of the powdered

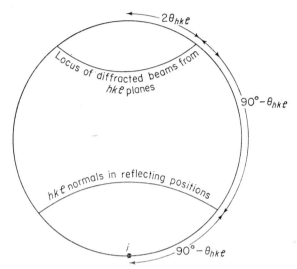

Fig. 10.1. The Bragg conditions for a reflection *hkl* for a random powder.

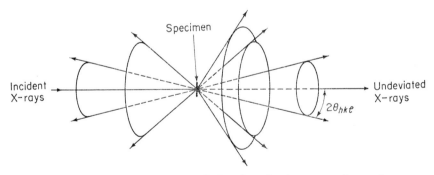

Fig. 10.2. Diffracted cones of radiation diverging from a powder specimen. Each cone corresponds to reflections for particular *hkl* planes, and has a semi-vertical angle $2\theta_{hkl}$.

specimen; in practice, a reasonable degree of approximation may be difficult to achieve. For bulk specimens, the crystallite size may be too large and there may be some restriction of the orientations present (see preferred orientation in Chapter 10.4 later). Even when there is some control over specimen preparation from crushed powders, there are certain snags in obtaining a suitably random aggregate. At first sight it might seem advantageous to have crystallites as small as possible; quite apart from any technical difficulties in producing very small grain sizes, there is a lower limit that is desirable for powder diffraction work. When the particle size decreases below about 10^{-5} cm, each diffracted cone is broadened to reflect over a range of Bragg angles, so that the pattern is less well defined and any measurements are consequently less accurate and significant; this broadening may be compared to that developed by the spectra from an optical diffraction grating as the number of lines in the grating is reduced. (Broadening may also be due to other causes such as imperfections and strains in the crystallites, and it is sometimes necessary to try to reduce these by annealing before X-ray examination.) The crushed material is usually sieved to give uniform crystallite dimensions, but often the grain size is large enough to prevent a sufficiently random array within the irradiated volume of the specimen; this results in the appearance of separated reflections from individual crystallites around the cones. Although 'spotty' cones are sometimes used in estimates of grain size (particularly for bulk specimens), most powder work relies on accurate measurements on smooth continuous diffracted cones; in practice, a sufficiently good approximation to the random array of the ideal powder pattern is ensured by continuously varying the orientation of the specimen with respect to the incident beam. For cemented aggregates a simple rotation is usually adequate, but for some bulk specimens with large grain sizes more complex motions may be essential.

10.2. Experimental methods

(a) Photographic

There are many different designs of camera used for particular experiments with various kinds of specimen; it would be inappropriate here to give an account of all these technical developments and we shall only outline the simplest cameras that are universally used for routine powder work. In these a cylindrical specimen (length ~ 2 cm, diameter ~ 0.5 mm) is usually made by rolling a paste of the crushed material and gum tragacanth (whose scattering does not interfere with the diffraction pattern); occasionally, if the paste method is unsatisfactory, the crushed powder is packed into a fine tube of borosilicate or silica glass. The rotation cameras mentioned in Chapter 9.3 can be used to record diffraction patterns from such powder specimens, but they have various disadvantages. With flat plate films in the transmission or back-reflection positions, diffracted cones will appear as concentric circles at radii dependent both on the specimen-film distance and on the Bragg angles, but only a very limited range of θ angles is recorded. On cylindrical films, the cones trace out complex curves around the loci of constant θ; in the equatorial plane of the film, however, it is possible to simply relate θ_{hkl} values to the intersections of the cones (as in Fig. 9.5(a)) over a very wide range of

Bragg angles; but in practice, the accuracy of measurement is not very high due to the relatively small diameters of single-crystal cameras. Moreover, collimation of the incident beam is normally circular, and this can lead to unnecessarily long exposures due to the weakness of the beams diffracted from powder specimens.

In purpose-built powder cameras larger diameters are used and collimation of X-rays, preferably from the line source of the tube, produces an incident beam of rectangular cross-section which bathes the cylindrical specimen; although this causes a set of cones displaced along the length of the specimen for each reflection, the width in the equatorial plane is not appreciably affected and the exposure time is much reduced. Usually a suitable filter is placed in the path of the incident beam, though occasionally crystal-reflected monochromatic radiation is necessary to define a complex pattern; this increases the normal exposure time (up to a few hours) by a factor of 5–10. The

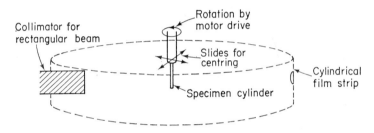

Fig. 10.3. The essential features of a powder camera.

specimen is on the central axis of a narrow cylindrical film strip placed around the equatorial plane containing the incident beam (Fig. 10.3); only a small sector of each diffracted cone is recorded, and it is customary to refer to the short arcs on the film as *powder lines* (Plate 4). Apertures must be provided to prevent incident and undeviated X-rays from striking the film, and this restricts the maximum and minimum Bragg angles that can be recorded. Many commercial instruments are of 9 cm or 19 cm diameter, the greater resolution of the larger camera inevitably demands longer exposures; more recently a radius of 5·73 cm has become popular, a value chosen so that a distance of 1 mm on the film corresponds to a change of 1° in 2θ angle. The specimen is carried on mutually perpendicular cross-slides to permit alignment in the incident beam and coincidence with the camera axis, about which it is rotated by motor drive. On a developed film line positions at the centres of the arcs are roughly measured with a steel scale, though greater accuracy, when necessary, can be obtained by some form of travelling microscope.

The three simplest common powder cameras differ only in the way in which the strip film is arranged (Fig. 10.4). In (a) and (b), patterns interchange the locations of low and high angle lines on the film; in the *Bradley–Jay* mounting, high angle lines with large values of θ_{hkl} (recognised by the doublet resolution which occurs at such angles) are at the outside of the processed film, whereas in the *van Arkel* setting they are in the centre. In most cameras

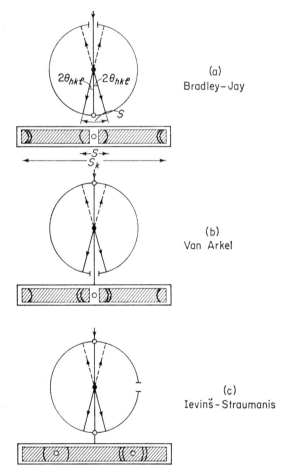

(a)
Bradley–Jay

(b)
Van Arkel

(c)
Ieviņš–Straumanis

Fig. 10.4. Common film mountings in powder cameras. A plan of the camera arrangement is shown in each case, with a sketch of a developed film beneath. In the upper diagrams low angle cones are indicated by full lines and high angle cones by broken lines; on the films low angle lines are single and high angle lines are sketched with resolution of a doublet.

screening the film from fogging by the incident and undeviated X-ray beam restricts the recording of diffracted radiation to θ angles between about 4° and 86°. The separation, S, of the two arcs due to the same diffracted cone is used to deduce the Bragg angle θ_{hkl} for the particular set of reflecting planes; for rough measurements the simple geometry of the arrangement is sufficient (e.g. $\theta_{hkl} = S/4R$ in the Bradley–Jay mounting, where R is the camera radius). For more accurate measurements, cameras are calibrated by means of 'knife-edges'; these are metal strips built into the instrument whose sharp straight edges parallel to the camera axis throw well-defined shadows at the low and high angle limits of the film by intercepting background radiation due to air scatter, Compton scattering, etc. A camera constant ϕ_K is obtained from the pattern of a standard specimen; it is usually defined as the Bragg angle for reflections which would fall on the knife-edge shadows. Values of S can then be related to θ in terms of ϕ_K and S_K, the measured distance between knife-edges on the film (e.g. $\theta_{hkl} = S\phi_K/S_K$, in the Bradley–Jay mounting). Assuming that linear distances on the film change uniformly during processing, a correction from calibrated knife-edge shadows will help to eliminate errors in line positions due to this cause. But, apart from film shrinkage, there are many other possible factors which can affect the measured values of S; these include absorption, failure to place the specimen at the centre of the camera, refraction, etc. All of these must be taken into account when the highest accuracy is required, and in Chapter 10.3 we will discuss extrapolation methods to eliminate or minimise systematic errors of this kind. Although θ_{hkl} values of quite high accuracy can be obtained from either of these mountings under favourable circumstances, rather more reliable and accurate data on line positions (particularly at low–medium Bragg angles) are found nowadays by direct-recording powder diffractometers, which will be described shortly. Photographic methods are confined mainly to rough measurements and surveys of patterns, for which the third mounting in Fig. 10.4(c) is well suited. *Ievinš and Straumanis* suggested that such an asymmetric film mounting would allow the positions corresponding to $\theta = 0°$ and $90°$ to be found at the mid-points between pairs of lines belonging to the same low or high angle cones; knife-edges are unnecessary, for the distance between these positions provides the calibration for film shrinkage. In practice, it is often difficult to define these positions with great accuracy, particularly from the high angle regions of complex patterns with rather weak lines; but such cameras are quite accurate enough for the majority of modern photographic powder work, and they have become increasingly popular.

Before the advent of the direct-recording methods which now dominate modern powder work, cameras were designed for specialised purposes such as low- and high-temperature experiments, the examination of block specimens, etc. Many of these photographic techniques are still important, particularly if they are not readily adaptable to diffractometry; we cannot pursue the design of such cameras here. Nor is it worthwhile to discuss other cameras in which increased resolution is attained by divergent beams and focusing geometry, although, as we shall see, most modern diffractometers use a similar technique. The past and present ramifications of photographic powder methods are very extensive, and they are well documented in the biblio-graphy.

(b) Diffractometers

An alternative to the photographic film in the recording of diffracted X-rays had long been desired, but it has only been with the development of reliable detectors such as Geiger, proportional or scintillation counters over the past twenty years that direct-recording of diffraction patterns has become a routine operation. Both because of their wider technical applications and their simpler experimental geometry, powder methods were the first to be adapted to these new techniques in commercial instruments; at the present time most modern crystallographic laboratories possess (or have access to) powder diffracto-meters, but the more complex and expensive instrumentation of single crystal diffractometers does not yet justify a similar widespread use.

All powder diffractometers employ some kind of focusing geometry to aid resolution, and most adopt a form of the *Bragg–Brentano system* illustrated diagrammatically in Fig. 10.5(a). Filtered X-rays diverge from a fixed line source at A to strike the specimen (in the form of a flat rectangular or circular disc of powder) at the centre C of a circular diffractometer table; reflected radiation is focused on to the detector slit B, when the normal to the plane of the specimen bisects $A\widehat{C}B$ (perfect focusing would be achieved only if the speci-men surface were curved so as to lie on the circle through A, C and B). Once the instrument has been adjusted to satisfy this condition, this approximation to perfect focusing is maintained at all angles of incidence by ensuring that the detector rotates about the central axis of the table in the same sense as the specimen but with twice the angular speed. Different gear systems allow a choice of scanning speed for the detector, usually in the range between 0·05° (θ)/min. and 1–2° (θ)/min.; faster speeds are used for general surveys of pat-terns, whilst slower speeds are reserved for accurate measurements within a limited range of Bragg angles. The detector converts an X-ray quantum into an electrical impulse, so that the intensity of X-rays (or the rate of arrival of quanta in the counter) is matched by a corresponding rate of output of pulses from the detector; with suitable electronic circuitry these pulses can either be counted individually over a unit time interval for every counter setting, or—as is more usual—averaged electronically over a few seconds whilst the detector is in continuous motion to give an output that activates a pen-recorder. When the detector output is coupled to a pen-recorder, the peaks of a powder pat-tern are displayed on a chart as in Fig. 10.5(b); the scale of this chart depends on the chosen relation between the scanning speed and the chart speed of the pen-recorder. Most commercial instruments also include a second pen to indicate the scale of angular intervals on the chart, and these suffice for rough measurements of line positions. Accurate determinations of line positions need a calibration of the pattern which is usually made by mixing a small quantity of a standard with the specimen; measurements on the chart are then made relative to the known positions of the standard lines.

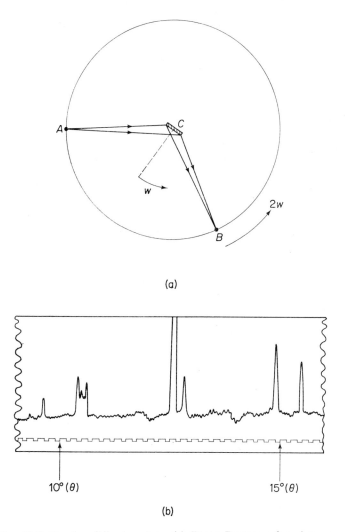

(a)

(b)

Fig. 10.5. Powder diffractometers. (a) Bragg–Brentano focusing system. (b) Part of a diffractometer chart showing peaks over the range of Bragg angles marked.

Around instruments of the kind outlined, a field of diffraction technology has grown up; this is concerned with improving the geometry of reflection, the counters and their associated electronic circuits, and the automation of diffractometers and their adaptation for specialised work, etc. It is still a fast developing area, and the more detailed accounts of diffractometry which will be found in the bibliography will probaby have to be changed as new techniques are developed. In our context we can recognise that, whenever the highest accuracy in relative line positions (and intensities) is demanded, this is most likely to be obtained from some sort of powder diffractometer; nevertheless, there will always remain many simpler routine tasks in which the humbler powder photographs provide all that is needed with relatively inexpensive equipment, not least in the field of very accurate determinations of cell constants from the positions of sharp, indexed high angle lines, as discussed in the next section.

10.3. Interpretation of powder records

However a powder pattern is recorded, its measurement will provide (i) values of θ_{hkl} for all lines, and (ii) the relative intensities of the various lines. Detailed studies of intensities are mainly relevant to structure determination, and the many factors that influence line intensities in complex patterns militate against the determination of accurate $|F(hkl)|$ values; whenever possible, this work is best carried out by single crystal methods, and structure determination from powder data alone is relatively rare. In this section, we shall address ourselves only to the interpretation of line positions and the dimensional values that can be derived from them.

Measured θ_{hkl} values can be converted into a set of interplanar spacings d_{hkl} for the specimen by the Bragg law; to use these d_{hkl} values to find cell constants, we must be able to associate particular values of h, k and l with the different diffracted cones, i.e. we need to index the lines. The problems of *indexing a powder pattern* are less formidable when rough values of the cell constants have already been found; indexing *ab initio* is particularly difficult in the lower symmetry systems where d_{hkl} depends on an increasing number of unknown cell parameters (see Table 7.2) and there are many more lines. For cubic crystals it is always relatively simple, for there is only one variable, the cell edge a. We can combine the relationship

$$d_{hkl} = a/(h^2 + k^2 + l^2)^{\frac{1}{2}}$$

with the Bragg equation to give

$$\sin^2 \theta_{hkl} = \frac{\lambda^2}{4a^2} (h^2 + k^2 + l^2)$$

For a particular pattern $\lambda^2/4a^2$ is a constant, so that $\sin^2 \theta_{hkl}$ values must be directly proportional to $h^2 + k^2 + l^2$; by examining our measured values we should be able to identify the constant factor ($= 4a^2/\lambda^2$) by which they must be multiplied to give a set of integers formed by summing the squares of the indices. In general $h^2 + k^2 + l^2$ can have all integral values except those given by $m^2(8n - 1)$, where m and n are integers, i.e. a sequence 1, 2, 3, 4, 5,

6, 8, 9, 10, 11, 12, 13, 14, 16, ... from which certain integers (7, 15, etc.) are excluded; not all lines in this sequence are necessarily observed on all cubic powder patterns, due to the accidental and systematic absences described in Chapter 8. For example, assuming no accidental absences, the sequence above would be that found for a cubic P-lattice; when the material has an I-cell, systematic absences modify the sequence to 2, 4, 6, 8, 10, 12, 14, 16, ..., and for a cubic F-lattice, the sequence becomes 3, 4, 8, 11, 12, 16, ... In fact, provided there are not too many accidental absences, the indexing of cubic powder photographs is a relatively straightforward exercise, and with a little experience can often be accomplished by an initial inspection of the sequence of low angle lines. In other systems the complexity of patterns increases with the degeneration of cell shape; in cubic crystals all planes of the form {100} have the same spacing (a) and contribute to one line of the pattern; but in tetragonal crystals there will be two lines corresponding to $d_{100}(= d_{010} = a)$ and $d_{001}(= c)$; in the orthorhombic system there will be three independent reflections for 100, 010 and 001, and so on; without some prior knowledge of the cell constants, the problem of indexing the observed $\sin^2 \theta_{hkl}$ values becomes more and more difficult. When there are only two unknown parameters—as in the tetragonal, trigonal and hexagonal systems—*ab initio* indexing is usually possible by trial and error methods; these can be systematised by graphical or numerical techniques. With three, four, or six undetermined parameters—as in the orthorhombic, monoclinic and triclinic systems —the difficulties of *ab initio* indexing are greatly magnified; various numerical procedures have been proposed, and these have had some success in restricted applications. But in general, in these three systems it is preferable, if possible, to obtain some idea of the cell constants by single crystal examination before attempting to identify powder lines by comparing observed d_{hkl} values with those calculated from the equations of Table 7.2; even so, there are often ambiguities in identification, particularly at larger Bragg angles where there can be considerable overlapping of line positions which are more sensitive to small changes in d_{hkl} values.

Once a powder pattern has been indexed, we can derive values for the *cell constants*. For example, a rough value of a for a cubic crystal can be obtained from any $\sin^2 \theta_{hkl}$ value from the equation above. But the powder method is capable of higher accuracy more readily than most other simple X-ray techniques; the greatest accuracies will be obtained by measurement of those lines which are most sensitive in position to small changes in d_{hkl} (and the cell constants). By differentiating the Bragg equation we find that

$$\Delta\theta_{hkl} = -\Delta d_{hkl}/d_{hkl} \cot \theta_{hkl}$$

so that for a given change in spacing (Δd_{hkl}), the change in angle ($\Delta\theta_{hkl}$) increases as $\theta_{hkl} \to 90°$. (It is for a similar reason that the closely spaced emission peaks of a $K\alpha$ doublet are resolved for high angle reflections.) The best values of cell constants will be those calculated from the measured positions of high angle lines, particularly when there is doublet resolution. When extreme accuracy is required, it is necessary to use extrapolation methods to minimise the effects of systematic errors (mentioned in Chapter 10.2(a)) by extrapolation to $\theta = 90°$; for a cubic crystal, values of a calculated independently for a number of high-angle reflections are plotted against a suitable angular function ($\frac{1}{2}(\cos^2 \theta/\theta + \cos^2 \theta/\sin \theta)$) to provide a linear extrapolation to the best value of a at $\theta = 90°$ (this does not allow for refraction, which must be subsequently corrected). Whenever high angle lines can be indexed, these extrapolation techniques can be adapted for use in any crystal system; they can provide cell constants with far greater accuracy (about ± 0.0001 Å) than is normally obtainable by any other method. But, as we remarked above, reliable indexing of such lines in the patterns of many crystals of the three lower symmetry systems is often impossible, and in these circumstances the best values of cell constants are calculated from very accurate diffractometer measurements of the positions of unambiguously indexed low-angle lines; cell dimensions are then fitted to the observed d_{hkl} values by a least-squares deviation method. This numerical process can be carried out by standard computer programmes, and although the quality of results is not comparable with that obtainable by high-angle extrapolation methods, the accuracies (usually about 0.001 Å and 0.005°) are sufficient for most purposes.

More detailed accounts of the indexing of non-cubic patterns and the derivation of accurate cell constants will be found in references in the bibliography.

10.4. Some applications of powder methods

In many ways powder methods are less valuable in providing fundamental crystallographic data on a substance than an examination of single crystals; but they form the basis of a plethora of important technical applications of X-ray diffraction. It is impossible to provide an exhaustive list of the ways in which powder techniques are applied, and this section contains a few illustrative examples selected for their universality and importance.

We will consider first the *identification of an unknown crystalline substance*. If every crystalline material has a characteristic structure with its own peculiar cell constants and contents, it should have a unique powder pattern by which it can be recognised. This kind of 'finger-print' identification has been systematised in the *X-ray Powder Data File* (published by the American Society for Testing and Materials with the assistance of many other learned societies); in its basic format, this file is a catalogue with card entries for each substance. Each card shows as its prime data the interplanar spacings and relative intensities of the three strongest lines of the powder pattern, and is identified by a file index number. Additional data on the substance are listed on each card; these include a complete set of d_{hkl} values and relative intensities (with indices where known), the experimental method by which the pattern was recorded, the name and chemical formula, space group and cell dimensions (if known), some account of physical properties (usually optics, density, colour) and any references to related or obsolete earlier data. Cards are arranged in the file in ascending order of prime interplanar spacings. In the original format each substance had three cards at the three positions in the file corresponding to the d_{hkl} values for each of its three strongest lines. Nowadays each substance has only a single card in the file, but the cards are accompanied by a catalogue index in book form; in this, each substance has three entries in positions corresponding to its three strongest lines. The file contains entries for many thousands of substances, and additional supplements are published regularly; in recent years these have been separated into organic and inorganic materials, and automatic sorting by punched cards or computer tape provides extra refinements.

In theory, identification of an unknown follows from the set of interplanar spacings calculated from measurement of the powder lines. Entries in the catalogue index with spacing values of the strongest lines are inspected; since each substance occurs in three positions, any ambiguity in selecting the most intense line of the pattern is unimportant. Each catalogue entry actually states the spacings (and rough relative intensities) of the six strongest lines, so that a fuller comparison of the d_{hkl} values and their intensities for very few cards extracted from the file should serve to identify the specimen. In practice, there are some obvious snags in identification by this method. Firstly, there is a possibility that two quite different substances fortuitously have indistinguishable powder patterns; fortunately this seems to be relatively rare. More importantly, it is clear that the efficacy of the method depends on the purity of the specimen; the recorded pattern of a mixture is the superposition of all the lines due to the various components weighted according to their relative proportions. When there are only two or three components, it is often possible for their separate patterns to be disentangled and identified, though the measure of success depends on the nature of the components; for complex multicomponent mixtures this is not usually practicable. A more fundamental difficulty relates to the basis of recognition by a pattern characteristic of each substance. Many of the structural configurations which determine the general character of a powder record allow the substitution of chemically different atoms in some, at least, of the cell sites; when this happens the overall powder pattern retains its general characteristics but the precise positions of lines (and the values of their relative intensities) may be affected by the different sizes (and scattering powers) of atoms introduced in the substitution. The crystallographic significance of substitutional solid solution of this kind will be mentioned again in Chapter 12, but its widespread existence promotes difficulties in identification by the *Data*

File which we can illustrate by a simple example. The metallic elements Cu ($a = 3·608$ Å) and Ni ($a = 3·517$ Å) have a similar structural arrangement with a simple cubic *F*-cell; the two elements can form binary alloys of any intermediate composition in which appropriate numbers of Cu and Ni atoms are arranged statistically on the face-centred pattern of sites. In this binary system at all compositions we observe powder lines corresponding to a cubic *F*-cell but with slight but significant shifts in position as the proportions of the small Ni and larger Cu atoms are varied. It is not possible for any data file to take account of variations in d_{hkl} values due to this cause in any satisfactory systematic manner; our simple example may not seem to pose many problems, but in the inorganic field, particularly with natural materials such as minerals, solid solutions (and their effects) can be extremely complicated. In summary, identification by powder methods is an excellent, non-destructive test, especially if only a small amount of a pure specimen (< 1 mg) is available; whilst unique solutions can be obtained, there are often limitations; and the value of identification by powder methods is greatest when it can be used in conjunction with other tests, such as chemical or electron-probe analysis, optical examination, etc.

Powder techniques are widely employed to investigate the *crystallisation of multicomponent systems* under various physico-chemical conditions, i.e. in the establishment of phase diagrams. Such diagrams often contain ranges of substitutional solid solution as, for example, in the Cu–Ni alloys above, where the change in cell dimensions is linear to a reasonable approximation and so the movement of a line can be used to estimate the composition of the alloy; in more complex substitution series the changes in cell constants (and line positions) are not necessarily linear, but once a variation has been established it can subsequently be used to estimate the extent of atomic substitutions, which otherwise might require long and tedious chemical methods. Moreover, each of the limited number of phases which crystallise within a system has a characteristic structural pattern, and in this restricted context each can be recognised from its powder pattern without difficulty despite any small changes due to solid solution; in a multicomponent system therefore, we can readily identify the various phases that occur at different points under varying conditions of temperature, pressure, etc., and establish their composition fields to construct the phase diagram. Powder methods are particularly suitable for this work, for patterns can be obtained from aggregates of grain sizes too small to be conveniently examined by most other methods.

Some physical properties of crystals are related to changes in interplanar spacings in the structure, and powder techniques can be adapted for the measurement of some physical constants. An obvious example is the *measurement of thermal expansion coefficients* where, from powder records of the same material at two different temperatures, we can determine the line shifts (and changes in interplanar spacings) caused by the temperature change; since the expansion coefficient normal to (hkl) planes (α_{hkl}) is defined as $\Delta d_{hkl}/d_{hkl} \cdot \Delta t$, where Δd_{hkl} is the spacing change caused by a temperature interval Δt, reasonably accurate values of α_{hkl} can be obtained quite simply. Moreover, as we shall see in the next chapter, thermal expansion (like many other physical properties) can vary with the direction of measurement in the crystal lattice; the different values of expansion coefficients for various planes in different lattice orientations are derived from the shifts of the relevant lines on the powder pattern.

Among the crystalline solids of technical importance are bulk materials which are neither single crystals nor randomly oriented aggregations of small crystallites; their textures show a preferred orientation in the arrangement of constituent single crystals, so that they are, in a sense, intermediate between the ideal single crystals and powders that we have already discussed. Although preferred orientation textures can be found in the crystallisation of natural mineral formations, they are more importantly developed in some of the common industrial processes for the production of metallic sheets or wires, synthetic fibres, etc. In this context they can have a significant effect on physical properties such as strength, ductility, etc., and study and control of the textures formed by different manufacturing processes is essential to the quality of products; X-ray methods, among others, can give the essential data on any preferred orientation within a material. Basically there are two different kinds of *preferred orientation texture*, those of fibres and sheets. In an ideal fibre texture, a particular zone axis in all crystallites is oriented parallel to a fixed external direction, though individual crystals are free to assume any orientation about this direction; for example, when a cast metal with randomly oriented crystallites is extruded through a die to produce a wire, not only is the size of the metal crystallites changed but also they tend to assume a fibre texture with a common zone axis parallel to the length of the wire, which develops a kind of cylindrical symmetry in crystallite orientation. In an ideal sheet texture two or more zone axes are parallel to external directions, so that if a texture were perfect all crystallites would be identically orientated to build up a single crystal; in practice, textures are imperfect, so that when, for example, a cast metal block is passed through a rolling mill to produce a sheet, there is only a tendency to a greater or lesser extent for the same mutually perpendicular zone axes of all crystallites to be parallel to the rolling direction and the normal to the sheet. For both fibre and sheet textures, diffraction patterns can be interpreted to find the particular zone axes parallel to external directions and an estimation of their misalignment from the ideal orientation; we will illustrate this by considering diffraction by a specimen with a fibre texture.

Let us suppose that a wire has been produced from a cubic metal with an F-lattice of known cell dimensions for which the process has developed an ideal texture with $\langle 111 \rangle$ zone axes for all crystallites parallel to the length of the wire; we will work out the appearance of a diffraction pattern from this specimen which is recorded on a flat-plate film in the transmission position when monochromatic X-rays are incident normal to the length of the wire. A powder specimen of the cubic metal has a first low-angle cone for which $h^2 + k^2 + l^2 = 3$; with a random orientation of the crystallites this would

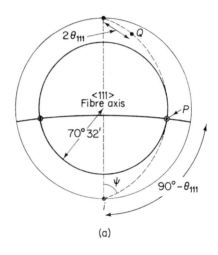

(a)

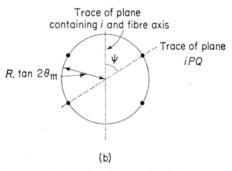

(b)

Fig. 10.6. Geometry of reflections from an ideal fibre texture. (a) Stereo-gram showing the distribution of {111} normals in the wire and the conditions for 111 reflections. (b) The appearance of spots on a 111 powder ring.

intersect the film as a continuous circle of radius $R \tan 2\theta_{111}$, where R is the distance of the film from the wire. But the ideal fibre texture of our wire considerably restricts the possible orientations of {111} reflecting planes in the specimen, and this limits the distribution of diffracted beams around the 111 powder ring. Using stereographic projection of the Bragg geometry, reflecting normals must lie on a small circle of radius $90° - \theta_{111}$ about the pole i of the incident beam; in cubic crystals the only possible angles between the fibre axis $\langle 111 \rangle$ and $\{111\}$ normals in the crystallites are $0°$, $70°\ 32'$, $109°\ 28'$ and $180°$. In an ideal fibre texture cylindrical symmetry about the fibre axis gives the possible positions of $\{111\}$ normals for crystallites within the wire as the centre and small circles of the stereogram of Fig. 10.6(a); there are just four positions at which crystallites are oriented to satisfy Bragg conditions for these planes. At any one of these points, say P, the direction of the diffracted beam (Q) is such that the plane iPQ is inclined to that containing i and the axis of the wire at an angle ψ; blackening will occur on the film at the point at

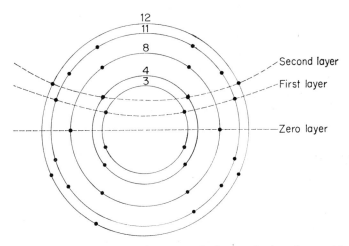

Fig. 10.7. The appearance of spots on the low angle rings for a cubic specimen with an ideal $\langle 111 \rangle$ fibre texture.

which the 111 powder ring is intersected by the plane iPQ, and the three other intersections on the stereogram will give other symmetrically disposed spots as shown in Fig. 10.6(b). (In practice, the texture is likely to be imperfect with some misorientation of $\{111\}$ normals from their ideal positions of Fig. 10.6(a). This would draw out the four spots into limited arcs around the powder ring; it might even cause additional regions of blackening, as, for example, if there were sufficient misorientation around the fibre axis at the centre of the stereogram to intersect the small circle locus of the Bragg condition. From such effects the extent of any departure from the ideal texture can be judged.) For the rest of the diffraction pattern, the same geometrical arguments can be adapted to locate ideal spot positions on the powder rings for other reflecting planes; the results for a few low-angle lines are shown in Fig. 10.7. The appearance of this pattern emphasises the regularity of the arrangement parallel to the fibre axis, for we can distinguish curved layer lines at positions corresponding to an S_{111} value of $\sqrt{3}a$, just as might be expected on a photograph of a single crystal rotated about [111].

Photographs from specimens with a sheet texture are generally similar in having a pattern of discontinuous arcs around powder rings; but the lack of cylindrical symmetry implies that the distributions of blackening on individual rings depend not only on any degree of misorientation from the ideal texture but also on the particular orientation of the incident beam relative to the external reference directions. To find both the texture and the extent of departures from ideal orientations it is necessary to combine data from a number of photographs taken at different inclinations of the X-ray beam to the sheet; for any one family of planes, e.g. {111}, the results are usually presented as a *pole figure*, a stereogram in which the external reference directions are marked together with shaded areas whose densities represent the populations of {111} poles in these orientations. The importance of pole figures in materials technology has led to the development of more complex cameras with moving films, diffractometers, etc. whose purpose is to obtain the densities of crystallites in various orientations from a single exposure; details of these instruments will be found in the bibliography.

10.5. Exercises and problems

1. A powder photographic technique will resolve two lines if their Bragg angles differ by not less than 1°. At what minimum Bragg angle will there be resolution of the doublet components of Cu$K\alpha$ radiation? ($\lambda_{\text{Cu}K\alpha_1} = 1\cdot5405$ Å; $\lambda_{\text{Cu}K\alpha_2} = 1\cdot5443$ Å.)

2. The following θ values were obtained for the first few lines on a powder photograph of Cu_2O taken with radiation of wavelength $1\cdot542$ Å:

$$14° \; 47', \quad 18° \; 13', \quad 21° \; 09', \quad 26° \; 14', \quad 30° \; 40', \quad 36° \; 46'.$$

What are the probable indices of these lines and the cell side of Cu_2O?

Why does the observation on a single-crystal rotation photograph about [111] that the height of the 3rd layer line above the equatorial plane is $2\cdot35$ cm (camera radius 3 cm; $\lambda = 1\cdot542$ Å) confirm this interpretation? If the powder camera can record Bragg angles up to 86°, what are the indices of the highest angle line which can occur on the photograph?

3. A powder photograph of VN (cubic) is taken on a Straumanis camera of radius $5\cdot73$ cm with Fe$K\alpha$ radiation. Measurement with a travelling microscope gave the following readings for line positions:

$2\cdot085$, $4\cdot655$, $5\cdot420$, $14\cdot630$, $15\cdot395$, $17\cdot965$, $19\cdot725$, $20\cdot315$, $22\cdot925$, $22\cdot975$, $25\cdot945$, $26\cdot075$, $29\cdot975$, $30\cdot105$, $33\cdot075$, $33\cdot125$, $35\cdot735$, $36\cdot325$ (all $\pm0\cdot005$ cm).

(The scale zero of the microscope is at an arbitrary position with the two holes in the film at readings ~ 10 cm and ~ 28 cm.) Index the lines, determine the lattice type and cell side. ($\lambda_{FeK\alpha_1} = 1\cdot936$ Å; $\lambda_{FeK\alpha_2} = 1\cdot940$ Å.)

4. A powder photograph of a specimen known to be a mixture of two cubic compounds is taken on a 9 cm diameter camera ($\lambda = 1\cdot542$ Å) with van Arkel geometry. Measurement of the lines on the left- and right-hand sides of the film are as follows:

   24·90, 23·46, 22·35, 22·29, 21·38, 21·32, 20·52, 19·72, 18·26, 18·16, 17·57, 16·90, 16·25, 16·12, 15·62, 15·47, 12·90, 10·91, 10·23, 7·10 (all in cm).

   Identify the lines belonging to each compound and determine the probable lattice types and cell sizes.

5. In the experimental syntheses of the mixed oxide $CaTiO_3$, other oxides can be formed. For a fine-grained product formed under particular conditions of synthesis, the results of a diffractometer scan with $CuK\alpha$ radiation ($\lambda = 1\cdot542$ Å) over a limited low angle region of the powder pattern are:

θ (degrees)	11° 38′	13° 05′	14° 55′	15° 39′	16° 18′
Intensity	70	50	10	20	60
θ (degrees)	16° 28′	16° 35′	16° 41′	17° 31′	
Intensity	100	120	100	30	

   From the data on $CaTiO_3$ and possible contaminants below, demonstrate that the product of synthesis is impure, and determine the most probable impurity.

CaTiO₃		Ca₃Ti₂O₇		Ca₄Ti₃O₁₀		CaTi₂O₄	
d(Å)	Intensity	d(Å)	Intensity	d(Å)	Intensity	d(Å)	Intensity
3·82	60	3·76	8	3·80	10	2·99	20
3·41	40	3·30	4	3·34	6	2·86	30
2·72	80	2·73	100	2·73	100	2·74	100
2·69	80	2·72	65	2·72	100	2·56	30
2·56	10			2·71	45		

6. A single crystal was set on an oscillation camera (radius 3 cm) with a prominent zone axis parallel to the rotation axis. Laue photographs showed identical patterns with symmetry $mm2$ at 90° intervals as the crystal is rotated; it was also observed that Laue symmetry $mm2$ is displayed by directions midway between those for which the first set of Laue photographs were taken, though the patterns were different from those of the first set. On a rotation photograph about this axis with radiation of wavelength 1·542 Å, the height of the first layer line is 0·493 cm. On the zero layer of this photograph, reflections are symmetrically disposed about the exit of the undiffracted X-rays, and measurements of pairs of corresponding reflections across the exit gave the following separations: 3·802, 5·488, 9·256, 12·964, 14·494 and 16·734 (all in cm). On a powder diffractometer

trace of the same material, the first few low angle lines occurred at spacings of 4·95, 3·27, 2·475, 2·447, 2·389, 2·193, 1·740, 1·719 (in Å).

Assuming that the only absences are systematic, what does this data imply about the crystallography of this material? How would you confirm your deductions?

7. In a solid solution series between Mg_2SiO_4 and Fe_2SiO_4, the interplanar spacing d_{130} varies linearly with composition from 2·7659 Å for Mg_2SiO_4 to 2·8328 Å for Fe_2SiO_4; members of this series have the space group *Pbnm*, and approximate cell dimensions $a = 4·78$, $b = 10·33$, $c = 6·03$ (in Å). For an intermediate member of the series, a diffractometer record gave the following values of $1/d^2 (\times 10^5)$ for low-angle lines:

3779, 6563, 7165, 8112, 11069, 12917, 14899, 15708, 16409, 17947.

Index these lines, and from the value of d_{130} find the composition of this specimen.

SELECTED BIBLIOGRAPHY

General reading

AZAROFF, L. V. and BUERGER, M. J. 1958. *The powder method in X-ray crystallography.* McGraw-Hill.

HENRY, N. F. M., LIPSON, H. and WOOSTER, W. A. 1960. *The interpretation of X-ray diffraction photographs.* Macmillan.

Practical methods and applications

KLUG, H. P. and ALEXANDER, L. E. 1959. *X-ray diffraction procedures.* Chapman and Hall.

PEISER, H..S., ROOKSBY, H. P. and WILSON, A. J. C. 1955. *X-ray diffraction by polycrystalline materials.* Institute of Physics, London.

Source of powder diffraction data

X-ray powder data file and index, a continuing publication by the American Society for Testing and Materials, Philadelphia.

11

SYMMETRY RELATIONSHIPS IN PHYSICAL PROPERTIES

11.1. Introduction

In most elementary treatments of physical properties of solids it is assumed that any relation between cause and effect is the same in all directions within the body; whilst such *isotropic behaviour* may occur for some crystals and some properties, the nature of atomic structures suggests that it is unlikely to be universally valid. For example, many materials are known to have layered structures in which sheets of relatively tightly bound atoms are stacked in a repetitive sequence, with rather weaker binding forces between the layers; for a crystal with this kind of structure we would expect that a property such as thermal expansion would vary with the direction of measurement, i.e. coefficients must be larger normal to the sheets than for any direction within them. Variations of this kind are typical of the *anisotropic behaviour* which is widespread in the properties of many crystalline materials; in elementary physics they are neglected, either because we are referring to a particular crystal or are describing a particular property which is not anisotropic, or because experiments are carried out on solids which are inherently isotropic by their nature (e.g. a non-crystalline glass, a polycrystalline metal block with a random orientation of grains so that average bulk properties mask any anisotropy of individual crystallites, and so on).

In a sense, there are two levels of enquiry into the physical properties shown by a crystal with a particular structure. One of these is concerned with the origins of properties; we must decide, for example, why one material is a good electrical conductor and another is not, why a given stress only causes elastic deformation of one body but permanent deformation of another, and so on. Discussion of this kind requires a fundamental investigation of the source of each property in so far as it depends on the atoms and their interactions in the crystal structure; with the exception of a brief description of the structural origins of optical properties in Chapter 11.5, this is not attempted here. We shall accept the occurrence of the different properties, but pursue an enquiry at a different level into their relationships to the symmetry of crystals in which they are found; even when a crystal is a good electrical conductor, the conductivity can be isotropic or anisotropic, and our concern is with the way in which this kind of behaviour is dependent on any symmetry the crystal may possess. The relationship between symmetry and the physical behaviour of a crystalline solid depends on the nature of the property in such a way that various properties may be grouped together for common discussion; in this chapter we shall not investigate all the different possible groups in detail but describe in simple terms one behavioural category so as to clarify the principles of this division into particular groupings with separate relationships to symmetry.

11.2. Properties relating two vectors

For some properties of isotropic bodies there is a simple linear relation between vectors representing cause and effect; in thermal conduction (some other properties of this kind are listed later in Table 11.2), a temperature

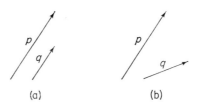

(a)   (b)

Fig. 11.1. Relations between cause $(\vec{q})$ and effect $(\vec{p})$ vectors in a general direction (a) in an isotropic body and (b) in an anisotropic body.

gradient causes a resultant heat flow, and both of these quantities can be represented in magnitude and direction by vectors. In elementary physics these two vectors are always taken to be parallel, whatever the direction of measurement, and they are related by a constant scalar quantity, the thermal conductivity coefficient of the material; this kind of relationship is shown in Fig. 11.1(a) where $\vec{p}$ is the effect (viz. heat flow) produced by the cause $\vec{q}$ (viz. temperature gradient) according to the equation

$$\vec{p} = T\vec{q}$$

in which T is the appropriate physical constant (viz. thermal conductivity). In crystal physics, experiments show that the situation is complicated by the possibility of non-parallelism of the cause and effect vectors as shown in Fig. 11.1(b); a temperature gradient applied in one direction produces a resultant heat flow in another. In any general direction, this disposition of the vectors is characteristic of an anisotropic body, in which, moreover, the precise relationship between $\vec{p}$ and $\vec{q}$ both in relative magnitude and inclination varies with direction. Our first problem is to decide how the kind of anisotropic relationship typified by Fig. 11.1(b) can be described, for clearly the simple vector equation above is not valid; this is most elegantly treated by tensor notation, but we will eschew such mathematical sophistication in favour of more rudimentary methods.

For this kind of property, whilst there is non-parallelism of $\vec{p}$ and $\vec{q}$ in general directions, there are always in any crystal at least three mutually perpendicular directions in which physical measurement shows the cause and effect vectors to be parallel. As we shall see shortly, some crystals exhibit this parallelism in more than the minimum of three directions; but for the moment we will treat the most general case and select these perpendicular directions as

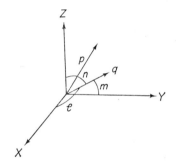

Fig. 11.2. General relations of $\vec{p}$ and $\vec{q}$ with respect to the orthogonal axes X, Y, Z.

a set of reference axes X, Y and Z (not necessarily related to crystallographic axes x, y and z). In these directions we can write the equations

$$\vec{p}_X = T_X\vec{q}_X; \quad \vec{p}_Y = T_Y\vec{q}_Y; \quad \vec{p}_Z = T_Z\vec{q}_Z$$

as for an isotropic body, except that the coefficients T_X, T_Y and T_Z are of different values, so as to express the essential anisotropy in the behaviour of the crystal. In such a crystal $\vec{q}$ applied in some general direction (specified by the direction cosines l, m and n with respect to the reference axes) will cause a resultant $\vec{p}$ in some other direction (Fig. 11.2). We can choose to define a *coefficient* T_{lmn} which relates the applied vector $\vec{q}$ to that component of $\vec{p}$ which is parallel to the direction lmn, and an expression for the coefficient T_{lmn} specified in this way can be simply derived. If $\vec{q}$ is resolved on to the reference axes, $\vec{q}_X = ql$, $\vec{q}_Y = qm$, $\vec{q}_Z = qn$, so that from the relations above

$$\vec{p}_X = T_Xql; \quad \vec{p}_Y = T_Yqm; \quad \vec{p}_Z = T_Zqn.$$

When these components of the resultant $\vec{p}$ are resolved back into the direction *lmn* of the applied vector $\vec{q}$, we get

$$\vec{p}_{lmn} = p_X l + p_Y m + p_Z n = (T_X l^2 + T_Y m^2 + T_Z n^2)\vec{q}$$

so that $T_{lmn} = T_X l^2 + T_Y m^2 + T_Z n^2.$

Thus, if the values of the *principal coefficients* T_X, T_Y and T_Z are known, other coefficients defined in this way can be calculated for all general directions; whereas in an isotropic body there is only one coefficient of thermal conductivity, an anisotropic material has at the most three independent values of the principal thermal conductivities. In general, we need to know at least the values of all the principal coefficients before we can calculate the components of the resultant heat flow normal and parallel to a pair of opposite crystal faces across which a temperature gradient is maintained.

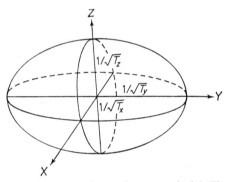

Fig. 11.3. A representation surface as a triaxial ellipsoid.

The general relationships between $\vec{p}$ and $\vec{q}$ in crystals can be conveniently visualised by a form of *representation surface*, like the triaxial ellipsoid of Fig. 11.3; this is constructed on the axes X, Y and Z, so that its principal semi-axes are of lengths $1/\sqrt{T_X}$, $1/\sqrt{T_Y}$ and $1/\sqrt{T_Z}$, i.e. it is the ellipsoid $T_X x^2 + T_Y y^2 + T_Z z^2 = 1$. A radius vector of length r in a direction *lmn* defines a point on the surface with co-ordinates (rl, rm, rn), so that by substitution

$$r^2 = 1\big/(T_X l^2 + T_Y m^2 + T_Z n^2) = 1/T_{lmn}.$$

Hence, when a radius of the representation surface is identified with the applied vector q, its length is a measure of the reciprocal of the square root of the coefficient in this direction. Moreover, since the tangent plane to the surface at the point (rl, rm, rn) is

$$T_X rl \cdot x + T_Y rm \cdot y + T_Z rn \cdot z = 1$$

the normal to the surface at the same point has direction cosines proportional to $T_X rl$, $T_Y rm$ and $T_Z rn$. We showed earlier that the components of a resultant

$\vec{p}$ produced by an applied vector $\vec{q}$ were $\vec{p}_X = T_X ql$, $\vec{p}_Y = T_Y qm$ and $\vec{p}_Z = T_Z qn$, so that the direction of p must be that of the normal to the surface at the end of the radius vector identified with q. Generalising and summarising this brief account of a representation surface for this group of properties:

(i) A representation surface is erected on mutually perpendicular principal axes X, Y, Z so that the lengths of the principal semi-axes are $1/\sqrt{T_X}$, $1/\sqrt{T_Y}$ and $1/\sqrt{T_Z}$, where T_X, T_Y and T_Z are the principal coefficients. In general such surfaces must be of a quadric form; principal coefficients are usually of the same (positive) sign to give ellipsoidal shapes, but occasionally they can have different signs to give a hyperboloid.

(ii) When a radius is drawn parallel to the direction of the applied vector $\vec{q}$ with direction cosines l, m and n, the direction of the resultant vector $\vec{p}$ is that of the normal to the surface at the end of this radius.

(iii) The length of this radius is $1/\sqrt{T_{lmn}}$, where T_{lmn} is the value of the coefficient by which q must be multiplied to give the component of p in the direction lmn.

Having established that properties within this grouping have variations described by such representation surfaces we can turn to the main problem of their relationship to crystal symmetry. For any physical property this is determined by *Neumann's principle* which is formally stated: 'Any kind of symmetry which is possessed by the crystallographic relations of the material is possessed by the material in respect of every physical quality'; more loosely, this may be interpreted as indicating that the symmetry shown by a physical property is greater than, or equal to, that of the crystalline point group. The symmetry of the representation surfaces for the present group of properties can only be of three distinctive types. When $T_X \neq T_Y \neq T_Z$, it is identical with that of the point mmm; when $T_X = T_Y \neq T_Z$, there is an axis of revolution normal to an m plane; when $T_X = T_Y = T_Z$, the surface is completely symmetrical; with the common positive principal coefficients, these three shapes are respectively a triaxial ellipsoid, an ellipsoid of revolution, and a sphere. A spherical surface must denote isotropic properties, for the radius is constant and the normal to the sphere is parallel to the radius at any point. Both ellipsoids must represent anisotropic behaviour, though of a different kind; in an ellipsoid of revolution, any direction within the circular principal section normal to the axis of revolution will show isotropic relations between $\vec{p}$ and $\vec{q}$, but any general direction inclined to this axis will be anisotropic; in a triaxial ellipsoid all principal sections are elliptical and indicate the most general form of anisotropic behaviour (though there are two inclined circular sections of such a surface which are of particular importance in the optical properties discussed in Chapter 11.4). From Neumann's principle, the sphere (and isotropic behaviour) is the only possible surface to be linked to the high and distinctive symmetries of the classes of the cubic system; the other two surfaces are not symmetrical enough. Grouping together the classes of the trigonal, tetragonal and hexagonal systems, their symmetries would be most in accord with surfaces of revolution in which the axis of revolution is parallel to the direction of the triad, tetrad or hexad axis; complete isotropy would not be expected, and the other anisotropic surface is not symmetrical enough. The most general surface must be reserved for the properties of the classes of the remaining orthorhombic, monoclinic and triclinic systems. In orthorhombic classes the relationship to symmetry is clear; a triaxial ellipsoid must be aligned so that its principal axes are parallel to the crystallographic axes (which are always diad symmetry directions). In monoclinic classes this control over the orientation of the triaxial ellipsoid is relaxed; the only essential symmetry axis is a diad (along the y-crystallographic axis) and, whilst one of the principal axes of the representation surface must be parallel to this, the ellipsoid is free to assume any orientation by rotation about this direction. While the other two principal axes must lie in the (010) plane, they need have no particular relationship to arbitrarily chosen x and z crystallographic axes in directions without symmetry significance; for the same crystal the orientation of principal axes in this plane could be different for different properties or even for the same property at different temperatures. In monoclinic crystals the complete physical relationships require a knowledge both of the magnitudes of principal coefficients and the orientation of principal axes in the xz plane. Extending this to triclinic crystals there is no symmetry control over the orientation of a triaxial representation surface whose principal axes may assume any inclination to arbitrarily chosen crystallographic axes. A summary of the relationships between properties represented by these kinds of surfaces and crystal symmetry is given in Table 11.1; the number of inde-

Table 11.1. Representation surfaces for properties relating two vectors (positive coefficients)

Crystal system	Surface and orientation	Independent constants
Cubic	Sphere	1 (coefficient)
Trigonal Tetragonal Hexagonal	Ellipsoid of revolution; axis of revolution parallel to principal axis of symmetry.	2 (coefficients)
Orthorhombic	Triaxial ellipsoid; principal axes parallel to crystallographic axes.	3 (coefficients)
Monoclinic	Triaxial ellipsoid; one principal axis parallel to diad (y-axis), with other two in general orientation in (010) plane.	4 (3 coefficients and 1 orientation parameter)
Triclinic	Triaxial ellipsoid; no necessary relation between principal axes and crystallographic axes.	6 (3 coefficients and 3 orientation parameters)

pendent constants is the minimum necessary to completely specify each property for each crystal system.

Representation surfaces of this kind can be used for a group of quite unrelated properties; some of the commoner ones are listed in Table 11.2.

Table 11.2. Common properties with representation surfaces of Table 11.1

Electric polarisation	= dielectric susceptibility × **electric field strength**
Electric flux density	= permittivity × **electric field strength**
Intensity of magnetisation	= magnetic susceptibility × **magnetic field strength**
Heat flow	= thermal conductivity × **temperature gradient**
Current density	= electrical conductivity × **electric field intensity**
Change in length per unit length	= thermal expansion × temperature interval

Whilst this tabulation is not exhaustive, in all cases except one the vectors involved in the physical relationship are readily identifiable (and in heavy type in the table); only in cubic crystals will there be isotropic behaviour for any of these properties; in all other systems, there is some kind of anisotropy with physical relationships described by an appropriate representation surface. An exception to the vector–vector relations is the last entry, which is an elementary statement about thermal expansion. The variations of thermal expansion coefficients in crystals may be represented by the same kind of surfaces as all the other properties, but an interpretation of the surface in terms of cause and effect vectors p and q is not possible, for the physical qualities involved are different from the rest of the list. This raises the fundamental question of how physical properties are to be grouped in regard to their relation to symmetry; so far we have established a form of representation surface common to one group; in the next section we try to disentangle the criteria for inclusion in this group so that we may see how the other groupings arise.

11.3. Order in properties of crystals

In thermal expansion, it is clear that the temperature change which causes expansion is a scalar quantity; it cannot be applied in a particular direction in the crystal as can vector quantities like a temperature gradient. In this sense, it cannot change the nature of the right-hand side of the equation for expansion in Table 11.2, and this suggests that the key to the constancy of behaviour which allows the grouping of physical properties in this Table lies in the significance of the coefficients T_{lmn}. Coefficients for the properties of Chapter 11.2 cannot be generally regarded as scalars or even vectors; in mathematical terms, they are the components of a particular kind of second-order tensor, whose properties are described by a certain type of representation surface. It is for this reason that the properties in Table 11.2 can all be grouped together as second-order; although such properties commonly arise in a general relation of two physical quantities expressed as vectors, the example of thermal expansion shows that there can be exceptions. When second-order coefficients are multiplied by a scalar property, the product must be a second-order property itself; in the case of expansion, this product, a change in length per unit length, is a strain (more accurately a homogeneous strain) and the implication must be that such strains can only be generally expressed in the form of a second-order tensor. In turn, this implies that we must first examine the nature of the physical quantities involved when trying to place a property in one of these general groups; they may be *scalars* (like a temperature interval) or *vectors* (like an electric field) or *second-order tensors* (like a strain).

The different orders of properties arise from the form of the independent coefficients that are necessary to relate the two physical quantities of a particular kind; again in mathematical terms, these coefficients are the components of various orders of tensor, and the properties that they describe are grouped accordingly into orders as in Table 11.3. In the preceding section we discussed

Table 11.3. The nature of the different orders of common properties

Order	Physical quantities related	Example
Zero	Two scalars	Heat capacity
First	A scalar and a vector	Pyroelectricity
Second	Two vectors	Thermal conductivity
	or a scalar and a second-order tensor	Thermal expansion
Third	A vector and a second-order tensor	Piezoelectricity
Fourth	Two second-order tensors	Elastic properties

the common form of second-order properties* and established their relationship to crystal symmetry through the representation surface; a systematic study of the physical properties of crystals requires a similar treatment of the different orders listed in Table 11.3. In each case we should find that the symmetry dependence of a property depends on the nature of the tensor and its order, which also affects any division between isotropic and anisotropic behaviour; for example, even cubic crystals need more than one independent modulus to specify their elastic properties. Representation surfaces change, becoming more complicated to visualise and less useful for the higher orders, which are better treated by mathematical analysis; appropriate references are given in the bibliography. It is sufficient for us to recognise that the symmetry dependence of all properties is not the same, so that in the converse problem (discussed in Chapter 11.6), in which physical measurements are used to

* There are some different second-order properties, notably optical activity, which due to their different qualities have different surfaces and symmetry relationships.

help with symmetry determination, we realise that different properties can give different data according to their order.

11.4. Optical properties

In the next two sections we shall look in rather greater detail at the optical properties arising from the transmission of light waves through a crystal; for optical techniques are the commonest form of physical examination in practice. Firstly, we shall briefly establish the general nature of the phenomena, neglecting the effects of absorption, optical activity etc., and omitting also any description of the practical methods; in the following section we shall give some account of the origins of optical behaviour and the way in which this is linked to structural arrangements within a crystal. In this section we are concerned only with optical properties as they arise from the propagation of

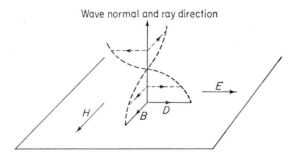

Fig. 11.4. Propagation of electromagnetic wave in an isotropic body.

electromagnetic waves through the general continuum of a crystalline solid; as the dielectric and magnetic properties of crystals are second-order, we would expect optical behaviour to have some resemblance to that of the second-order properties described in Chapter 11.2; nevertheless, there are important dissimilarities due to the essentially different character of optical phenomena.

When light waves pass through an isotropic dielectric medium, at any given point the electric field of the radiation $\vec{E}$ causes an electrical displacement $\vec{D}$; under isotropic conditions these two vectors are parallel. At the same time the periodic waxing and waning of the electric field at this point is associated with a similar magnetic field $\vec{H}$ perpendicular to $\vec{D}$ and $\vec{E}$; this in turn causes a magnetic induction $\vec{B}$, parallel to the direction of $\vec{H}$ in the isotropic medium. The disposition of these vectors for a point on a planar wavefront advancing through the medium is shown schematically in Fig. 11.4. Under these conditions, solution of Maxwell's equations, which control propagation, shows that the velocity of the wavefront is $c/\sqrt{K}$, where c is the velocity of light in vacuo and K is the dielectric constant of the medium; the *refractive index n* is therefore $\sqrt{K}$, a constant for an isotropic material. The wave-form shown in the figure is linearly polarised and conventionally its *vibration direction* is taken to be that of the electrical vectors; similar planar wavefronts would be propagated in this medium with the same speed (and refractive index) in any direction of travel and for any vibration direction. In propagation of this kind the direction of movement of a bounded part of a wavefront (i.e. the direction of the flow of energy)—often called in optics, the *ray-direction*—is parallel to the normal to the wavefront—often called the *wave-normal direction*. Moreover, unpolarised light will not be changed in character by its passage through the medium, for it can be regarded as a succession of linearly polarised pulses at random intervals with random vibration directions; apart from a reduction in velocity, each of these will move unmodified through the medium to constitute a transmitted unpolarised beam. These, then, are the elementary conditions of electromagnetic wave transmission in isotropic media and are the basis of ordinary isotropic optics; they depend on isotropic dielectric properties and we know that these occur only in cubic crystals.

For all other crystalline matter dielectric (and magnetic) behaviour is anisotropic; although $\vec{B}$ remains parallel to $\vec{H}$ to a high degree of approximation for most transparent or non-ferromagnetic solids owing to their very small magnetic susceptibilities, there can be significant departures from parallelism of the electrical vectors $\vec{D}$ and $\vec{E}$. For a particular point on a planar wavefront defined by $\vec{B}$ and $\vec{D}$, the general relations between the vectors are shown in Fig. 11.5; although $\vec{D}$ and $\vec{E}$ are in a plane normal to $\vec{B}$ (and $\vec{H}$), $\vec{E}$ is inclined to $\vec{D}$ out of the plane of the wavefront. As the wavefront advances, the ray direction (along the flow of energy perpendicular to $\vec{E}$ and $\vec{H}$) is different from the wave-normal direction; the vibration direction $\vec{D}$ of the light is contained in the wavefront which moves forward in a crab-like manner through the medium. The refractive index is related to the velocity of advance of the wavefront, and this is determined by the dielectric constants relating $\vec{D}$ and $\vec{E}$; in an anisotropic body the values of dielectric constants vary with the direction of $\vec{D}$, so that the refractive index will depend on the vibration direction of the transmitted light waves. A full solution of Maxwell's equations under these conditions (see Appendix H, Nye) shows that propagation is restricted, in the sense that only two mutually perpendicular directions of the displacement $\vec{D}$ are possible for a general wave-normal direction; if light is travelling along this direction, only two waves (of the type in Fig. 11.5) are permissible; each is linearly polarised but their vibration directions are mutually perpendicular. Naturally, the two waves travel with different speeds, i.e. they have refractive indices determined by the two permitted vibration directions; they will also have separate ray directions which are, in general, different from their common wave-normal direction. In contrast to isotropic behaviour, unpolarised light cannot be transmitted unchanged along such a direction but must be resolved into the two linearly polarised disturbances.

This short account of electromagnetic wave propagation in an anisotropic dielectric medium outlines the complexities of crystalline optical properties; in practice, it is essential to describe optical behaviour by some kind of representation surface. The most convenient represents the variation of refractive index with vibration direction and is known as the *optical* (or *Fletcher*)

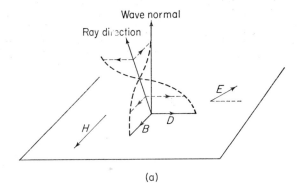

(a)

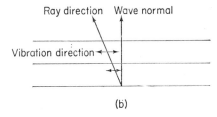

(b)

Fig. 11.5. Propagation of electromagnetic wave in an anisotropic body. (a) The relations between the electric and magnetic vectors. (b) Successive positions in the advance of a planar wavefront.

indicatrix. As might be expected, it corresponds closely to the related surface for dielectric properties; it has the same directions of principal axes but the lengths of the semi-axes are $\sqrt{K_X}$, $\sqrt{K_Y}$ and $\sqrt{K_Z}$, i.e. they are determined by values of principal refractive indices. Like the second-order dielectric surface, the shape of the indicatrix may be a sphere, an ellipsoid of revolution, or a triaxial ellipsoid; in the relations between shape and orientation and crystal symmetry the principles set out in Table 11.1 are followed; but in its properties and use the indicatrix is somewhat different from the normal second-order surfaces of this table. Its most important properties may be stated thus:

(i) By its method of construction, a radius parallel to the vibration direction of light waves in the crystal has a length which is the refractive index for these waves.

(ii) The two permitted perpendicular vibration directions for wave propagation along a given direction are parallel to the principal axes of a central section of the indicatrix perpendicular to the wave-normal.

A justification of these (and other) properties of the indicatrix can be found in the references cited in the bibliography, and they are the foundation of its practical utility in describing and predicting the common optical properties of crystals. Let us suppose that the shape and orientation of the indicatrix is known; for propagation in a general direction along a wave normal such as OP, we draw the central section of the figure perpendicular to OP (Fig. 11.6).

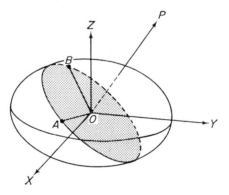

Fig. 11.6. The use of the optical indicatrix. The indicatrix is an ellipsoid whose principal semi-axes along OX, OY and OZ are the principal refractive indices n_X, n_Y and n_Z. With a general wave normal OP, we consider the central shaded section of the ellipsoid perpendicular to OP; the semi-minor and semi-major axes of this section, OA and OB, indicate the permitted vibration directions and, by their magnitudes, the associated refractive indices for waves propagated along OP.

In general, this will be elliptical and we can identify the semi-major and semi-minor axes of the ellipse, OA and OB. From (ii) above, these denote the permitted vibration directions for waves travelling along OP; from (i) the lengths of OA and OB represent the two refractive indices associated with such waves. Numerical computation of vibration directions and refractive indices is not too tedious, and constructions can be adapted to stereographic projection; often qualitative consideration of the indicatrix is sufficient to solve practical optical problems. The surface can also be used to predict the two different ray directions associated with waves travelling along OP which are responsible for the two images found in the well-known phenomenon of double refraction.

Whatever the shape of an indicatrix, there are certain directions of OP for which a central section is circular with no distinction between OA and OB. In such directions of propagation all possible vibration directions within the plane of the circular section are equally favourable and can be found in waves travelling along OP; in these directions the crystal behaves as if it is optically isotropic. In cubic crystals, any central section of the spherical indicatrix is circular, as is expected for completely isotropic behaviour. For trigonal,

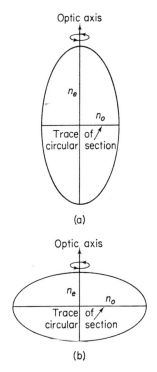

Fig. 11.7. Indicatrices for optically uniaxial crystals. Both figures are ellipsoids of revolution for which a central section containing the optic axis has been drawn (a) for an optically positive crystal with $n_e > n_o$ and (b) for an optically negative crystal with $n_e < n_o$.

tetragonal and hexagonal crystals, the circular section of an ellipsoid of revolution is normal to the triad, tetrad or hexad axis; when light travels along these symmetry axes, the optical behaviour is isotropic. Directions of isotropy in an anisotropic body are known as *optic axis directions*, and crystals of these three systems are said to be *optically uniaxial* for they have only one optic axis; uniaxial crystals are further sub-divided by *optic sign*, positive or negative, depending on whether the principal refractive index along the axis of revolution (n_e) is greater or less than the second principal refractive index (n_o), which is the radius of the circular section (Fig. 11.7). Less symmetrical crystals in the orthorhombic, monoclinic and triclinic systems have optical properties described by the general indicatrix, a triaxial ellipsoid. Such a surface has two circular sections symmetrically disposed about the direction of the greatest principal refractive index (n_γ); the normals to these sections lie in the plane containing the greatest (n_γ) and least (n_α) principal refractive indices

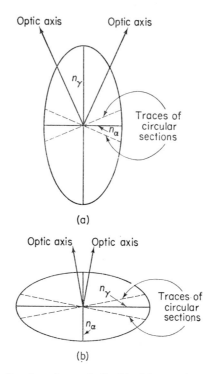

Fig. 11.8. Indicatrices for optically biaxial crystals. Both figures are triaxial ellipsoids for which a central section containing the greatest (n_γ) and least (n_α) principal refractive indices has been drawn, i.e. a section perpendicular to the third principal refractive index (n_β). (a) For an optically positive crystal in which the acute bisectrix of the optic axes is n_γ (or ($n_\gamma - n_\beta$) > ($n_\beta - n_\alpha$) (approx.)), and (b) for an optically negative crystal in which the acute bisectrix of the optic axes is n_α (or ($n\gamma - n_\beta$) < ($n_\beta - n_\alpha$) (approx.)).

at an inclination which depends on the relative magnitudes of all three principal refractive indices n_γ, n_β and n_α. Since they have two optic axis directions, crystals of these three systems are said to be *optically biaxial*; like uniaxial crystals, biaxial substances may be sub-divided by optic sign in terms of the inclination of the optic axes to the direction of n_γ (Fig. 11.8).

Further developments concerning the propagation of light in uniaxial and biaxial crystals and the practical techniques for the determination of optical constants (principal refractive indices, orientation of the indicatrix, optic sign, double refraction, optic axial angle, etc.) are fully described elsewhere. For our purpose, these preceding elementary remarks are sufficient description of optical properties for the discussion which follows of their general dependence on structural features.

11.5. Optical properties and crystal structure

It might be thought that it is possible to relate any property of a crystalline material to the structural arrangement of its individual atoms; but some properties depend more on the imperfections which characterise a given crystal than its atomic structure (see Chapter 12); and even when this is not true, detailed quantitative calculations are difficult owing to the complexities of crystal structures. Optical data are more widely known than any other physical property of crystalline material for they are easier to obtain in practice and are part of the routine investigation for many crystalline solids. Since they are effectively constant for a given structural arrangement and do not appear to be significantly changed by any imperfections which distinguish one crystal from the next, there has been some attempt to relate them both quantitatively and qualitatively to crystal structure; in this section there is a short account of the important features of this work. The optical properties with which we are concerned originate in the dielectric behaviour of the solid, and it is with this in mind that we can try to separate (i) those aspects of crystal structures which cause the general range of values of refractive indices shown by crystals, and (ii) those which are responsible for the isotropic and anisotropic behaviour implied by optical indicatrices.

In the passage of light through matter the effect of the fields represented by the electric vector will be to displace centres of positive and negative charge for an atom so that it becomes polarised and behaves as an alternating dipole. The frequency of light waves is about 10^{14} cycles per second; at such frequencies the nucleus is too heavy to respond, so that polarisation depends on the distortion of the extranuclear electronic charge clouds. For a given atom the dipole moment induced by a particular field strength is measured by its *polarisibility*; this will depend on the charge cloud system both in its size and looseness of binding, and will vary from one atomic species to the next. For a particular atom, even though these extranuclear clouds may be affected by any binding forces linking it to neighbouring atoms, we can regard its polarisibility as constant (at least to a first approximation) over the range of frequencies in visible light; if atomic polarisibilities are known, the total polarisation (or electric moment per unit volume) produced by an electric field can be calculated from the numbers and kinds of atoms in unit volume; this enables us to deduce the dielectric constant (and hence the refractive index). Such a calculation can only be valid for gases in which atoms are relatively distant from one another, for in the other states of aggregation we must expect neighbouring atomic dipoles to have mutual induction effects; values of refractive indices for gases are very different from those of liquids and solids. The effects of mutual induction in liquids and solids will depend on the arrangement of neighbours; with the random distribution of closely packed atoms of a liquid or amorphous solid, statistical electrostatic methods show that the polarisation of each atom is increased by mutual induction to give much larger refractive indices than those of gases. For a given atom, this treatment leads to the *Lorentz–Lorenz* equation

$$\frac{n^2 - 1}{n^2 + 2} = \frac{4}{3}\pi\alpha N'$$

which relates the refractive index n to α, the atomic polarisibility and N', the number of atoms per cc; this can be adapted for polyatomic systems and has been used with some success in predicting the refractive indices of liquids.

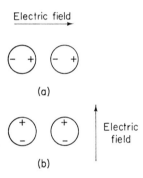

Fig. 11.9. Induction effects for a pair of isolated atoms.

In crystalline matter, although atoms are in close proximity their spatial distribution is often far from random; we might reasonably expect 'mean' refractive indices comparable to those of liquids, etc., but the mutual induction effects of particular spatial arrangements of near neighbours may act so as to decrease or increase the polarisation of an atom from any mean value that it would have in a random array. Such changes in polarisation affect dielectric constants, and in both magnitude and sign they depend on the orientation of the electric vector (or vibration direction) of the light wave; they are responsible for any variations about a mean refractive index for a particular atomic assemblage. We can demonstrate this by considering the two atoms of Fig. 11.9, which are close enough to have marked mutual induction but far enough from any other atoms to allow all other induction effects to be neglected. In (a) the electric vector is parallel to the line joining the atoms; at a given instant the dipoles induced by the field can be as indicated, and the moment of each will be increased by induction due to its neighbour. In (b) the electric vector is normal to the line of atomic centres; the dipoles due to the field at a given instant are again shown, and their moments will be reduced by mutual induction. For (a) the effective dielectric constant (and refractive index) has been increased by mutual induction, whereas in (b) their values have been decreased. In most crystals we can therefore expect the refractive index to vary about a mean value in a manner which depends on the asymmetries of their atomic patterns; the departure of the refractive index from its mean value is determined by the vibration direction of the light waves, and it is, of course, this variation which is plotted by the optical indicatrix. In theory, this approach to crystalline optical properties permits detailed calculation of the indicatrix for a known structure; an average refractive index can be computed rather as for liquids, and then the induction effects can be evaluated for different orientations of the light vector within the atomic configuration. In practice, this has been done with reasonable success in a few cases (e.g. see Chapter IV, Hartshorne and Stuart for a simple account of methods used by Bragg for $CaCO_3$); in complex structures induction effects due to more distant dipoles are difficult to compute accurately, and calculations that have been attempted are not entirely satisfactory.

Nevertheless, qualitative treatments on this basis can often provide an order of magnitude and sign for the maximum difference between principal refractive indices (or *double refraction*) to be expected for a particular crystal. In many structures there are complexes (molecules or ion groups) that dominate optical properties because of their greater polarisibilities; in the calculations on $CaCO_3$ mentioned above, Ca (and C) atoms make very little contribution to the average refractive index and even less to the optical anisotropy, so that the essential optics are controlled by the more numerous oxygens arranged in triangular planar groups (around the C atoms). If we restrict discussion to the extent of optical anisotropy as measured by the double refraction, this must depend primarily on the form of any highly polarisible atomic groups and their mutual arrangement in the structure; we can distinguish three different general shapes for these groups, roughly spherical, rod-shaped or plate-like. For a spherical group, there will be little or no anisotropy and so, whatever their mutual dispositions in any structure, the optical properties of the crystal will be isotropic or nearly so. Individual rod-like groups will behave similarly to the diatom of Fig. 11.9 and develop the greatest polarisation when the electric vector is parallel to the length; the optical properties of a crystal containing rod-like highly refracting groups, however, will be determined by the mutual orientations of the rods in the structure. If these are effectively random, the crystal must appear isotropic despite the anisotropy of separate groups; if there is preferred orientation, the anisotropy of individual groups will become apparent in the crystalline optics. When the rods are all parallel in the structure, the refractive index associated with light vibrating parallel to this common direction will be very much greater than for any other vibration direction, i.e. one principal refractive index is very much larger than the other(s) and the indicatrix has a shape associated with a high positive double refraction. When the rods are all parallel to a plane but not to each other, refractive indices for any vibration direction within this plane will be much the same but far larger than the value for light waves vibrating normal to the plane, i.e. one principal refractive index is very much less than the other(s) and the optical indicatrix is that for high negative double refraction. Plate-like refracting groups (as the O groupings in $CaCO_3$) can be considered in the same way. Individual groups are more highly polarised by fields within their planes than by fields across their planes; the optical properties of a crystal then depend on the mutual orientations of the planes of the plates in the structure. A summary of these qualitative conclusions is given in Table 11.4; in a practical context the measured optical

Table 11.4. Qualitative relationships between the optical properties of crystals and the shape and arrangement of highly refracting complexes in their structures

Shape and arrangement of groups	Qualitative optical properties of crystals	Structural inferences from optical data
Roughly spherical (e.g. SO_4, ClO_4, $PtCl_6$ groups)	Isotropic or weakly anisotropic; double refraction about 0·01 or less	None possible
Rod-shaped (e.g. paraffins, CO_2 molecules)		
(a) All parallel to one direction	Strongly anisotropic; large positive double refraction greater than about 0·05	Rods parallel to vibration direction of greatest refractive index
(b) All parallel to a plane but not to each other within the plane	Strongly anisotropic; large negative double refraction greater than about 0·05	Plane containing rods normal to vibration direction of least refractive index
(c) Inclined in all directions	Isotropic or weakly anisotropic; double refraction about 0·01 or less	Cannot be either of the arrangements of (a) or (b)
Plate-like (e.g. NO_3, CO_3 groups, aromatic molecules)		
(a) Planes parallel to one another	Strongly anisotropic; large negative double refraction greater than about 0·05	Planes of plates normal to vibration direction of least refractive index
(b) Planes all parallel to one direction but not to each other	Strongly anisotropic; large positive double refraction greater than about 0·05	Plates all parallel to vibration direction of greatest refractive index
(c) Inclined in all directions	Isotropic or weakly anisotropic; double refraction about 0·01 or less	Cannot be either of the arrangements of (a) or (b).

properties for a crystal of unknown structure can infer the mutual arrangements of any highly refracting groups known to be present; these are given in the last column.

11.6. Symmetry determination by physical properties

A key point in the study of the crystallography of any material is its assignment to one of the 32 crystal classes; once this is done, its ultimate classification by the methods of Chapter 8.5 into one of the 230 space groups is usually possible. As we saw in Chapter 6, the determination of point group symmetry from crystal morphology is unambiguous only when certain diagnostic forms are present; if they are not, the material may belong to any one of several classes within a system. But often we have to work with specimens without well-developed crystalline faces, so that symmetry must be found by physical examination. This can start with the X-ray methods described in the past few chapters; but in the majority of cases, these can only establish membership of one of the Laue symmetry groups of Table 9.1, which embrace two or three possible crystal classes. To separate the classes within a Laue group, we must use other physical tests for symmetry and those most commonly employed are described in this section. Earlier in this chapter we saw how physical properties vary in their relationship to symmetry with their order. Some properties, like those of the second-order described in Chapter 11.2 and the optical properties of Chapter 11.4, are not suitable for point group determination, for the subdivisions that they establish are far too large; their value in this work is indirect as when, for example, optical examination recognises important directions (such as an optic axis) in the specimen along which X-ray and the other symmetry tests can be carried out. But some other properties are more informative in their symmetry relations.

Since less symmetrical classes within a Laue group are not differentiated by X-ray methods due to the addition of a centre of symmetry in the diffraction pattern, tests for this element are particularly valuable; the most common involve the properties of pyroelectricity and piezoelectricity. *Pyroelectricity* is the development of an electrical dipole due to a distortion of the charge-cloud systems of the atoms when the crystal is heated (or cooled); in terms of Chapter 11.3, it is a first-order property in which a vector (the dipole moment) is related to a scalar (the temperature change). It should occur only in those classes in which there is a unique direction along which the separation of charge can take place; these are the ten polar classes described in Chapter 4.4. *Piezoelectricity* is the term used to describe either the development of an electric dipole on the crystal by an applied stress (the direct effect) or the development of a strain within a crystal subjected to an electric field (the indirect effect); in either case, it must be regarded as a third-order property which relates a vector to a second-order tensor and its relationship to symmetry should be different from that of the first-order property pyroelectricity. Analysis of the direct effect shows that it can occur in all twenty-one acentric classes except 432 when tensile or shear stresses are applied. Now these twenty classes include the ten polar classes in which pyroelectricity is possible, and this can cause difficulties in the interpretation of simple experiments. The temperature change of a pyroelectric experiment can cause strains due to anisotropic expansion within a crystal; these in turn lead to dipole formation by the direct piezoelectric effect; thus a piezoelectric class in which there is no unique direction can show a separation of charges due to this coupling of pyroelectric conditions and piezoelectric phenomena. A fuller discussion of the relationships between the two effects is given in Chapter X of Nye, and practical details of pyroelectric and piezoelectric tests are described in Chapters 6 and 7 of *Experimental crystal physics* by Wooster and Breton. For our part, we must recognise that, in the absence of any special precautions to distinguish true and false pyroelectricity, the only safe conclusion to be drawn from the formation of an electric dipole when a crystal is stressed or heated is that it belongs to one of the acentric classes (other than 432). We must emphasise that not only must the crystal have the requisite symmetry but, more importantly, its atomic structure must have certain other features necessary for these properties; so that a positive test eliminates centric classes but a negative test is without any significance. For example, in Laue group *mmm* the development of a dipole is not compatible with point group symmetry *mmm*; but the absence of piezoelectricity or pyroelectricity must be interpreted to allow the crystal to belong to any of the three classes *mmm*, *mm2* and 222 within this Laue group.

From this example we see that tests for other elements are often required even when electrical dipoles are found under stress or on heating. One of these relates to *optical activity*, which in its simplest form may be described as a rotation of the vibration direction of linearly polarised light as it advances along an isotropic direction in the crystal (such behaviour is also observed in certain solutions, where it has a different but related origin). Although optical activity can be thought of as a second-order property, its nature is rather unusual and its relationship to symmetry is unlike that of the ordinary second-order and optical properties that we discussed earlier. A detailed analysis shows that it is possible in all acentric classes except 4*mm*, $\bar{4}3m$, 3*m*, $\bar{6}$, 6*mm* and $\bar{6}m2$.* As with other properties, optical activity depends on the existence of certain kinds of structural configurations which may or may not

---

* It has often been said that optical activity is associated with enantiomorphism; but this is not strictly true, for it can occur in the non-enantiomorphic classes *m*, *mm2*, $\bar{4}$ and $\bar{4}2m$. In these classes there are experimental difficulties in detecting optical activity and examples are rare, so that it is only recently that its existence has been demonstrated for the uniaxial classes; from this it is probable that a practical association between optical activity and enantiomorphic classes has grown up.

be present in crystals belonging to one of the fifteen possible point groups; a positive test allocates the crystal to one of these classes, but a negative test is without significance. The elementary form of optical activity is an aspect of wider problems in the propagation in crystals of light with general polarisations; accounts of the theory of optical activity and its relationship to symmetry will be found in the books by Nye and Shubnikov, whilst practical aspects of observation are described elsewhere in the bibliography.

There are a few further physical symmetry tests, which are so complicated or so restricted in use that their inclusion in an introductory text cannot be justified; but there is one other procedure (of rather different character) which has been sufficiently regularly used for it to be mentioned briefly. It is restricted to those crystals which have well developed faces or which can be cut parallel to known planes, and it involves a controlled solution of the specimen by a suitable solvent. This solvent must not attack the crystal too violently to produce rounded-off shapes; the rate of solution should be slow and controllable to leave the specimen faces flat but with pitted regions. *Etch pits* so formed are polygonal with a shape and orientation determined by the differential rates of solution in different directions; they might reasonably be expected to be related to the underlying structure in such a way that they reflect the symmetry of its arrangement. In fact, the symmetry of etch figures can be regarded as a projection of the point group elements on to the plane of the face on which the pits are produced. However, in their interpretation we must accept that under a given set of experimental circumstances solution might take place at the same rate in two or more directions unrelated by symmetry; significant conclusions can only be related to any lack of symmetry shown by the figures.

The study of etch pits is well suited to crystals whose shapes are composed of forms which may be characteristic of several different classes; we will illustrate this by a short example. Let us suppose that our specimen always crystallises as a cube, a special form in all five classes of the cubic system; X-ray examination indicates the Laue group $m3m$, which contains the possible classes $m3m$, $\bar{4}3m$ and 432, but other physical tests are negative. If we find a suitable solvent, we can attempt to establish the crystal class by examining the shape and orientation of etch pits formed on the {100} cube faces; possible etch figures are shown diagrammatically in Fig. 11.10. The outline of the pits is either rectangular or square; rectangular shapes cannot be consistent with the operation of a rotation tetrad normal to the face but can occur with inversion tetrad (or diad) symmetry. When rectangular pits are formed as in (b) we may conclude that the crystal belongs to class $\bar{4}3m$; notice, too, that the pits are oriented on the face so that their symmetry lines are consistent with the diagonal symmetry planes that occur in this class. The observation of square pits as in (a) or (c) is not conclusive evidence that the crystals must belong to one of the other two classes. Such shapes might be developed in $\bar{4}3m$ by the particular conditions of solution; the repeated formation of square pits with other reagents might be interpreted as reinforcing the elimination of $\bar{4}3m$, but they do not provide incontrovertible proof. Similar considerations apply to the orientation of pits; in 432 they should be in a general

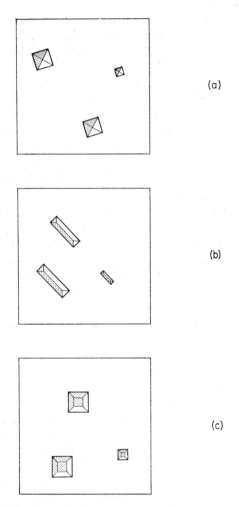

Fig. 11.10. Etch pits on the faces of a cube with Laue symmetry $m3m$.

orientation as in (a), but in $m3m$ we would expect their symmetry lines to be contained in the symmetry planes of a crystal belonging to this class as in (c). However, the only firm conclusion from the pits of (c) is that the maximum symmetry is $m3m$; they might be developed in the less symmetrical cubic classes by particular solution effects. In some cases we can also make use of a lack of orientation between pits on two adjacent faces related by different symmetries in the different classes; but the limitations of etch pits in symmetry determination are clear from our example.

Problems of symmetry determination can be resolved in many cases by physical tests, but we must recognise that the best that can be done with some specimens is to establish a limited choice of related alternative point groups (and space groups). Ultimately their space group (and point group) symmetry could be found by a determination of crystal structure; structures based on the alternative space groups are considered to find that which gives the best agreement between the observed and calculated intensities of X-ray reflections. As may be imagined, this is a complicated and tedious process so that it would be undertaken only with the objective of determining the atomic structure; the resolution of point and space group symmetry would be incidental to the main purpose.

11.7. Exercises and problems

1. The principal coefficients of thermal conductivity of a certain tetragonal crystal are: $k_X = k_Y = 100 \times 10^{-4}$, $k_Z = 400 \times 10^{-4}$ cal. cm^{-1} sec.$^{-1}$ °C^{-1}.

   (i) On a suitable scale, draw a section of the conductivity surface parallel to the YZ plane.

   (ii) Use this drawing to find the coefficient of thermal conductivity for the direction cosines $(0, \sqrt{3}/2, \frac{1}{2})$.

   (iii) Determine the components of heat flow parallel to the principal axes when a temperature gradient of 100°C cm^{-1} is established along the direction in (ii).

   (iv) Use these to find the magnitude and direction of the resultant heat flow. Compare this direction with the normal to the surface at the extremity of the radius in (ii). Check the value of the coefficient found in (ii) by determining the heat flow produced by this temperature gradient in the direction $(0, \sqrt{3}/2, \frac{1}{2})$.

2. A trigonal crystal $(c/a = 1.50)$ has principal coefficients of thermal conductivity $k_X = 4.0$, $k_Z = 12.0$ $(\times 10^{-6}$ cal. cm^{-1} sec^{-1} °C$^{-1})$; it has a good $\{10\bar{1}1\}$ rhombohedral cleavage. One face of a cleavage rhombohedron is coated with a low melting point wax, and a point source of heat applied. What is the shape of the melted wax a short time later?

3. The principal coefficients of thermal expansion of silver iodide (hexagonal) are: $\alpha_X = 0.65$, $\alpha_Z = -3.97$ $(\times 10^{-6}$ °C$^{-1})$.

   (i) Calculate coefficients of thermal expansion in directions inclined at angles 0°, 10°, 20°, ..., 80°, 90° to the hexad axis, and use these to draw a principal section of the representation surface for a plane containing the hexad axis.

(ii) Determine the angle between the hexad axis and the direction of invariant length; notice the significance of this direction in the section of the representation surface.

4. (i) Rutile (a form of TiO_2) belongs to class 4/mmm with cell dimensions $c = 2.96$, $a = 4.59$ (in Å) and has principal refractive indices $n_{o} = 2.61$, $n_e = 2.90$. Determine the permitted vibration directions and associated refractive indices for light travelling normal to the face (011).

(ii) A crystal of class mmm with cell dimensions $a = 8.878$, $b = 5.450$, $c = 7.150$ (in Å) has good {001} and {201} cleavages; its optic properties are described as $n_\alpha = 1.637$, $n_\beta = 1.639$, $n_\gamma = 1.649$, $\alpha = z$, $\beta = y$, $\gamma = x$. Determine the permitted vibration directions and associated refractive indices for normal incidence for the two kinds of cleavage flake.

(iii) The following crystallographic and optical data are given for a substance:

Class 2/m, $a:b:c = 1.4:1:3.2$, $\beta = 150°$; $n_\gamma = 1.60$, $n_\beta = 1.59$, $n_\alpha = 1.56$; $y = \beta$, $\widehat{zy} = 25°$ in the acute $\widehat{xz}$ angle, with $2V_\gamma = 120°$; good {110} cleavages. Draw sketches of (100) and (010) faces showing the trace of the cleavages; on each indicate the permitted vibration directions and associated refractive indices for normal incidence.

5. What conclusions can be drawn about the orientations of refractive complexes from the following data:

(i) Planar groups; Laue symmetry mmm; $n_\alpha = 1.531$, $n_\beta = 1.682$, $n_\gamma = 1.686$, $\alpha = z$, $\beta = x$, $\gamma = y$.

(ii) Planar molecules; Laue symmetry 2/m; $n_\alpha = 1.561$, $n_\beta = 1.590$, $n_\gamma = 1.594$, $\beta = y$, $\alpha = z$ (approx.).

(iii) Linear molecules; Laue symmetry 6/mmm; $n_o = 3.00$, $n_e = 4.04$.

6. (i) A crystal has Laue symmetry $\bar{3}m$. What conclusions about its possible crystal classes can you draw if (a) it exhibits piezoelectricity, (b) it also shows optical activity.

(ii) A crystal has Laue symmetry 4/mmm. The following data on classes of X-ray reflections were obtained:

$0kl$ reflections present only if $k + l = 2n$
hhl „   „   „   „ $l = 2n$

It is found that it displays piezoelectricity. What is the space group?

(iii) Crystals of a certain cubic material only exhibit the octahedral form {111}. A Laue photograph with X-rays normal to one of the faces of the octahedron shows the symmetry 3m. It does not exhibit pyroelectricity, piezoelectricity, or optical activity. Etch pits formed by a suitable reagent on the faces are triangular in shape, but in a general orientation with respect to the edges of the octahedral faces. What is the probable crystal class? If subsequent X-ray examination revealed that the only systematic absences occurred in $h00$ when $h = 2n + 1$, $0k0$ when $k = 2n + 1$ and $00l$ when $l = 2n + 1$, how would this confirm your deductions?

SELECTED BIBLIOGRAPHY

General reading
BROWN, F. C. 1967. *The physics of solids.* Benjamin.
NYE, J. F. 1957. *Physical properties of crystals.* Clarendon Press.
WOOSTER, W. A. 1938. *A textbook on crystal physics.* Cambridge University Press.

Experimental methods
WOOSTER, W. A. and BRETON, A. 1970. *Experimental crystal physics.* Clarendon Press.

Optical properties
GAY, P. 1967. *An introduction to crystal optics.* Longmans.
HARTSHORNE, N. H. and STUART, A. 1970. *Crystals and the polarising microscope.* Arnold.
SHUBNIKOV, A. V. 1960. *Principles of optical crystallography.* Translated by Consultants Bureau, New York.

Source books for optical data
TRÖGER, W. E. 1959. *Tabellen zur optischen bestimmung der gesteinsbildenden minerale.* Schweitzerbartsche Verlag, Stuttgart.
WINCHELL, A. N. and WINCHELL, A. 1964. *The microscopic characters of artificial inorganic solid substances or artificial minerals.* Academic Press.
WINCHELL, A. N. 1964. *The optical properties of organic compounds.* Academic Press.

12

IMPERFECTIONS IN REAL CRYSTALS

12.1. The nature of imperfections

In most of the discussion in earlier chapters we have assumed a perfection of atomic structure consequent upon the precepts of classical crystallography. Any crystal, whether of macroscopic or sub-microscopic dimensions, is considered to be composed of a strictly ordered and regular arrangement of constituent atoms in a repetitive three-dimensional continuum; identical atoms are situated in the same structural sites in every one of the equivalent cells that are stacked together to form the crystal. To be realistic, we must recognise that the locations of these structural sites represent only mean positions for the atoms. At any finite temperature an atom has a thermal motion whose amplitude and direction depend on its electronic configuration and the restoring forces determined by its binding into the general structural arrangement. Sometimes these restoring forces can be overwhelmed by thermal energies in localised volumes so as to bring about a continuous free rotation of some atom groups (such as molecular or ionic complexes) without total disintegration of the overall structural pattern; for example, at temperatures slightly below their melting points some paraffins go into free rotation about the lengths of molecules; but neither this nor any other effect of thermal atomic vibrations conflict seriously with the concepts of classical perfection. Nor indeed do most of the modern generalised theories of binding which suggest that similar atoms behave quite differently in close proximity and in relative isolation; in solids charge cloud distributions are modified by the presence of neighbouring atoms, even to the extent that some valency electrons cannot be regarded as attached to particular atoms, but these changes are not out of harmony with a strict structural regularity in crystalline materials. Again on a realistic view, we must expect some departure from ideal uniformity near the surface of a crystal, due to the asymmetry of the binding forces acting upon atoms, but the effects will be restricted to surface layers, except for crystals of very small dimensions, and the main body of the crystal remains perfect.

In fact, real crystals cannot have such perfection of atomic arrangement, and in the past fifty years it has become increasingly clear that for some properties structural imperfections can be just as important as the overall atomic pattern; it is these defects which give a degree of individuality to crystals with the same structure and constitution. We have already described in Chapter 8.4 how the observed intensities of X-ray reflections are profoundly affected by the textural perfection of a specimen, and there are many other important properties* (such as mechanical deformation, strength, ionic and electronic conductivity, diffusion, etc.) which depend critically on the micro-textures formed by the imperfections within a specimen. In much of the earlier work the presence of many different kinds of defect was inferred in order to account for particular crystalline properties, but recent technical improvements in the resolution of electron microscopes, etc. have confirmed many of the imperfections by more direct methods. We shall not explore the origins of properties except in so far as this is necessary to introduce a description of those defects commonly found in any real crystal; no description of crystalline matter would be complete without some account of these. They are usually classified according to the geometrical nature of the lattice imperfection, e.g. a point (or zero dimensional) imperfection has a centre of disruption at a point in the structure; line (or one-dimensional) imperfections cause disruptions along a line, surface (or two-dimensional) imperfections along a surface, and so on.

12.2. Point imperfections

Some types of point defect are essential to any understanding of the chemical constitution of much of crystalline matter. In compounds an exact identity in every unit cell is possible only within the restriction of *stoichiometric combination* in which elements of the appropriate valencies are present in fixed proportions according to the chemical formulae; but in many crystalline compounds, particularly in the inorganic and metallic fields, the chemical combination is non-stoichiometric. This is often due to some kind of *solid solution* in which a parent structure is able to take in foreign atoms (or ions) without its structural pattern being significantly changed; in one simple form, this was studied by early crystallographers in 'mixed' crystals grown from solutions containing a mixture of pure end members, e.g. the hydrated mixed sulphates known as the alums, still much used in elementary demonstrations of crystal growth. Non-stoichiometric mixed crystals are no different in general character from those of pure end members with stoichiometric combination; indeed, the concepts of classical crystallography were often based on many observations and measurements on crystals of naturally occurring minerals, which usually contain variable amounts of impurity elements. If the

* These properties are sometimes referred to as 'structure-sensitive', a term which has unsatisfactory connotations; the use of 'texture-sensitive' might be preferable.

overwhelming evidence that stoichiometric combination (and chemical for-mulae) are of no fundamental importance in the chemical constitution of crystalline matter is accepted, we must replace our present abstraction of ideal perfection by some more realistic conception of atomic structure. In effect, the structure must be regarded as a three-dimensional pattern of sites available for occupation by the atoms of the crystal; subject to the rules of crystal chemistry for the occupation of particular kinds of site, we may find chemically different atoms in crystallographically equivalent sites in different parts of the structural continuum, and, sometimes, some sites may be unoc-cupied. It is the distinctive pattern of sites which distinguishes one structure from another and determines its basic crystallographic properties; to this extent the particular atoms which occupy individual sites are secondary in their influence on any averaged properties by which a given structural pattern can be recognised. In these terms point defects are natural and quite permis-sible in the detailed atomic arrangement in any real crystal.

Impurity atoms which define some of the point defects can be accom-modated in two different ways in the structural pattern. In *substitutional solid solution* they replace atoms of an ideal perfect structure on certain kinds of site; the particular foreign atoms which are acceptable on given sites and the range of solid solutions brought about by their substitution are dependent on a wide variety of factors, including the relative sizes and chemical natures of solute and solvent atoms, the binding forces which are operative, the physico-chemical conditions of formation, etc. In the simple Cu–Ni alloy system (mentioned in Chapter 10.4) there is a complete range of substitutional solid solution, as there is at high temperatures for mixed crystals in which Ba^{2+} ions replace Sr^{2+} ions on the divalent cation sites of the strontium carbonate structure; but in another alloy, formed from Ag and Cd, the amount of Cd atoms that can be taken into solid solution by Ag is limited and varies with the temperature, and other carbonates have very restricted solid solution even at high temperatures. Certain kinds of impurity atoms can also be taken into solution if they are small enough to fit into cavities between atoms of the parent structure without causing too much distortion; this leads to *interstitial solid solution* in which the foreign atoms are in positions that are not sites of the original structural pattern. Interstitial solid solutions are particularly important in metallic systems where small atoms, like H, B, C and N, introduced to fill to a greater or lesser extent the interstices of a parent metal framework can cause marked changes in properties; the technology of steels is vitally concerned with the behaviour of C atoms in various metallic structures at different temperatures. Solid solution ranges can also be due to the absence of atoms from some available sites. For ex-ample, analyses of the mineral pyrrhotite (FeS) have an irregular sulphur content which is always greater than the stoichiometric 50 at.%; this ori-ginates in lattice vacancies on structural sites which should be occupied by Fe atoms. But *lattice vacancies* are possible even in strict stoichiometric combinations, and they are essential to modern explanations of the colour, luminescence, etc. of many ionic solids. Various kinds of defects have been postulated; the simplest are the Frenkel and Schottky imperfections. In the former the lattice vacancy is created when a small cation leaves its structural site but is retained in an interstitial position in the structure, whilst in the latter a vacant cation site is accompanied by a vacancy on a neighbouring anion site; in most structures, simple Frenkel defects (cation vacancy and intersitial cation) are less probable than simple Schottky defects (cation and anion vacancies) due to the lack of sufficient empty space to accommodate a displaced cation. However, both kinds of defect are present in ionic crystals at normal temperatures, either as simple or complex clusters of vacancies (sometimes with captured electrons which form colour centres).

Other kinds of point defect cause changes in the electronic configurations of particular atoms, as, for example, when excited states are produced by bombardment of the crystal by radiation of a suitable frequency. We must expect all real crystals of elements or compounds of stoichiometric or non-stoichiometric proportions to contain point imperfections of some kind; their distribution is statistical, and their presence must exclude the strict identity of all unit cells.* Their effect on the continuity of a structural pattern will be to cause localised regions of distortion, rather as knots in a piece of wood distort, but do not interrupt, the flow of the grain. Lattice and symmetry operators are only strictly valid in a statistical sense over the crystal as a whole; since we observe the behaviour of many hundreds of thousands of unit cells in most experimental techniques, we shall only detect the averaged effects of individual structural irregularities, e.g. in substitutional solid solution in the Cu–Ni system we shall see a gradual decrease in the averaged cell edge as the larger Cu atoms are randomly replaced by increasing numbers of smaller Ni atoms.

12.3. Line imperfections

A realisation that real crystals must contain line (as well as point) imperfections grew from explanations advanced for the mechanism of slip, one of the important processes in the plastic deformation of crystals. It had long been known that when a crystal is permanently deformed by an applied stress, plastic flow takes place by the movement of lamellae of the crystal relative to each other; this movement occurs on particular parallel planes spaced at

---

* For some compounds and some defects, the random distribution can sometimes be induced to become more regular. At certain compositions for some mixed crystals, for example, a change in the physical conditions of formation can lead to a regular (ordered) occupation of certain structural sites by the foreign atoms as opposed to the random (disordered) occupation of solid solution; this is the basis of the widespread order-disorder phenomenon. Regular distributions of interstitial atoms and lattice vacancies can also occur.

relatively narrow intervals by the translation of whole blocks of structure between the active planes. Early studies of crystalline deformation showed that
displacements always take place on crystallographic planes (known as *slip* or
glide planes) along a lattice direction within the plane (known as a *slip* or *glide
direction*); the particular slip system that is operative is related to the crystal
and its structure and the method of deformation, and the effect of structural
movements during deformation are often visible as slip lines formed by the
intersection of slip planes with the surface of the crystal (Fig. 12.1). At first

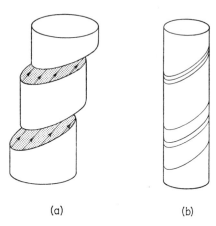

(a)   (b)

Fig. 12.1. Plastic deformation by slip. (a) The displacement of blocks of
a crystal by movements on active slip planes in a glide direction indicated
by the arrows. (b) Slip lines appearing on the surface of a single crystal
wire of a metal during extension.

sight, the atomic mechanism of a process in which blocks of structure glide
over one another on well-defined slip planes in certain slip directions presents
no real difficulties; most slip systems describe atomic movements which are
reasonable from the viewpoint of the structure within which they take place,
and the shearing effect of these movements, like that of a pack of playing
cards, can generate the macroscopic slip lines. Unfortunately, however, this
simple picture encounters difficulties when any attempt is made to calculate
the stresses necessary for the onset of slip; if perfect blocks of structure are
required to move bodily over one another, the initiation of movement requires stresses that are 10^3–10^4 greater than those actually measured, i.e. the
observed plastic yield point of most materials is far smaller than it ought to
be on this basis.

The dilemma of an acceptable slip mechanism involves retaining those
features of simple block movement that account for the observed geometry
while resolving the discrepancy between the observed and theoretical yield
points. It was recognised that theoretical yield points would be reduced if the

atomic forces acting across the slip plane could be overcome one at a time rather than simultaneously over the whole area, i.e. the movements involved in the slip process should spread consecutively across the slip plane. Eventually a mechanism was postulated in which slip is caused by the movement through the structure of line defects, known as *dislocations*; this is illustrated schematically in Fig. 12.2, which shows the passage of one kind of dislocation through a crystal; the diagrams depict a small section of a plane of atoms

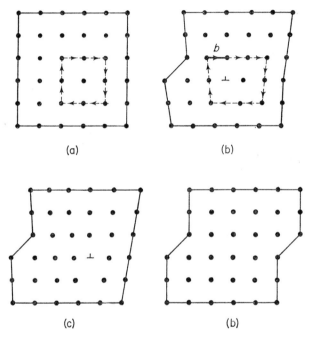

Fig. 12.2. Passage of a dislocation through a block of crystal. (a) The perfect block. (b) The block containing the dislocation centred at ⊥. (c) The block containing the dislocation centred at ⊥. (d) The block after the passage of the dislocation.

normal to the line of disruption, the dislocation line. In (b), the imperfection has moved from the left into the perfect region shown in (a); its centre is marked, and around this line there is compression of the atoms in the upper half of the crystal and tension in the lower half, i.e. across the slip plane on which the imperfection is moving there are $(n + 1)$ atoms in the upper half opposite n atoms in the lower half. In (c) the dislocation line has moved farther to the right (notice the small atomic adjustments required by this movement), whilst in (d) the imperfection has passed out of this section of the crystal; the visible effect of its passage is in the shearing of one half of the crystal over the other, as in slip. The mobility of this type of imperfection must be quite high, for there are only small atomic adjustments between (c) and (d), and restoring forces acting across the glide plane close to the dislocation line are almost exactly balanced; it can be set in motion by relatively small applied stresses. Provided that crystals contain line imperfections of this kind, the motion of dislocations allows a slip mechanism that is compatible both with the observed geometry and the measured plastic yield points.

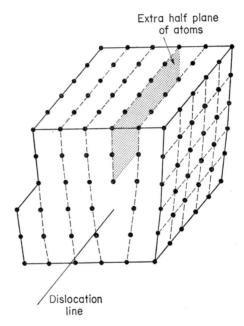

Fig. 12.3. An edge dislocation. Fig. 12.2 (c) is a section of this block normal to the dislocation line.

Initially this hypothesis was slow to be accepted, but gradually the favourable evidence became overwhelming, and it is now recognised that many important crystalline properties are closely related to the presence of line imperfections; the field of dislocation studies has developed into one of the most active branches of solid-state work. In this advance it has become clear that the particular dislocation shown in Fig. 12.2 is really a special case of a more generalised form of line imperfection. Fig. 12.3 shows a three-dimensional representation of the imperfection in Fig. 12.2(c), and from this we see that the line of the dislocation is at the terminating edge of an extra half plane of atoms in the upper portion of the crystal;* for this reason this imperfection

* This extra half plane could be in the lower portion of the crystal, when its movement to the left under an applied stress would produce exactly the same slip as in Fig. 12.2(d); for convenience, these two imperfections are described as positive (as in Fig. 12.2) and negative.

is usually known as an *edge dislocation*. Its geometry (and that of other dislocations) is best described in terms of the Burgers vector which defines both the magnitude and direction of atomic displacements. If we make an atom-to-atom circuit around the dislocation line, it is incomplete in the sense that a corresponding circuit in a perfect crystal returns to the initial atom; the circuit marked in Fig. 12.2(b) is incomplete when compared with the equivalent circuit in Fig. 12.2(a). The *Burgers vector* $\vec{b}$ is required to close the circuit in the crystal containing the imperfection, i.e. in Fig. 12.2(b) it has a length of one atomic spacing in a direction perpendicular to the dislocation line. A

Dislocation line

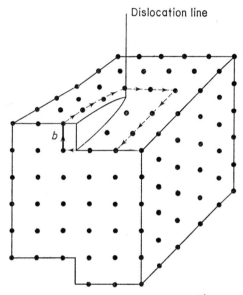

Fig. 12.4. A screw dislocation. The Burgers vector $\vec{b}$ is now parallel to the dislocation line.

Burgers vector must always have a magnitude which is a multiple of the atomic spacing so as to ensure continuity across slip planes, but its inclination relative to the dislocation line can be varied; it is this inclination that distinguishes various line imperfections. An edge dislocation always has a Burgers vector normal to the dislocation line, but Fig. 12.4 shows an imperfection in which $\vec{b}$, still with a magnitude of one atomic spacing, is taken parallel to the dislocation line. We see that successive traverses around the dislocation line provide a continuous descent through the crystal, which can be regarded as a single spiral ramp of atoms extending through the whole volume; for this reason such an imperfection is known as a *screw dislocation*.* When a screw

* Again it is convenient to describe screw dislocations of different signs according to the sense of the Burgers vector along the dislocation line.

dislocation is set in motion by an applied stress, its passage through a crystal can also produce the visible displacements of slip. In fact, the line of any dislocation may be conveniently considered as the boundary between slipped and unslipped regions of the structure; in any real crystal these lines of imperfection must wander through the structure to intersect the surface or form closed loops; their movements under applied stresses are responsible for the slip process. A closed dislocation loop is of particular interest, for it illustrates the relationship between the edge and screw dislocations that we have described, and establishes the existence of mixed (or hybrid) dislocations which must have edge and screw components. Fig. 12.5(a) shows schematically a closed dislocation loop on a slip plane around the boundary between a central portion of the crystal which has slipped and an outer portion which has not; Fig. 12.5(b) depicts a simple arrangement of atoms in part of this region. At all points around the loop, the magnitude and direction of atomic displacements denoted by the Burgers vector are the same, but their inclinations relative to the line of the imperfection are variable. At A, we may describe the imperfection as a pure edge dislocation (also at A', though with opposite sign); at B there is a pure screw dislocation geometry (also at B' though again with opposite sign); but at all other positions the Burgers vector is at some general inclination to the dislocation line to give an imperfection of mixed character with edge and screw components.

The original slip hypothesis asserted that some dislocations must be present in any real crystal; later experimental work showed that even those crystals which appear to be highly perfect still have dislocation densities of 10^3–10^6 lines intersecting a square centimetre of a random section. Naturally, the micro-texture of individual crystals depends on their conditions of formation, but further imperfections may be generated by subsequent treatments. The increasing stress essential for successive deformations of the same crystal is interpreted as due to an increasing density of dislocations which causes mutual hindrance to easy movement and an increase in the stress necessary to initiate motion; an increase in density requires some sources of dislocations within the crystal, and these must become active during mechanical deformation. Certain combinations of dislocations produce atomic distortions which are less mobile than the simple systems we have described; some of these anchored systems can generate further imperfections under applied stresses so that densities can be as high as 10^{12} dislocation lines per square centimetre in heavily worked materials. The motion of dislocations can be more complex than the simple movements that we have discussed, and their role in mechanical behaviour and other properties is so complex that it has developed a language of its own; dislocations may dissociate into partials, can climb or cross-slip to different slip planes, can contain jogs, and so on. We cannot here examine the enormous volume of work over the past few decades which has been devoted to interpreting various crystalline properties of technically important materials in terms of dislocation systems and their interactions; further reading on this aspect of crystallographic studies will be found in the bibliography.

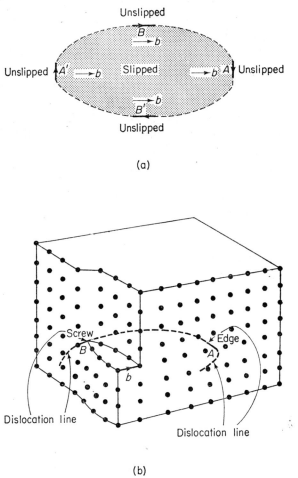

(a)

(b)

Fig. 12.5. A closed dislocation loop. (a) The line of the dislocation marks the boundary between a central region all parts of which have slipped with the same Burgers vector and an outer region which has not slipped. (b) Representation of an atomic arrangement across a segment of the loop.

12.4. Other imperfections

Individual point and line imperfections cannot account for inhomogeneities in any crystalline solid in which ideal structural perfection is disrupted over continuous surfaces; though, as we shall see shortly, atomic distortions in these boundary regions can be interpreted in terms of arrays of defects of simpler geometry in some cases. Surface imperfections have the effect of separating blocks of structure within the solid; their existence is implicit in the occurrence of polycrystalline solids whether their crystallites are in random array or have some sort of preferred orientation, and also in the majority of monocrystalline solids which behave as if they were constructed of a mosaic of slightly misorientated blocks (as in Chapter 8.4). Boundary regions have different characters; they can be non-structural, as in ferromagnetics where domain walls separate regions of different magnetisation, but we will consider only those which relate structural changes across the surface. In general we can divide such boundaries according to the misorientation of the blocks of structure that they separate; the greater any misorientation between the blocks, the more easily they are distinguished due to the more severe atomic distortions within the boundary regions.

Some surface imperfections, however, separate regions in which there are no orientational differences. In a sense the process of slip produces such a boundary, but there is another rather more distinctive surface imperfection for which orientational registry is maintained. The atomic structures of many materials can be built up by stacking layers according to a particular repetitive sequence; thus in many simple cubic metals the perfect crystal structure can be described as a stacking sequence *ABCABCABCA*... in which identical layers of atoms are stacked with particular lateral displacements according to the letters of this repetitive pattern. In some crystals, due to an accident in the growth process or dislocation movements, the sequence might become *ABCABABCABC*... so that a particular layer (the sixth in this example) is out of step with the ideal repetitive pattern (it is in the *A* position rather than the *C* position); subsequent layers continue in the normal sequence. This produces a thin boundary region with an incorrect alternative stacking pattern separating perfect regions in parallel orientation with the correct stacking pattern; this type of surface defect is known as a *stacking fault*.

In boundaries which mark orientational changes between relatively perfect structural regions we can separate those which lead to constant angular differences (large or small) determined by the particular nature of the atomic structure from those which give rise to the variable orientational changes found in the different textures of individual crystalline solids. Most important in the former category are boundaries between the sub-components of twinned crystals (described in Appendix D); these are regions of structure which are consistent (at least as far as neighbouring atoms are concerned) with the atomic arrangements in both adjacent sub-components of the twin, so that, whatever the angular change, there is a high degree of structural coherence across the boundary. In textural boundaries the continuity of structure between adjacent blocks depends on the extent of their misorientation; with large misorientation, as in a polycrystalline aggregate, the boundary regions, although only a few cells thick, must be highly distorted with little or no coherency with the structures of the regions on either side of them; with small misorientation (up to a few degrees), the atomic arrangement in the boundary is to some extent coherent with that of the adjacent crystalline blocks. *Low angle boundaries* are often formed by local concentrations of dislocations to relieve the strain energy of the crystal. Fig. 12.6 shows a low-angle tilt boundary formed by a wall of similar edge dislocations; there is a small rotational misorientation about an axis in the boundary wall to an extent

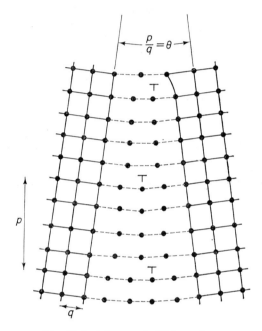

Fig. 12.6. A low angle tilt boundary.

which depends on the Burgers vectors and average separation of the dislocations. A wall of at least two sets of parallel screw dislocations produces a low-angle twist boundary in which the small rotational misorientation is about an axis normal to the boundary wall. Congregations of dislocations can form the boundary regions (part tilt, part twist) which delineate the mosaic texture of imperfect single crystals.

There are other forms of crystalline imperfection (even volume imperfections achieved by thermal excitation, etc.) that are important in the context of certain properties and certain materials, but these must be omitted here. As we remarked earlier, our aim is not to become deeply involved in the highly important and diverting topic of imperfections but only to dispel any idea of strict classical perfection in the atomic structures of real crystals that may have developed in the earlier chapters. For most crystallographic purposes a single crystal may be regarded as a regular and repetitive three-dimensional continuum in an averaged statistical sense, but detailed inspection on an atomic scale reveals that it must contain many flaws; at best, its texture will consist of small relatively perfect regions (still containing some defects) linked together through distorted boundary regions in which there are higher concentrations of imperfections of all kinds.

SELECTED BIBLIOGRAPHY

General reading

COTTRELL, A. H. 1967. *Introduction to metallurgy.* Arnold.

HAYDEN, H. W., MOFFATT, W. G. and WULFF, J. 1965. *The structure and properties of materials:* Vol. 3, *Mechanical behaviour.* Wiley.

KELLY, A. and GROVES, G. W. 1970. *Crystallography and crystal defects.* Longmans.

MOFFATT, W. G., PEARSALL, G. W. and WULFF, J. 1964. *The structure and properties of materials:* Vol. 1, *Structure.* Wiley.

Further reading on dislocations and other imperfections

COTTRELL, A. H. 1953. *Dislocations and plastic flow in crystals.* Clarendon Press.

FRIEDEL, J. 1959. *Dislocation interactions and internal strains, internal stress and fatigue in metals.* Elsevier.

READ, W. T. 1953. *Dislocations in crystals.* McGraw-Hill.

VAN BUEREN, H. G. 1961. *Imperfections in crystals.* North-Holland.

APPENDIX A

PROJECTIONS AND THEIR PROPERTIES

A.1. Stereographic projection

The principles of construction for this projection and practical manipulations in conjunction with the stereographic net have already been described in Chapter 3. In this section we shall demonstrate the validity of fundamental propositions that were assumed there and describe those geometrical constructions which can be used to draw the great and small circle loci on a stereographic net.

In Chapter 3.2, three main virtues of stereographic projection were listed: the representation of all directions in space, the preservation of angular truth, and the ease of projection of great and small circles which appear as arcs of geometrical circles; the first of these properties is self-evident but the other two require some justification.

(a) Angular Truth on the Stereogram

In Fig. A.1 two great circles on the surface of a sphere intersect at the spherical pole P. Tangents to these great circles are drawn at P to cut the tangent plane to the sphere at the south pole in U and V; W and X are respectively the points at which these tangents intersect the equatorial plane of stereographic projection. Q is the stereographic pole obtained by joining P to the south pole of the sphere.

In this diagram UP = US, for both are tangents to the sphere from the same external point; by the same argument VP = VS. Hence $\widehat{UPV} = \widehat{USV}$, and, since the triangles USV and WQX are parallel, it follows that $\widehat{WQX} = \widehat{USV} = \widehat{UPV}$. But WQ and XQ are projections on to the equatorial plane of the two tangents to the great circles through P; in this projection they must represent tangents to the projected great circles which intersect at Q, and, as we have just shown, their mutual inclination is identical to that of the tangents UP and VP constructed to the two great circles on the sphere. In this way a stereogram preserves angular relationships exhibited by a spherical projection of the subject.

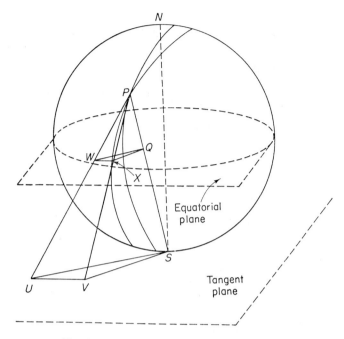

Fig. A.1. Angular truth on the stereogram.

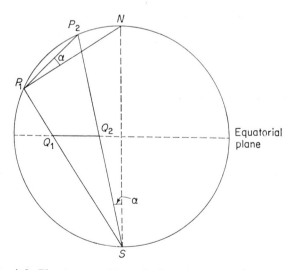

Fig. A.2. The stereographic projection of small (and great) circles.

(*b*) *Projection of Great and Small Circles on the Stereogram*

Fig. A.2 shows a central vertical section of a circumscribing sphere which contains a diameter P_1P_2 of a circle (great or small) on the surface of the sphere; Q_1 and Q_2 are stereographic poles formed by joining P_1 and P_2 to the south pole of the sphere.

The volume generated by joining all points on a circle on a spherical surface to the south pole is a cone with apex at S; P_2P_1S is a central section of this cone. Any right section of this general cone will be elliptical, but inclined sections can be circular; the inclination of the plane which intersects the sphere to give a circle of diameter P_1P_2 on its surface defines one set of circular sections. There is however a second set of planes which will also give circular cross-sections; these are symmetrically disposed with respect to the axis of the cone, i.e. in our central section they are inclined at the same angle to the second generator (SP_2) as P_1P_2 is to the first (SP_1). But from the figure $P_2\widehat{P_1}S = 90° + \alpha = Q_1\widehat{Q_2}S$, for the angle subtended by the arc NP_2 is the same at all points around the circle; thus Q_1Q_2 and P_1P_2 are symmetrically inclined to the axis of the cone, and both must be circular sections. Hence a circle with diameter P_1P_2 on the sphere must project on to the stereogram as a circular locus with Q_1Q_2 as diameter.

All constructions on the stereogram described in Chapter 3.2 using a net can be carried out by graphical methods alone. Such procedures are less convenient and rather tedious, and with the ready availability of nets of all sizes they are rarely used in practice. We shall describe only two geometrical constructions, those needed to draw the loci of great and small circles inscribed on a stereographic net. In both of these, as in many other graphical constructions on the stereogram, a simple stratagem is employed. The plane of the stereogram is visualised as a vertical section of the circumscribing sphere and a diameter drawn through the relevant stereographic pole P (Fig. A.3); this diameter is regarded as the equatorial projection plane with 'north' and 'south' poles placed on the primitive at the ends of a perpendicular diameter. By this device we can construct the position of another stereographic pole Q at an angular distance α from P on the same diameter. P is projected from the 'south' pole to cut the section of the sphere (the primitive) at P'; P' can be thought of as a spherical pole formed by the radius of the sphere CP'. CQ' is then drawn inclined at α to CP'; the point Q' (on the primitive) is again thought of as a spherical pole which is subsequently projected to the 'south' pole to give the stereographic pole Q on the same diameter as P. The angular separation of P and Q on the stereogram must now be α. (In construction (iii) of Chapter 3.2, this device, with $\alpha = 180°$, is used to re-project poles in different hemispheres into positions on the same diameter outside the primitive, i.e. as the points A' and B' at the bottom of Fig. 3.6.) We can use this method to construct any of the circles on a stereographic net.

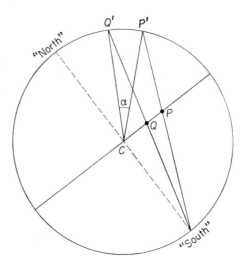

Fig. A.3. Device for use on the stereogram to construct a pole a particular angular distance (α) from a given pole.

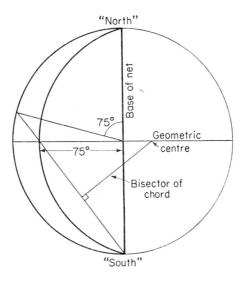

Fig. A.4. The construction of a great circle inclined at angle α ($=75°$) to
the centre of the projection.

(c) Great Circles of All Inclinations

Apart from the vertical circle at the base of the net, there are arcs of circles which pass through the ends of the base diameter. To construct any one of them we need only find another point on the appropriate circular arc; the device described above will locate a pole at the required inclination to the centre of the projection, as illustrated in Fig. A.4. The geometric centre of the circular arc of this projected great circle is at the intersection of bisectors of its chords.

(d) Small Circles about a Point on the Primitive

These are circular arcs whose geometric radii are determined by the angular radius α of the small circles they represent (if $\alpha = 90°$, we have a vertical

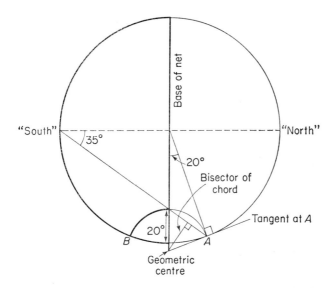

Fig. A.5. The construction of a small circle of radius α ($=20°$) about a point on the primitive.

great circle, an arc of infinite radius, which appears as a radius perpendicular to the base of the net). To construct any small circle, we must find a third point on the appropriate circular arc, in addition to its two intersections with the primitive at A and B which are inclined at α to the angular centre; Fig. A.5 shows how such a point on the base of the net can be found. The geometric centre of the circular arc for this small circle is at the intersection of bisectors of its chords; notice that by the geometry of the figure it is also the intersection of the tangent to the primitive at A (or B) with the base diameter produced.

A.2. Gnomonic projection

As an alternative to stereographic projection, this method is employed
sufficiently often in certain crystallographic work to merit a brief description
here. It has some features in common with stereographic representation and
is also developed after a projection subject has been replaced by a distribution
of spherical poles on a circumscribing sphere as in Chapter 3.1; it differs in
that subsequent projection of these poles into two dimensions takes place on
to a non-equatorial plane, conveniently a tangent plane to the sphere. The
gnomonic projection of a spherical pole is obtained by extending the radius
of the sphere through the pole to cut the projection plane (Fig. A.6); each

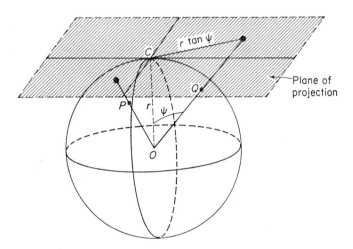

Fig. A.6. Gnomonic projection. Gnomonic poles are obtained by producing radii through P and Q to cut the tangent plane at the north pole
of the sphere.

gnomonic pole will be at a distance of $r \tan \psi$ from the centre of the projection,
where r is the radius of the sphere and ψ, the inclination of the radius defining
the spherical pole to the N-S axis.

Comparing its qualities with those of the stereogram, it is obvious from the
figure that the directions in each hemisphere that can be represented on a
single projection are restricted; as $\psi \rightarrow 90°$, gnomonic poles become farther
and farther from the centre of the projection. A particular virtue, however, is
that all great circles project as straight lines on the gnomonogram; this is
readily apparent, for all those that pass through the centre of the projection,
and a little consideration will show that all straight lines, whether they pass
through the origin or not, may be interpreted in the same way. But small
circles present some difficulties; their loci will be formed by the intersection

of the projection plane with the cone generated by radii of the sphere to all points of a circle on its surface. When the angular centre of a small circle is at the origin of projection, these loci will be circular; but for any other angular centre, loci will be conic sections in the form of ellipses when all points are in one hemisphere and hyperbolae when the circle lies in both hemispheres. In practice, this can be a disadvantage which more than offsets the simplicity of great circle construction; more importantly, it means that principles of angular measurement are more complex than on the stereogram.

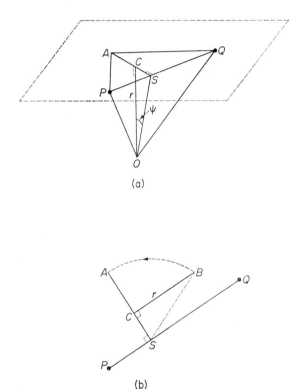

Fig. A.7. Angle point construction. (a) The location of the angle point. (b) Its construction on a gnomonogram.

Most graphical measurements are carried out by an *angle point construction*, i.e. the location of a point in the plane of projection at which a pair of gnomonic poles subtend the true angle between the two directions that they represent. In Fig. A.7(a), we wish to determine the true angle between gnomonic poles P and Q on a general zone line in a projection whose origin is at C; O is the centre of the circumscribing sphere of radius r, and another line SA in the projection plane is drawn perpendicular to PQ to pass through C so that AS = OS. The true angle is subtended at the centre of the sphere as $\widehat{POQ}$, but this must be equal to $\widehat{PAQ}$ subtended by the line PQ at an angle point A. In constructing the position of A in the gnomonogram of Fig. A.7(b), there is no difficulty in drawing the line SC produced, but the precise position

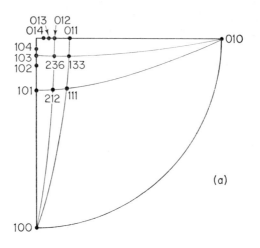

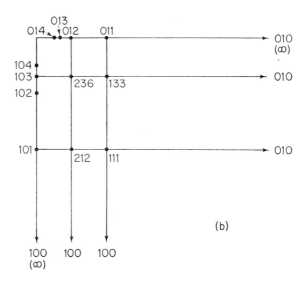

Fig. A.8. Projections of an orthorhombic crystal. (a) Stereogram. (b) Gnomonogram. Both diagrams show the same section of the projections made from the same circumscribing sphere with the same angular relationships between the poles (*hkl*).

of the angle point A along this line is not immediately obvious. It is best found by constructing CB of length r perpendicular to SC, when SB = SA; in Fig. A.7(a), OS = $r \sec \psi$ = AS, but from Fig. A.7(b), SB = $(SC^2 + BC^2)^{1/2}$ = $(r \tan^2 \psi + r^2)^{1/2}$ = $r \sec \psi$. The angle point A is then located on the extension of SC, when $\widehat{PAQ}$ is measured to give the angular separation of the two gnomonic poles. When available, gnomonic nets (used in a similar way to stereographic nets) are very helpful in manipulations on the projection; they usually show, for a particular value of r, the loci of perpendicular great circles at small intervals as a square mesh together with the loci of small circles at similar angular intervals in the form of the hyperbolic curves that appear when the angular centre is on a diameter of the sphere parallel to the projection plane.

The merits of gnomonic projection are best utilised in certain specialised circumstances. For example, in some morphological studies (or indeed in other work where it is necessary to project a group of approximately similar directions), a gnomonogram can avoid the crowding of poles which can occur on a stereogram particularly in the central regions; Fig. A.8 shows both gnomonic and stereographic projection of some normals to planes of an orthorhombic crystal. Gnomonic projection is also often used in the interpretation of Laue patterns when indexing of individual spots is required; there is a close connection between a gnomonic projection and the reciprocal lattice (which arises in the Ewald treatment of the Bragg conditions mentioned in Chapters 8 and 9) in that the gnomonic poles of plane indices can be formed by allowing the projection plane to be intersected by lines from the origin to the corresponding reciprocal lattice points.

Further reading
See the Selected Bibliography for Chapter 3.

APPENDIX B

CALCULATIONS ON THE STEREOGRAM

B.1. Spherical trigonometry

Angular measurements on an accurate stereogram are often insufficiently precise, and in many cases it is more convenient to proceed by calculation from a sketch of the distribution of poles. These calculations involve the solution of *spherical triangles*, which are areas on the surface of a sphere bounded by the intersections of three great circles (Fig. B.1). Each triangle has six angular elements, three angles between tangents to the great circles at their points of intersection $(\widehat{A}, \widehat{B}, \widehat{C})$ and three angles subtended at the centre of the sphere by the circular arcs forming the sides (a, b, c); clearly these angular elements of any spherical triangle are related in a manner which is independent of the size of the sphere. We can express relationships between various elements in forms which are often analogous to those found in planar trigono-

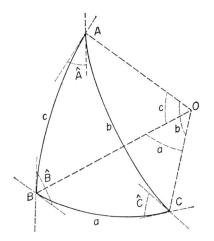

Fig. B.1. The angular elements of a spherical triangle on the surface of a sphere centred at O.

metry; there are, however, some important differences (in particular that the sum of the included angles $\widehat{A}$, $\widehat{B}$ and $\widehat{C}$ in a spherical triangle is not constant). For those who are unfamiliar with the elements of spherical trigonometry required for stereographic calculation, some of the important formulae are derived in this section.

Fig. B.2 shows a general spherical triangle ABC on the surface of a sphere

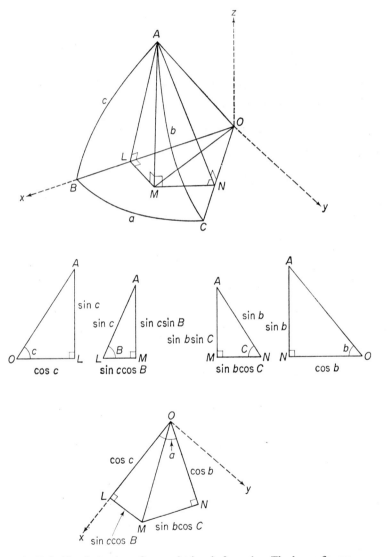

Fig. B.2. The derivation of general triangle formulae. The lower figures show the plane faces of the pyramid *LMNOA*.

of unit radius whose centre is at O; a set of orthogonal reference axes is chosen with the x-axis placed for convenience along the radius OB. AM is drawn perpendicular to the xy plane, with LM and NM normal to OB and OC respectively. For clarity the planar faces of the pyramid LMNOA are drawn out separately in the lower diagrams of the figures; the values of the sides and angles of this pyramid are expressed in terms of angular elements of the spherical triangle. We see from these that

$$AM = \sin c \sin B = \sin b \sin C$$

i.e. $$\frac{\sin c}{\sin C} = \frac{\sin b}{\sin B} \left(= \frac{\sin a}{\sin A}, \text{ by a similar construction} \right)$$

a general relation of spherical trigonometry, usually referred to as the *sine formula*. Moreover in Fig. B.2

$$\overrightarrow{OM} = \overrightarrow{OL} + \overrightarrow{LM} = \overrightarrow{ON} + \overrightarrow{NM}$$

and the x and y components of vectors in this identity are

$$\overrightarrow{OL} = (OL, 0), \overrightarrow{LM} = (0, LM), \overrightarrow{ON} = (ON \cos a, ON \sin a) \text{ and } \overrightarrow{NM}$$
$$= (MN \sin a, -MN \cos a).$$

Equating x components of the vectoral identity and substituting values for OL, LM, ON and MN from the figure, we get

$$\cos c = \cos a \cos b + \sin a \sin b \cos C$$

a relation often called the *cosine formula*. (Other forms of the cosine formula, e.g. $\cos b = \cos a \cos c + \sin a \sin c \cos B$ would, of course, be obtained by similar constructions from the other apeces of the spherical triangle.) Finally equating y components of the identity, another relationship is obtained

$$\sin c \cos B = \sin a \cos b - \cos a \sin b \cos C$$

the five element formula, which can again be expressed in terms of related combinations of elements. Manipulation of these formulae will also produce

$$\cot b \sin a = \cos a \cos C + \sin C \cot B, \text{ etc.}$$

the *cotangent-sine formulae*, and

$$\tan \frac{A}{2} = \left(\frac{\sin (s - b) \sin (s - c)}{\sin s \sin (s - a)} \right)^{\frac{1}{2}} \quad \text{etc.}$$

where $s = (a + b + c)/2$, the *half-angle formulae*.

In effect when three elements of a spherical triangle are known, the values of the remainder can be found by application of the most suitable trigonometrical formulae; in practice it is the first two, the sine and cosine formulae which are most used.

For some purposes, it is convenient to consider a *polar triangle* A'B'C' of the original triangle ABC, defined in such a way that A' is the pole of the great circle through B and C which lies on the same side of the plane of BC as A, B' is the pole of CA on the same side of the plane of CA as B, and C' is the pole of AB on the same side of the plane of AB as C. Under these conditions the sides and angles of the polar triangle are the supplements of the angles and the sides, respectively, of the original triangle (i.e. $a' = 180° - A$, $A' = 180° - a$, etc.). By applying to the polar triangle A'B'C' any of the general formulae established above, we can obtain a supplemental formula for the triangle ABC involving sides and angles opposite to these of the original formula; for example, since $c' = 180° - C$, $C' = 180° - c$, etc. the cosine formula quoted above becomes

$$\cos C = -\cos A \cos B + \sin A \sin B \cos c$$

in its supplemental form. Such variations are often invaluable in practical calculations.

In many crystallographic calculations, particularly in higher symmetry systems, we are concerned with the solution of right-angled triangles; naturally the formulae above are considerably simplified, e.g.

if $\widehat{B} = 90°$, $\sin c = \sin b \sin C$ or if $c = 90°$, $\cos C = -\cot a \cot b$

$\cot b = \cot a \cos C$, etc. $\sin B = \sin b \sin C$, etc.

A convenient method of memorising the manifold reductions of general triangle formulae was devised by Napier; it can be applied to obtain the complete solution of any right-angled spherical triangle (often called a *Napierian triangle*) in which two elements, other than the right angle, are known. Fig. B.3 shows the use of this device for Napierian triangles in which $\widehat{B} = 90°$ (on the left) and $c = 90°$ (on the right). Beneath the triangle a five-compartmented diagram is drawn with three sectors to the left of a vertical line and two on the right divided by a horizontal line; angular elements of the triangle are written into the five compartments in sequence as they are encountered in moving round the triangle whose right angle is set on the right-hand horizontal line. The sense, clockwise or anti-clockwise, of movement around a triangle is immaterial, but elements must be in the correct sequence from the position of the right angle *and* those in the three left-hand sectors must be subtracted from 90°. All Napierian solutions involve the use of this diagram; in solving a particular problem, two compartments must contain known elements, and, together with the unknown that we seek to calculate, they form three sectors that are either adjacent or have two sectors opposite the third; an example of both kinds of sector distribution is given for each triangle in the figure. All the necessary formulae are then summarised by the statement for alternative dispositions of the sectors:

sine of a middle part = product of tangents of adjacent parts

or

product of cosines of opposite parts.

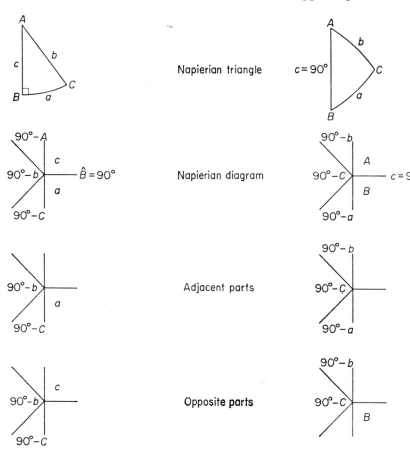

Fig. B.3. The Napierian device for the solution of right-angled spherical triangles.

Applying this to the examples in the figure for the left hand triangle, we get

$$\sin (90° - C) = \tan a \tan (90° - b)$$

$$\text{i.e.} \quad \cot b = \cot a \cos C,$$

and

$$\sin c = \cos (90° - b) \cos (90° - C),$$

$$\text{i.e.} \quad \sin c = \sin b \sin C,$$

which are reductions of general triangle formulae for $\widehat{B} = 90°$ already quoted. For the right hand triangle, in which one side is a right angle, the procedure is similar, except that when the included angle opposite the right angle is involved products are negative unless this angle is one of the opposite parts. For the examples in the figure we have

$$\sin (90° - C) = -\{\tan (90° - a) \cdot \tan (90° - b)\}$$
$$\text{i.e.} \quad \cos C = -\cot a \cdot \cot b$$

and

$$\sin B = \cos (90° - b) \cos (90° - C),$$
$$\text{i.e.} \quad \sin B = \sin b \sin C$$

which are two of the general triangle reductions for $c = 90°$ given earlier. (Alternatively such solutions can be developed by taking the sector opposite the side which is a right angle as $C - 90°$ (instead of $90° - C$); the correct sign is then derived during the evaluation of the appropriate product.)

B.2. The rational sine ratio

In any crystal the angular dispositions of planes (or faces) around the same zone can be related to their indices defined in terms of any choice of unit cell. In its most convenient form, this relationship is known as the *Law of Rational Sine Ratios* (occasionally as *Miller's Law*) and is expressed in terms of four non-parallel planes with indices $(h_1k_1l_1)$, $(h_2k_2l_2)$, etc. inclined at angles θ_{12}, θ_{13}, etc. around the same zone as shown in Fig. B.4(a). This law states that the ratios of the sines

$$\frac{\sin \theta_{12}}{\sin \theta_{13}} = \frac{U_{12}}{U_{13}} = \frac{V_{12}}{V_{13}} = \frac{W_{12}}{W_{13}} = \frac{p}{q}$$
$$\frac{\sin \theta_{42}}{\sin \theta_{43}} \quad\quad \frac{U_{42}}{U_{43}} \quad\quad \frac{V_{42}}{V_{43}} \quad\quad \frac{W_{42}}{W_{43}}$$

where p and q are integers, and $[U_{12}V_{12}W_{12}]$, $[U_{13}V_{13}W_{13}]$, etc. are the zone symbols formed by cross-multiplication of the corresponding plane indices as in Chapter 5.1. This expression can be manipulated into

$$p \cot \theta_{12} - q \cot \theta_{13} = (p - q) \cot \theta_{14}$$

a form which is less cumbersome and more convenient for certain types of calculation. A proof of these expressions is not particularly difficult but is rather too lengthy and tedious to merit inclusion here; it will be found in many of the older books on mathematical crystallography. Since the four faces are tautozonal, zone symbols produced by the cross-multiplication of any pair of indices must indicate the same lattice row direction, but in deriving the numerical value of the sine ratios, symbols are not divided out and are allowed to contain common factors; indeed to avoid incorrect solutions great care must be taken to construct zone symbols in strict conformity with angular subscripts. The equations are of most practical value in solving two kinds of problems, as we shall illustrate by example.

(*a*) *When the angular positions of four planes in a zone are known as well as the indices of three of them, to find the indices of the fourth:*

Fig. B.4(b) shows the angular positions of four planes together with their

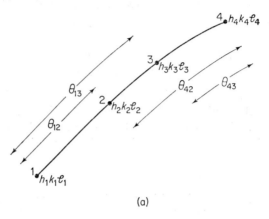

(a)

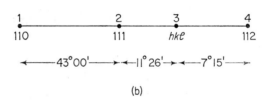

(b)

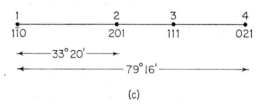

(c)

Fig. B.4. The rational sine ratio.

known indices; the problem is to find (*hkl*). The sine ratio in this example is

$$\frac{p}{q} = \frac{\dfrac{\sin 43^\circ\,00'}{\sin 54^\circ\,26'}}{\dfrac{\sin 18^\circ\,41'}{\sin 7^\circ\,15'}} = \tfrac{1}{3}$$

and the corresponding zone symbols are

$$12:\ \overline{\begin{matrix}110110\\111111\end{matrix}}\qquad 13:\ \overline{\begin{matrix}110110\\hklhkl\end{matrix}}\qquad 42:\ \overline{\begin{matrix}112112\\111111\end{matrix}}\qquad 43:\ \overline{\begin{matrix}112112\\hklhkl\end{matrix}}$$

$$[1\ \bar{1}\ 0]\qquad [l,\ \bar{l},\ k-h]\qquad [\bar{1}\ 1\ 0]\qquad [l-2k,\ 2h-l,\ k-h]$$

so that from the sine law

$$\cfrac{\dfrac{1}{\bar{l}}}{\dfrac{\bar{1}}{l-2k}}=\cfrac{\dfrac{\bar{1}}{\bar{l}}}{\dfrac{1}{2h-l}}=\tfrac{1}{3}$$

which implies that $3(2h-l)=l$, for $h=k$ in this zone or

$$6h=4l$$

to give (hkl) as (223).

(b) When the indices of four planes in a zone are known as well as the angular positions of three of them, to find the position of the fourth:

Fig. B.4(c) shows the four planes in a zone together with their known positions; the problem is to find the position of (111). Firstly we can compute the value of p/q from the zone symbols

$$12:\ \overline{\begin{matrix}1\bar{1}01\bar{1}0\\201201\end{matrix}}\qquad 13:\ \overline{\begin{matrix}1\bar{1}01\bar{1}0\\111111\end{matrix}}\qquad 42:\ \overline{\begin{matrix}021021\\201201\end{matrix}}\qquad 43:\ \overline{\begin{matrix}021021\\111111\end{matrix}}$$

$$[\bar{1}\ \bar{1}\ 2]\qquad [\bar{1}\ \bar{1}\ 2]\qquad [2\ 2\ \bar{4}]\qquad [1\ 1\ \bar{2}]$$

so that $p/q=\tfrac{1}{2}$. Using the alternative cotangent relationship, we have

$$\cot\theta_{12}-2\cot\theta_{13}=-\cot\theta_{14}$$

$$\cot\widehat{(1\bar{1}0)(111)}=\cot\theta_{13}=\tfrac{1}{2}(\cot\theta_{12}+\cot\theta_{14})$$
$$=\tfrac{1}{2}(\cot 33°\ 20'+\cot 79°\ 16')=0{\cdot}8565$$

to give $\widehat{(1\bar{1}0)(111)}=49°\ 28'$.

These two examples give some illustration of the practical application of rational sine ratios; there is a high probability of tedious computational error unless a systematic and orderly procedure is followed, and it is often preferable to avoid their use if at all possible by employing less direct solutions involving spherical trigonometry. Nevertheless, many zones in more symmetrical crystals have planes at intervals of $90°$; and if problems can be formulated so that $\theta_{14}=90°$, the sine ratio is simplified to

$$\tan\theta_{12}=\frac{p}{q}\tan\theta_{13}$$

a relationship which can be particularly useful in calculation.

B.3. Euler's construction

This is concerned with the permissible combinations of rotational symmetry operators passing through a point, and, as mentioned in Chapter 4.2(a), can be used in a systematic derivation of the point group symmetry associated with crystal classes.

On the stereogram of Fig. B.5, we show two inclined rotational operators, one acting through P with angular rotation α and the other through Q with angular rotation β. The points R and R' are on diameters symmetrically inclined to PQ at an angle of $\alpha/2$, i.e. on great circles related by the operator

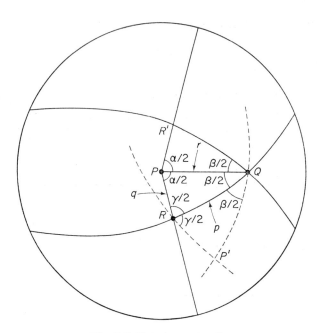

Fig. B.5. Euler's construction.

through P; moreover the positions of R and R' on these diameters are such that great circles through their poles and Q are symmetrically inclined to PQ at an angle of $\beta/2$, i.e. they are also on great circles related by the operator through Q. If we consider the effect of both operators on the point R, rotation about P will place it at R' from which subsequent rotation about Q will restore it to R. Thus combination of the two operators leaves R unmoved, so that the net movement of any other pole implies a rotation about R; there must be another rotational operator present acting through R. P is moved to P' after rotation through β about Q, when subsequent rotation about R must restore it to P; hence the angular rotation through R is γ where $\widehat{PRQ}$ in the figure is $\gamma/2$. The spherical triangle PQR expresses angular relationships between the three rotational operators at its corners in terms of their angular rotations and mutual inclinations, and after spherical trigonometric manipulation in this triangle we can write

$$\cos r = \frac{\sin \gamma/2 + \cos \alpha/2 \cos \beta/2}{\sin \alpha/2 \sin \beta/2}$$

together with other similar equations.

So far the construction has been quite general and has shown that a rotational operator (α) can be combined with another rotational operator (β) at an intersection angle r to give a third rotational operator (γ) in another specific orientation. But in crystalline matter, the only possible values of γ are those associated with monad, diad, triad, tetrad or hexad rotations (see Chapter 2.1); thus values of the intersection angle r are restricted. From the relations above we can derive all the permissible combinations of crystallographic rotational operators passing through a point by eliminating impossible and trivial combinations. For example, the combination 236, with $\alpha/2 = 90°$, $\beta/2 = 60°$, $\gamma/2 = 30°$ gives $r = 0°$, a trivial solution, whereas 246 is also rejected because it gives an impossible value for $\cos r$. Systematic investigation shows that apart from 211, 311, etc. the only permissible combinations of crystallographic rotations are 222, 223, 224, 226, 233 and 234, each with its own set of interaxial angles demanded by the equations above. A complete derivation of the crystal classes by this method requires us to examine all possibilities for each of these six combinations including both the proper and improper rotational operators described in Chapter 4.1; after establishing equivalences and decomposing improper elements (where necessary) into more conventional symmetry operators, we shall be left with the point group symmetries of the thirty-two crystal classes.

B.4. Examples

Calculations on a stereogram of the angular relations between crystallographic directions are needed for many purposes and can be undertaken in many different ways. The art of such calculations lies in expeditiously combining the techniques of this appendix with the basic relations between cell constants and geometry that have already been established for the various crystal systems in Chapter 5.4. In this section we present the outline of a solution for a problem with each different cell shape. No brief discussion of this kind can be in any way representative of the wide range of stereographic calculations, and it is not suggested that the solutions (using the Napierian triangles only) are unique; our only purpose is to introduce the reader by example to this aspect of crystallographic calculations.

(a) *Triclinic system:* $CuSO_4.5H_2O$ *belongs to class* $\bar{1}$ *and has cell constants* $a = 6\cdot11$, $b = 10\cdot67$, $c = 5\cdot95$ *(in Å),* $\alpha = 97°\ 35'$, $\beta = 107°\ 10'$, $\gamma = 77°\ 33'$. *Determine the angle* $[00\bar{1}](001)$ *implied by this description, an angle sometimes required in single crystal X-ray studies.*

Fig. B.6 shows a sector of a sketch stereogram with the poles (100)(010) and (001); the required angle is part of a radius drawn through the centre at z (or [001]) to cut the primitive at P (the pole P is unlikely to have any crystallographic significance). An angle ϕ is marked between this vertical great circle and the zone (001)(010) at their point of intersection (001); the other angles correspond to the relations of Chapter 5.4(a).

(i) In the triangle P(010)(001), $\cos(180° - \beta) = \cos(00\widehat{1)}P \cdot \sin\phi$

(ii) In the triangle P(100)(001), $\cos(180° - \alpha) = \cos(00\widehat{1)}P \cdot \sin(\phi + \gamma)$.

(iii) Eliminating $\cos(00\widehat{1)}P$ from these two equations and substituting numerical values for α, β and γ, we obtain $\phi = 77°\ 18'$.

(iv) From the equation of (i), $(00\widehat{1)}P = 72°\ 19'$.

(v) Since $[00\widehat{1}]P = 90°$, $[001\widehat{](0}01) = 90° - 72°\ 19' = 17°\ 41'$.

(b) *Monoclinic system:* *In a crystal of class* $2/m$, *three angles were measured as* $(12\widehat{1)(\bar{1}}2\bar{1}) = 103°\ 46'$, $(00\widehat{\bar{1})(\bar{1}}\bar{1}\bar{1}) = 37°\ 03'$ *and* $(10\widehat{0)(1}\bar{1}1) = 28°\ 27'$. *Determine the axial constants for this crystal, a problem that can arise in morphological studies.*

The sketch stereogram of Fig. B.7 shows these three angles, but from the symmetry of this class we see that

$$\tfrac{1}{2}(12\widehat{1)(\bar{1}}2\bar{1}) = (01\widehat{0)(1}21)$$
$$(00\widehat{\bar{1})(\bar{1}}\bar{1}\bar{1}) = (00\widehat{1)(1}11)$$
$$(10\widehat{0)(1}\bar{1}1) = (10\widehat{0)(1}11)$$

(i) From a multiple tangent relationship in the zone (010)(121) (111)(101), $\tan(01\widehat{0)(1}11) = 2\tan(01\widehat{0)(1}21)$ so that $(01\widehat{0)(1}11) = 68°\ 35'$ and $(10\widehat{1)(1}11) = 21°\ 25'$.

(ii) In the triangle (001)(111)(101), $\cos(00\widehat{1)(1}11) = \cos(00\widehat{1)(1}01) \cdot \cos(10\widehat{1)(1}11)$ and $\sin(10\widehat{1)(1}11) = \sin\widehat{00}1 \cdot \sin(00\widehat{1)(1}11)$, to give $(00\widehat{1)(1}01) = 31°\ 00'$ and $\widehat{00}1 = 37°\ 18'$.

(iii) In the triangle (100)(111)(101), $\cos(10\widehat{0)(1}11) = \cos(10\widehat{0)(1}01) \cdot \cos(10\widehat{1)(1}11)$ and $\sin(10\widehat{1)(1}11) = \sin\widehat{10}0 \cdot \sin(10\widehat{0)(1}11)$, to give $(10\widehat{0)(1}01) = 19°\ 12'$ and $\widehat{10}0 = 50°\ 01'$.

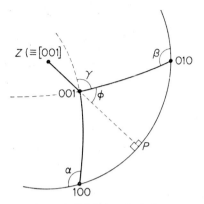

Fig. B.6. Triclinic calculation.

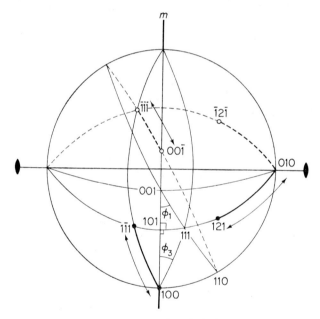

Fig. B.7. Monoclinic calculation.

(iv) From the relations of Chapter 5.4(b)

$$(100)\widehat{(101)} + (10\overline{1})\widehat{(001)} = (100)\widehat{(001)} = 180° - \beta$$

$$\widehat{00\overline{1}} = \phi_1 = \tan^{-1}(a/b) \quad \text{and} \quad \widehat{1\overline{0}0} = \phi_3 = \tan^{-1}(c/b)$$

so that for this crystal

$$\frac{a}{b}:1:\frac{c}{b} = 0\cdot762:1:1\cdot193, \beta = 129°\ 48'.$$

(c) *Orthorhombic system: A crystal of class mmm has cell dimensions $a = 4\cdot82$, $b = 11\cdot08$ and $c = 6\cdot37$ (in $\mathring{A}$). On a fragment of this crystal there are only three recognisable faces p, q and r; the angles between them were measured as $\widehat{pq} = 105°\ 46'$, $\widehat{qr} = 86°\ 43'$ and $\widehat{pr} = 31°\ 42'$. Physical examination suggests that the z-axis is the bisector of $\widehat{pq}$, and these faces lie in the (010) plane. Determine the possible indices of p, q and r, again a problem in morphological description.*

If the faces p and q lie in the (010) plane and are symmetrically disposed about the z-axis, their indices must be of the type $(h0l)$ and $(\bar{h}0l)$, so that $(00\bar{1})\widehat{(h0l)} = \frac{1}{2}\widehat{pq}$.

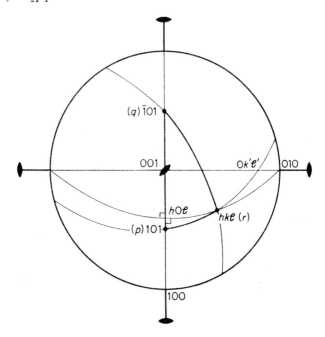

Fig. B.8. Orthorhombic calculation.

(i) From Chapter 5.4(c) $\tan \phi_5 = \dfrac{c/l}{a/h} = \tan (00\bar{1})\widehat{(h0l)}$, so that on substitution we obtain $\dfrac{h}{l} = 1$ to give the indices as (101) and ($\bar{1}$01). They are so marked on the sketch stereogram of Fig. B.8, together with the position of r as (hkl); zones are drawn through (010) and (hkl) to locate $(h0l)$ and through (101) and (hkl) to a pole $(0k'l')$.

(ii) In the triangle $(hkl)(101)(h0l)$, $\cos (10\bar{1})\widehat{(hkl)} = \cos (10\bar{1})\widehat{(h0l)} . \sin (hk\bar{l})\widehat{(h0l)}$.

(iii) In the triangle $(hkl)(\bar{1}01)(h0l)$, $\cos(\widehat{\bar{1}01)(hkl}) = \cos(\widehat{\bar{1}01)(h0l}) \cdot \sin$ $(\widehat{hkl)(h0l})$.

(iv) Eliminating $\sin(\widehat{hkl)(h0l})$ from these two equations and substituting numerical values, we obtain $(\widehat{101)(h0l}) = 19° 26'$.

(v) Hence $(\widehat{001)(h0l}) = (\widehat{001)(101}) - (\widehat{101)(h0l}) = 33° 27'$.

(vi) Using the relation quoted in (i), $\dfrac{h}{l} = \frac{1}{2}$, i.e. $(h0l) = (102)$.

(vii) In the triangle $(hkl)(101)(102)$, $\cos 1\widehat{0}1 = \tan(\widehat{101)(102}) \cdot \cot$ $(\widehat{101)(hkl})$, from which $1\widehat{0}1 = 55° 00'$.

(viii) In the triangle $(101)(0k'l')(001)$, $\sin(\widehat{001)(101}) = \tan(\widehat{001)(0k'l'}) \cdot \cot$ $1\widehat{0}1$ from which $(\widehat{001)(0k'l'}) = 48° 43'$.

(ix) From Chapter 5.4(c) $\tan \phi_3 = \dfrac{c/l'}{b/k'} = \tan(\widehat{001)(0k'l'})$, so that on substitution we obtain $\dfrac{k'}{l'} = 2$ to give $(0k'l')$ as (021).

(x) r is at the intersection of the zones $(102)(010)$ and $(101)(021)$; by cross-adding, its indices must be $(hkl) \equiv (122)$. The data is, of course, consistent with solutions involving negative axes, e.g. $(1\bar{2}2)(1\bar{2}\bar{2})$, etc.

(d) *Trigonal and hexagonal systems: A cleavage rhombohedron of a crystal of class $\bar{3}m$ was indexed in morphological examination as $\{10\bar{1}1\}$ and the angle* $(\widehat{10\bar{1}1)(\bar{1}101})$ *measured as $74° 56'$. Later X-ray examination showed that a minimum volume hexagonal cell had dimensions $a = 4·99$, $c = 17·06$ (in Å); systematic absences in reflections indexed on this cell revealed that the structure has a rhombohedral lattice. Give a description of the cleavage rhombohedron in terms of (a) the minimum volume hexagonal cell, and (b) the minimum volume rhombohedral cell.*

Fig. B.9 shows part of a stereogram with poles p and q representing the measured faces of the cleavage rhombohedron.

(i) The point r on the zone containing p and q would be indexed by the morphologist as $(01\bar{1}2)$. In the triangle $(0001)pr$,

$$\sin(\widehat{0001)r} = \tan \tfrac{1}{2}(\widehat{pq}) \cdot \cot 0\widehat{00}1$$

to give $(\widehat{0001)r} = 26° 16'$.

(ii) From Chapter 5.4(f) $\tan(\widehat{0001)(01\bar{1}2}) = \dfrac{c}{2a \cos 30°}$ so that the morphological axial ratio is $\dfrac{c}{a} = 0·855$.

(iii) The axial ratio for the minimum volume hexagonal cell of the X-ray examination is $\dfrac{c}{a} = 3\cdot418$, a value four times larger than that assigned in the earlier morphological examination. In this class (Chapter 6.3(e)) there can be no ambiguity in the choice of axial directions, so this discrepancy must correspond to an incorrect selection of parametral plane in the morphological study.

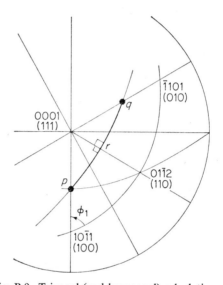

Fig. B.9. Trigonal (and hexagonal) calculation.

(iv) The poles $(10\bar{1}1)(01\bar{1}2)(\bar{1}101)$, etc. corresponding to the X-ray cell are inserted on the sketch stereogram at positions different from p, q and r, e.g. using the relation of (ii) the angle $(0001)\widehat{}(01\bar{1}2)$ for this cell is $63°\,08'$. r must now be indexed as $(0h\bar{h}2l)$, so that

$$\frac{\tan (0001)\widehat{}(0h\bar{h}2l)}{\tan (0001)\widehat{}(01\bar{1}2)} = \frac{h}{l} = \frac{1}{4};$$

the indices of r now become $(01\bar{1}8)$.

(v) Reversing the procedure of (i), the cleavage faces p and q would be indexed as $(10\bar{1}4)$ and $(\bar{1}104)$, so that the morphological description of the cleavage rhombohedron on Miller-Bravais axes corresponding to the X-ray cell should state that it is the form $\{10\bar{1}4\}$ with axial ratio $c/a = 3\cdot418$.

(vi) Indexing in terms of the smallest rhombohedral cell transforms $(10\bar{1}1) \rightarrow (100)$, $(01\bar{1}2) \rightarrow (110)$, $(0001) \rightarrow (111)$, etc. as described in Chapter 7.1(e). On this basis $(10\bar{1}4) \rightarrow (211)$.

(vii) From Chapter 5.4(f), the axial constant α, the interaxial angle of the cell edges is $180 - 2\phi_1$; ϕ_1 is calculated from $(111)(110)(100)$ as $\cos \phi_1 = \sin \widehat{1\bar{1}1} \cdot \cos \widehat{(1\bar{1}0)(111)}$ to give a value of $66° 59'$. The morphological description of the cleavage rhombohedron on Miller axes corresponding to the X-ray cell should state that it is the form $\{211\}$ with $\alpha = 46° 02'$.

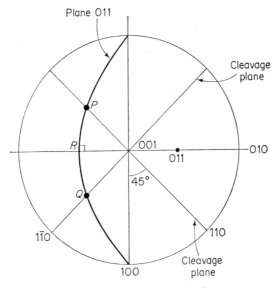

Fig. B.10. Tetragonal calculation.

(e) *Tetragonal system: A section of a crystal of class 4/mmm with a = 4·59, c = 2·96 (in Å) is cut parallel to the plane (011). When light is incident normally on this section determine the angles between the permitted vibration directions and the traces of {110} cleavages, a problem which can occur in practical optical work.*

From Chapter 11.4, this crystal is optically uniaxial with optic axis in the [001] direction; the indicatrix is an ellipsoid of revolution, and the permitted vibration directions are the principal axes of the central elliptical section perpendicular to the common wave normal direction. Fig. B.10 shows a sketch stereogram with the poles (001), (110), $(1\bar{1}0)$, etc.

(i) The great circle of which (011) is the pole denotes the orientation of the appropriate indicatrix section for light travelling along the normal to

(011) planes. One of the principal axes of this ellipse must be normal to the optic axis, i.e. at the pole (100); the other must be perpendicular, i.e. at the pole R in the same plane.

(ii) Great circles of which (110), $(1\bar{1}0)$, etc. are the poles represent the cleavage planes, and these will intersect the plane of the section in directions given by P and Q.

(iii) In the triangle (001)RP, $\sin (001)\widehat{\,}R = \tan \widehat{RP}.\cot \widehat{001}$; but

$$(00\widehat{1})R = 90° - (001)\widehat{(}011),$$

and from Chapter 5.4(d), $\tan (001)\widehat{(}011) = c/a$. With $(00\widehat{1})R = 57° 11'$ and $\widehat{001} = 45°$, $\widehat{RP} = 40° 03'$.

(iv) The permitted vibration directions for light incident normally on the section are the internal and external bisectors of the traces of {110} cleavages which are inclined at 80° 06′ in the plane of the section.

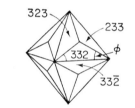

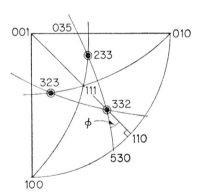

Fig. B.11. Cubic calculation.

(f) *Cubic system: A crystal of class m3m shows the form {332}. Determine the plane angles of one of its faces; these faces could then be assembled to build a cardboard model of this form.*

The form {332} in this class is a triakisoctahedron; a sketch of this shape and part of a sketch stereogram are shown in Fig. B.11. The plane faces are isosceles triangles whose edges are zone axes formed by the intersections of pairs of faces. The plane angles of a face are therefore the angles between the appropriate zone axes, i.e. the angles between corresponding zones at their points of intersections.

(i) One angle of a triangular face ϕ is defined by zones containing (332) $(33\bar{2})$ and (332)(233). The former will pass through the pole (110) and the latter through (530) and (035).

(ii) From Chapter 5.4(e), $\tan{(00\bar{1})\widehat{\ }(035)} = \frac{3}{5} = \tan{(100)\widehat{\ }(530)}$. In the triangle (035)(010)(530), $\sin{(010)\widehat{\ }(530)} = \tan{(010)\widehat{\ }(035)} \cdot \cot{\widehat{530}}$ to give $\widehat{530} = 62° 47'$.

(iii) In the triangle (110)(530)(332), $\cos{\widehat{332}} = \cos{(110)\widehat{\ }(530)} \cdot \sin{\widehat{530}}$, from which $\widehat{332} = \phi = 30° 24'$.

(iv) The plane angles of a triangular face of {332} in class $m3m$ are therefore 30° 24', 30° 24' and 119° 12'.

Further reading

Spherical trigonometry
DONNAY, J. D. H. 1945. *Spherical trigonometry*. Interscience.
MACROBERT, T. M. and ARTHUR, W. 1938. *Trigonometry: part IV, Spherical trigonometry*. Methuen.

Crystallographic calculations
DUCROS, P. and LAJZEROWICZ-BONNETEAU, J. 1967. *Problèmes de cristallographie*.
TERPSTRA, P. 1952. *A thousand and one questions on crystallographic problems*. Wolters, Groningen.

APPENDIX C

THE GROWTH OF CRYSTALS

C.1. Crystallisation

When conditions within a medium (solution, melt or vapour) at a given supersaturation, temperature, pressure, etc. are such that it is energetically favourable for the constituent atoms (or ions) to form permanent links in a regular and orderly manner, crystallisation becomes possible. In practice, the nucleation and growth of crystals is much more complicated than this might imply; conditions may change too rapidly for the constituent atoms to assume the regularity of the crystalline state and we may be left with a glass or an amorphous solid; or sometimes, even when a solid is formed slowly under conditions which are more or less constant, spontaneous nucleation does not occur and the medium crystallises only after seeding. Anyone with experience of elementary growth experiments will have encountered these or other difficulties, and extensive studies have revealed the sensitivity of the crystallisation process to a wide variety of factors (rate of cooling, concentration and temperature gradients, impurities, etc.); the crystallisation of a particular medium under various experimental conditions becomes a complicated study whose results all too often remind us of the unpredictability of many natural processes. Nevertheless, there are some remarks on nucleation and subsequent growth as they should occur in ideal controlled crystallisation that are relevant and they are set down briefly here.

The *formation of nuclei* requires atoms to form attachments after the manner of their arrangement in the appropriate crystal structure, at least over small volumes of the medium. In the medium, the atoms are in constant motion and these random movements must from time to time provide the essential aggregations of atoms held together by attractive forces developed by close proximity; in the flux of atomic movements such aggregates appear and disperse locally as random events. The size (and frequency) of such embryonic nuclei will be determined by conditions within the medium; for crystallisation to occur these must allow nuclei larger than a certain critical volume to be formed spontaneously. Just as drops of liquid evaporate unless they have achieved a critical radius, so the embryonic nuclei are broken up by thermal agitation unless aggregates of sufficiently large size are developed by random atomic movements. (It is by overcoming difficulties in the spontaneous formation of adequate nuclei that the seeding of supersaturated solutions with a small crystal is often successful in stimulating crystallisation.) Once suitable nuclei are present they will grow by assimilating more and more atoms from the medium, and by joining together. In this latter case of the conjunction of nuclei, energy considerations favour their orientation into a single structural continuum, though sometimes it is thought that this mechanism might be responsible for a slightly less favourable arrangement which develops into the different sub-components of a twinned crystal (see growth twins in Appendix D.2). The frequency of nuclei of suitable size will also depend on crystallisation conditions; if the number per unit volume is large, the crystalline texture will be that of a fine grained aggregate; but when spontaneous

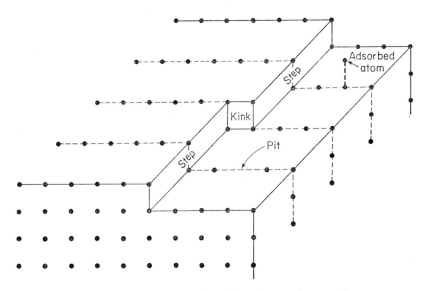

Fig. C.1. Irregularities on the surface of a growing crystal.

nucleation is rare (or crystallisation requires the introduction of a seed), the texture is coarse and large single crystals have an opportunity to grow.

Another important factor in any crystallisation sequence is the *growth rate*, and this, too, must be related to conditions within the medium. In the assimilation of material on to a growing crystallite, it is reasonable to assume that any growth surface will be irregular on an atomic scale; there will be pits due to atomic vacancies and projections in the form of incomplete atomic layers (Fig. C.1). In the process of adding the atoms required for growth, some will form permanent attachments immediately or shortly after collision with the crystallite, but some will merely be adsorbed temporarily to return to the medium after an interval on the growth surface; in growth there must be a net flow from the medium to the crystal and this will determine the growth rate. Clearly some sites on an irregular growth surface are more favourable for permanent attachment than others; as illustrated in the figure, these are in the pits, or adjacent to the kinks within steps or edges of steps where atoms will be more securely bound. A few atoms will be deposited directly in such sites, but the majority will be adsorbed elsewhere on the surface and must migrate to such positions within the short period which elapses before they are dislodged and returned to the surrounding medium by thermal agitation. Under such circumstances, the density of inhomogeneities on a growth surface is an important factor in determining growth rates; but we must realise that such a mechanism tends to eliminate the pits, steps and kinks as extra atoms are added to the growing crystal. The step in Fig. C.1 will be removed as the extra layer of atoms advances across the surface, and for growth to continue at the same rate similar irregularities must be created on the new surface. Vacancies will be generated as occasional atoms are lost to the medium by thermal agitation, but steps and kinks can only be formed by clusters of adsorbed atoms produced by their random motions; such *surface nucleation* can maintain the the irregularities which are an essential part of the growth process. The problem presented is similar to that of the original spontaneous nucleation in that only if an aggregation of a critical minimum radius is possible can it remain on the surface to provide further growth points for the crystal; if conditions in the medium do not allow enough surface nuclei of sufficient size to develop before the adsorbed atoms are removed by thermal agitation, the crystal cannot continue to grow at the original rate.

For various conditions of crystallisation, statistical treatments can be used to predict growth rates; densities of surface inhomogeneities can be calculated, the probability of the formation of suitable surface nuclei can be assessed, and eventually theoretical growth rates can be derived to compare with those measured in carefully controlled experiments on growing crystals. In general, such comparisons are fairly satisfactory at medium and high supersaturations, and it is justifiable to assume that under such conditions the processes that we have outlined represent a reasonable approximation to a possible growth mechanism. But at lower supersaturations, predicted and observed growth rates are very different; in some cases the divergence is so great that a crystal which should not perceptibly change in size over millions of years is seen to be growing by a human observer. At these supersaturations, it is the assumption of surface nucleation which causes the discrepancies; the probability of random aggregations of a critical size is so low that this mechanism could not possibly produce the density of inhomogeneities needed to maintain a detectable growth rate. However, the stage of surface nucleation could be omitted if the crystal contained a suitable form of inhomogeneity which was not eliminated by the addition of atoms during growth; this could be provided by one of the imperfections described in Chapter 12.3, the screw dislocation. When the Burgers vector is parallel to the dislocation line, a step, whose height depends on the magnitude of this vector, is formed on the surface of the crystal. We can regard the crystal as a single spiral ramp of atoms, so that, although atoms added to the side of this surface step will cause the crystal to

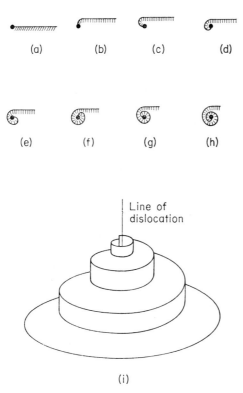

Fig. C.2. Formation of a growth spiral. (a)–(h) Show successive stages in the advance of the surface step due to a screw dislocation normal to the paper; the surface on the unshaded side of the line is lower than that on the shaded side; the step changes position as atoms added to the unshaded side but is never eliminated. (i) Shows the appearance of the surface after growth in this manner.

grow, the step will not disappear. The result of growth from such an imperfection will be to form spiral patterns on the surface of the crystal, as shown in Fig. C.2; a simple pattern will produce a surface which is like the slide of a helter-skelter tower. In practice, the surface patterns may be more complicated with spirals and closed loops of various kinds due to interactions of growth at various points of the network of line imperfections contained in any crystal. Initially the role of dislocations in the growth process was hotly disputed, and definitive evidence in the form of residual growth spirals and other patterns was hard to obtain, for the height of steps is usually small (of the order of a few lattice translations for $\vec{b}$). Eventually direct observations of simple growth spirals on paraffins by electron microscopy were decisive, and since that time growth patterns have been found regularly on the surfaces of an extensive range of crystalline materials; the steps are often detected by interferometry and occasionally are large enough to be seen with an ordinary optical microscope. The importance of crystalline imperfections in the growth process, at least at lower supersaturations, is now universally accepted.

C.2. The external shapes of crystals

In crystallisation, the opportunity for individual crystals to develop characteristic external shapes depends on the processes of nucleation and growth that we have just described. Rapid crystallisation from a large number of nuclei is unlikely to allow the formation of any recognisable shapes in the crystallites, and the texture of the fine-grained aggregate will resemble an interlocking mosaic of irregular fragments. But limited nucleation and slow growth will permit the formation of single crystals of appreciable size; under these conditions there is a wealth of different morphological forms and habits to be observed, often within the same point group and even among crystals of the same material. Naturally, the particular conditions of crystallisation must be expected to play some part in determining external shape, but even in undisturbed and controlled growth there are wide variations in form and habit between crystals with the same point group symmetry but different atomic structures. From this it seems probable that features of the atomic structure must control growth forms to some extent, and in this section we shall consider their role in determining crystal shape.

In a general way, morphological data support the view that the assembly of constituent atoms into a crystal structure takes place in the most economic and efficient manner; if the structure contains dominating extensible complexes, it is easier to continue to add atoms to one of these than to start to build another. For example, when a crystal structure is dominated by long, continuous, tightly bound, parallel chains of atoms, it is simpler for extra atoms to continue existing chains than to start new ones; the growth rate will be faster along a chain than normal to it, and any crystal develops a fibrous elongated habit. By similar arguments we might expect crystals with well-developed layered structures to have platy, tabular habits in their growth forms. But apart from these reasonable but rather vague generalisations, it should be possible to offer some explanation of the predominance of specific growth forms for different structures, even if the atomic arrangements do not contain such obvious dominating features. Such problems exercised earlier crystallographers, particularly in France, who sought first to establish the precedence of forms shown by crystals of a particular substance and then to provide some kind of structural basis for their observations. For example, they might have found that the order of precedence (roughly the frequency of occurrence) might be cube, rhombic dodecahedron, octahedron for one material of class $m3m$ but octahedron, cube, rhombic dodecahedron for another; they then tried to relate such a change in precedence to facets of the structural patterns of the two materials.

In treatments of this problem much work was carried out by Friedel by applying the *Law of Bravais*, which asserts that crystalline forms which tend to occur must frequently are those whose faces are parallel to lattice planes containing the greatest density of points; the general validity of this statement is borne out by the low values of plane indices for most natural crystalline faces which are therefore parallel to lattice planes with high concentrations of points; as atoms are added, such planes show the largest increases in surface area and will eliminate any other planes with higher indices which attempt to develop. An illustration of this approach is provided by Fig. C.3, in which the distributions of lattice points on (100), (110) and (111) planes for the *P*, *F* and *I* cubic lattice types are drawn; the smallest area is marked in each case, and this can act as a measure of the density of lattice points. From this diagram we see that the relative dominance of the three shapes derived from these planes depends on the lattice type so that:

P-lattice has cube, rhombic dodecahedron, octahedron
F-lattice has octahedron, cube, rhombic dodecahedron
I-lattice has rhombic dodecahedron, cube, octahedron.*

Experimentally there does seem to be some correlation between lattice types and distributions of these forms; the dominant form for $NaClO_3$ (*P*-lattice) is the cube, for diamond (*F*-lattice) the octahedron, for garnets (*I*-lattice) the rhombic dodecahedron, and many other examples can be found to support the predictions of this illustration, and those for the different cells and lattices in other systems to which similar arguments can be applied. Promising though this is, there are many anomalies even among the simplest materials, as when NaCl and KCl (both with *F*-lattices) commonly crystallise as cubes. Clearly, a treatment which seeks to differentiate various atomic structures solely on grounds of their different lattice types and cell shapes cannot be more than the first step in a theory of growth forms.

Later work by Donnay and Harker takes into account the presence of associated symmetry elements (particularly the glide planes and screw axes of space groups). These authors have worked out lists of forms in expected order of precedence for all space groups, and have shown that these account for many of the anomalies of the earlier treatment; for the various space

* Actually on these arguments two other forms the icositetrahedron {211} and the tetrahexahedron {310} have precedence over the octahedron for an *I*-lattice.

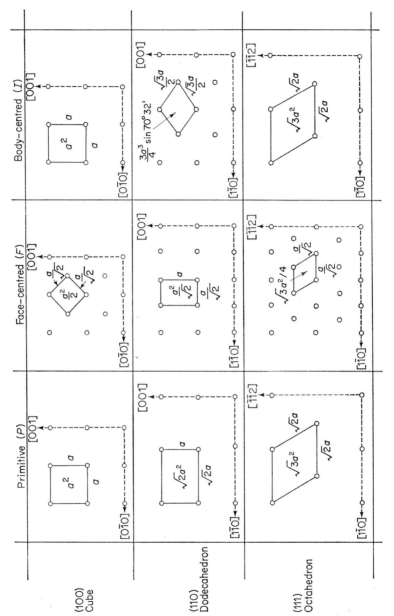

Fig. C.3. Distributions of lattice points on planes parallel to faces of the cube, rhombic dodecahedron and octahedron for the three cubic lattice types.

groups of the cubic system, for example, the dominant form can be {100} or {110} or {111} for P and F lattice types and {100} or {110} or {211} for I lattice types, depending on the particular symmetry elements of the group. They even suggest that, for a particular material, a list of forms in an observed order of precedence can be used to resolve any ambiguities in space group determination which remain after X-ray examination and other physical tests. Despite the successes of these refinements, there are still, unfortunately, many examples which confound what is essentially a geometrical approach. Undoubtedly some of these discrepancies may be discounted as due to the many and various extraneous factors that are known to be important in experimental crystallisation; but at best geometrical considerations can only represent an aspect of the atom-by-atom assemblage of a crystalline solid. Ultimately, explanations must be sought in an assessment of the detailed energy changes in nucleation and growth; this requires calculations based on the positions of atoms, the attractive forces between them, etc., at present a formidable undertaking for even the simplest structure. Until this is possible, we must accept that the less sophisticated geometrical arguments that have been illustrated provide the best available framework for an understanding of the influence of structural factors on external shapes of crystals; we should not be surprised, however, by the unexpected forms and habits shown by natural and synthetic crystals; and the important technical field of the controlled growth of crystalline materials in varying shapes and sizes must remain something of an art.

Further reading

BUCKLEY, H. E. 1951. *Crystal growth.* Chapman and Hall.

STRICKLAND-CONSTABLE, R. F. 1968. *Kinetics and mechanism of crystallisation.* Academic Press.

VAN HOOK, A. 1961. *Crystallisation.* Chapman and Hall.

VERMA, A. R. 1953. *Crystal growth and dislocations.* Butterworth.

APPENDIX D

TWINNING IN CRYSTALS

D.1. The nature of twinning and its geometry

The aggregates of separate crystals which form most polycrystalline solids can be randomly oriented or they can have some degree of preferred orientation; in either case, the angular orientations between a grain and its neighbours are variable from one crystal to the next. There are, however, some

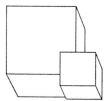

Fig. D.1. Cubes in parallel growth.

composite crystalline solids which have sub-components related to one another in a fixed orientation according to the geometry of their crystallographic lattices. Such solids may be large and recognisably composite, as when the sub-components display morphological shapes, or they may be small, as when sub-components are contained within only a single irregular grain of a polycrystalline aggregate; whatever their dimensions and shape, the structure within each sub-individual is the same, but they are distinguished by boundaries across which there is a specific orientation change. Composite crystals of all kinds were encountered in the earliest morphological studies, which showed that, apart from simple random aggregation, there are often types of parallel growth in which all recognisable sub-components are strictly parallel to one another; Fig. D.1 depicts two cubes in parallel orientation (there are often many more) but there must be structural continuity throughout the entire solid. More significantly, however, composite crystals with sub-individuals in different but fixed angular positions were found so regularly for the same material that they could not be dismissed as accidents of crystallisation; the two cubes of Fig. D.2 are intergrown in different orientations, but the angular dispositions of the two sets of faces are always the same whenever such an intergrowth is found. Composite crystals of this kind are described as *twinned* and their shape and angular relationships were specially studied by the morphologist.

As a result of this work, the nature of the *twin laws* which describe the relations between the orientations of the lattices of sub-components was discovered; in simple geometrical terms it was found that every twin law could be formulated either:

(a) in such a way that the orientation of one sub-individual to the next is described by rotation about the normal to a lattice plane (*hkl*) through $360°/n$. In most common *rotation twins* $n = 2$, and, although the twin axis can be a lattice row in more symmetrical systems, it cannot be assigned a simple symbol [*UVW*] in every case,

or (b) in such a way that the orientation of one sub-individual to the next is described by mirror reflection across a lattice plane (*hkl*); such combinations are known as *reflection twins*.

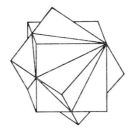

Fig. D.2. A cube twinned by rotation about a triad axis.

From this it is clear that the twin axes which relate sub-components by rotation cannot be symmetry axes of even degree and that the twin planes of a reflection relationship cannot be mirror-symmetry planes, for these would always produce sub-individuals in parallel orientations. Moreover, the distinction between reflection and rotation twins is significant only in acentric crystal classes. Fig. D.3(a) shows schematically the development of a composite crystal by both types of operation in a centric class; an identical twinned crystal is produced by reflection in a plane and rotation (through 180°) about the normal to this plane so that the composite shape could be described as a crystal of class *mmm* (with certain axial ratios) showing the forms {001} and {*hk*0} twinned by reflection in (*hk*0) or twinned by rotation about the normal to (*hk*0). In Fig. D.3(b), however, the centre of symmetry has been destroyed by hatching the ($\bar{h}k$0) face; the arrangement of hatched faces is now different in the composite shapes produced by the two types of twinning operation, and we must recognise this distinction in any description.

During these morphological studies, twinning acquired an extensive descriptive terminology; apart from names specific to twins of particular substances (after a locality, viz. a Brazil twin of quartz or after the shape of the composite crystal, viz. a butterfly twin of calcite), there are some more general terms in common use that merit some explanation. Twins are *simple* when only two sub-individuals are present, and *multiple* when there are more than two orientations; they are said to be *interpenetrant* (or *penetration*) twins

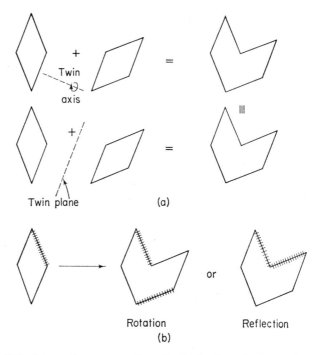

Fig. D.3. Schematic representation of twinning by reflection and rotation. (a) Equivalence in a centric crystal. (b) Non-equivalence in an acentric crystal.

when sub-components are intimately embedded in one another (as in Fig. D.2) or *contact* twins when sub-components are joined along a definite plane (as in Fig. D.3); in this latter case the *composition* (or *junction*) plane may, or may not, be the same as the twin plane and will need description. A crystal is described as having *lamellar* twinning when it is divided into thin plates of sub-components formed by multiple twinning on parallel twin planes: or as having *mimetic* twinning when it is multiply twinned so that, considered as a homogeneous solid, the arrangement of faces is apparently indicative of

higher symmetry than the material actually possesses. Rotation twins are sub-
divided into: (i) normal, in which the twin axis is perpendicular to the composition plane; (ii) parallel, in which the twin axis is a possible zone direction
in the composition plane (which is not necessarily a possible crystallographic

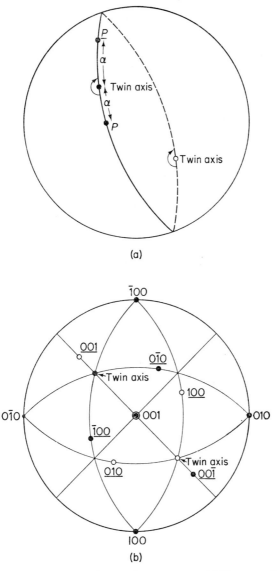

(a)

(b)

Fig. D.4. Twinning by rotation on the stereogram. (a) The general movement of a pole. (b) The stereogram of a cube twinned by rotation about
[$\bar{1}\bar{1}1$].

plane); and (iii) complex, in which the twin axis is a direction in the composition plane normal to a possible zone axis for the crystal.

The angular relationships between lattice directions in sub-components related by a particular twin law are readily displayed on a stereogram. Rota-

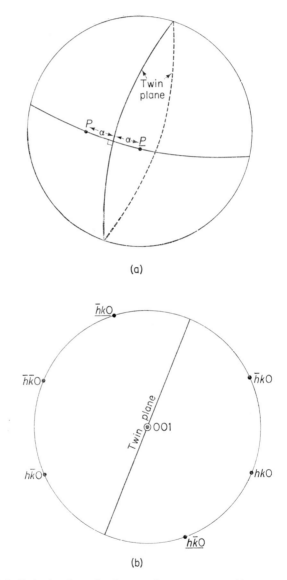

(a)

(b)

Fig. D.5. Twinning by reflection on the stereogram. (a) The general movement of a pole. (b) The stereogram of an orthorhombic crystal showing the forms {001} and {hk0} twinned by reflection in (hk0).

tion twinning requires twinned and untwinned poles to be related by the stereographic operation of rotation axes, which has already been described in Chapter 3.3. Unless otherwise stated, the rotation may be taken as 180°, and this causes all great circles through the poles of the twin axis to be brought into coincidence with themselves in a reversed position; in other words, as illustrated in Fig. D.4(a), an untwinned pole P is repeated in a twinned position $\underline{P}$ by transporting it along the great circle to the pole of the twin axis and on to a point which is angularly equidistant on the other side; Fig. D.4(b) plots the poles for the interpenetrant twin shown in Fig. D.2 developed when the cubic {100} form is twinned about [$\bar{1}\bar{1}1$]. Reflection twinning relates twinned and untwinned poles by the operation of a mirror symmetry plane; in other words, as illustrated in Fig. D.5(a), the untwinned pole P is carried along a great circle, which intersects that of the twin plane at right angles, to a symmetrical twinned position $\underline{P}$ on the other side of the point of intersection; Fig. D.5(b) plots the poles for the contact twins shown in Fig. D.3 produced when orthorhombic forms {001} and {hk0} are twinned by reflection in (hk0). Whatever twin law is operative, it is important to realise that the crystallographic directions represented by twinned poles must be indexed in terms of the twinned orientations of lattice translations defining the cell; their indices must, therefore, be the same as those of the corresponding untwinned pole, but they are distinguished from these by underlining; thus in Fig. D.4(b), ($\underline{100}$) must be interpreted as the position of the (100) pole when twinned by rotation about an axis [$\bar{1}\bar{1}1$].

D.2. The genesis of twins

We have introduced the concept of twinning and its geometry from morphological studies in which certain crystals, by some quirk of the growth process, develop as regular composite solids; such crystals are at all times visibly twinned (i.e. there is no detectable stage in growth at which the solid is a homogeneous single crystal) and are often called *primary* (or *growth*) twins. There are, however, other origins of twinned crystals in which sub-components are formed in a homogeneous single crystal after growth; such crystals are often known as *secondary* twins. In this section we discuss probable atomic mechanisms responsible for both primary and secondary twins.

Growth twinning is very widespread and is found in some form in about a quarter of crystalline substances; its incidence is variable, and some materials rarely, if ever, develop single crystals of any appreciable size, whilst some are never twinned; observed twin laws are many and diverse as are the forms and arrangements of the sub-individuals that are produced. Undoubtedly, particular conditions of crystallisation must play some part in the nucleation of growth twins, but the ability to grow twinned associations must also be related to the internal atomic structure and its geometry. Early in the study of twins it was found that twin axes and planes were often the directions of pseudo-elements of symmetry in a particular material; this suggests that twinning is most likely when the underlying structure has a pseudo-cell which approximates in symmetry and shape to a higher crystal class. When such a structure is twinned, this pseudo-cell is maintained throughout all the sub-individuals, which are separated by boundary regions with atomic arrangements in conformity with the structural orientations adopted by adjacent

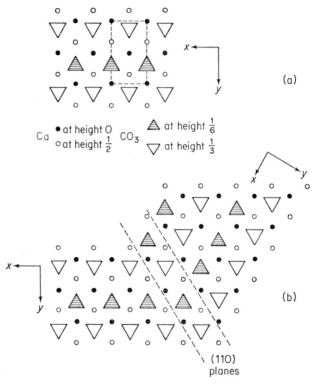

Fig. D.6. Growth twinning in aragonite ($CaCO_3$). (a) The structure projected on (001); only the contents of the lower half of the cell are shown. (b) The structure of an aragonite crystal twinned by reflection in (110).

sub-components. It is possible to provide structural explanations for most twin laws in this way, and twin formation is then taken to be an accident in the growth process when some atoms in the boundary region are attached in incorrect sites. These ideas are best demonstrated by example, and one of the most suitable is provided by aragonite, a variety of $CaCO_3$. Aragonite is orthorhombic but pseudo-hexagonal ($a = 4.96$, $b = 7.97$ (in Å), i.e. $a \sim \sqrt{3}b$), and shows twins on $\{110\}$ which may be repeated to give mimetic twins in crystals of approximately hexagonal cross-section. Fig. D.6(a) shows an idealised representation of the aragonite structure projected on to (001); for clarity, only the contents of the lower half of the cell are included (in the upper half of the cell the planar CO_3 groups are rotated through 180° and are at heights of $\frac{2}{3}$ (above the shaded triangles) and $\frac{5}{6}$ (above the unshaded triangles)). The hexagonal pseudo-symmetry is obvious in the arrangement of Ca ions, and a truly orthorhombic cell is defined only by the different heights and orientations of CO_3 groups. When twinning takes place by reflection on (110), the structural pattern of Fig. D.6(b) will be formed; a hexagonal pseudo-cell in the arrangement of Ca ions is continuous through the composite crystal and the transition layer (between the dashed lines) may be thought of as part of the structure of either of the adjacent sub-components. As an accident in growth to initiate twin formation, we may imagine that as the lower crystal grows from left to right, the atoms of a CO_3 group continuing one of the rows parallel to the x-axis are added in the wrong orientation and at the wrong height, i.e. as a shaded triangle instead of an unshaded one or vice versa. This mistake will raise the energy of the crystal above the minimum which it has when the CO_3 group is correctly oriented to extend the original single crystal, and the magnitude of this energy change determines the probability of such a mistake; when it is too large, even if incorrect siting of the atoms occurred there would be little chance of permanent attachment, and the group would evaporate from the growing surface to be re-deposited in a much more energetically favourable correct position. Under the conditions of structural conformity in boundary regions, as in aragonite, the energy increase due to a mistake cannot be very large; for with the continuity of first nearest neighbours in both sub-components, any energy difference between correct and incorrect locations is determined by the effects of more distant neighbours, which must necessarily be quite small. Once the growth mistake has been made, it will play its part in influencing the orientation of further atoms (in CO_3 groups) as they are added, and the second sub-component will begin to grow in the twinned orientation.

Moving from growth twins to secondary twins, these may be divided into two categories according to their origins; but we should perhaps emphasise first that all twins (primary or secondary) have the same kinds of geometrical relationships (twin laws, etc.) and it is only in their modes of formation that they differ. *Transformation twins* are secondary twins which can develop in certain types of transition between alternative structural arrangements of the same constituent atoms; polymorphism (or the occurrence of alternative crystal structures with the same chemical constitution) is very common in many crystalline solids. In one kind of structural transformation, atoms have the same first nearest neighbours in both polymorphs which differ only in the arrangement of second nearest and more distant neighbours; sometimes the transition does not involve the breakage of any permanent bonds; for the structures change from one to the other by simple geometrical movements in the linked framework of atoms (often called a *displacive change*). Fig. D.7 depicts a model for such polymorphic change; a shaded square represents an atom and its first nearest neighbours (such as the co-ordination polyhedron of anions surrounding a cation). A segment of relatively open and symmetrical structure is represented by B, and A_1 and A_2 represent similar segments of less symmetric collapsed structures based on a similar linkage of shaded squares; we notice that transitions between all three structures can be effected without breaking any links within or between the square structural units, and that the energies of A_1 and A_2 are identical for they are mirror images of one another. The more open symmetrical structure of polymorph B is likely to be stable at

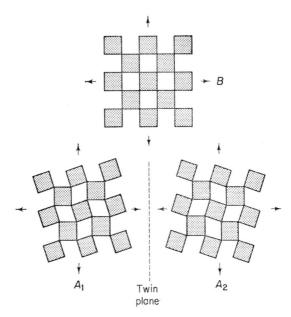

Fig. D.7. A model for transformation twinning.

more elevated temperatures when the atom sites linking the units together at the corners of squares will represent average positions of atoms which have considerable thermal motion. In fact, these mean positions are midway between the corresponding atomic sites for the alternative polymorphs A_1 and A_2, and an interpretation of structure B is that it is the pattern of sites that we shall observe when the thermal motion is sufficient to provide atoms with the energy necessary to oscillate between the positions of A_1 and A_2. As the temperature is decreased, the amplitude of atomic thermal vibrations will become smaller, until the motion of the atoms is insufficient to pass through the neutral point between A_1 and A_2; the less symmetrical collapsed structure will be developed and we shall recognise that our material has undergone a polymorphic transformation. But as this happens some regions of the homogeneous crystal of B will nucleate and grow the low temperature structure in orientation A_1 and others will produce a similar structure in orientation A_2; the crystal will become composite with differently oriented structural regions related by mirror symmetry, i.e. we have formed a twinned crystal. Two of the many polymorphic modifications of SiO_2, high quartz (hexagonal) and low quartz (trigonal), both with structures based on linked frameworks of Si-O tetrahedra, undergo such a transition at 573°C; transformation twinning by this mechanism is considered to be the origin of sub-components twinned according to the Dauphiné law which are found in many low quartz crystals and which may be removed permanently to give homogeneous single crystals of low quartz after suitable controlled heat treatments.

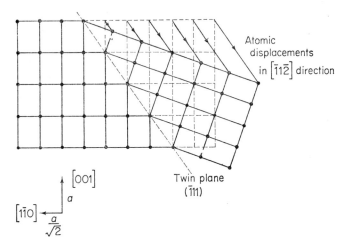

Fig. D.8. Atomic movements on a (110) plane of a cubic metal with an F-lattice to produce glide twinning.

Secondary twinning can also occur as part of the process of plastic deformation of single crystals; it is particularly important in metals for which the atomic mechanism has been widely studied but is also found in many other inorganic materials and minerals (such as calcite $CaCO_3$, sphalerite ZnS, etc.). The twinned orientations are produced by atomic movements comparable to those of slip (Chapter 12.3) though they differ in the form of movement and do not necessarily have the same glide planes or directions for a given structure. In slip there is uniform displacement of an entire block of structure across the glide plane, whereas in *glide twinning* the twinned orientation is produced by atomic displacements which are proportional to their distance from the twin plane. Fig. D.8 shows the atomic movements within a (110) plane of a face-centred cubic metal which produce sub-components twinned by reflection in a {111} plane; in this twinning the glide planes are {111} and the glide direction is ⟨112⟩, but, as the diagram shows, the atomic displacements of planes are those of a uniform shear strain with movements related to distances from the twin plane. Notice, too, that the glide movements must be made with the correct sense, for a simple reversal will not produce the twinned orientation; in slip the direction of movements can be in either sense. Stresses necessary to initiate twinning movements tend to be larger than those required for slip, and twins are often formed during deformation so as to reorient a structure into a position where slip on an alternative system is favourable. The general similarity of the atomic movements of deformation twinning to those of slip suggests that they may also be caused by dislocation motion (Chapter 12.3); any dislocations involved, however, would have to have Burgers vectors which are not units of the atomic repeat and would have to be active on all the successive glide planes on one side of the boundary between sub-components formed by the twin plane. Systems of dislocations which can account for the necessary atomic movements in relatively simple metallic structures have been proposed, but direct confirmation of such mechanisms has not yet always been obtained. In more complex inorganic structures the atomic movements in deformation twins are usually capable of a rational structural explanation, but any dislocation systems involved in their propagation are less well understood.

Further reading

HALL, E. O. 1954. *Twinning*. Butterworth.
KELLY, A. and GROVES, G. W. 1970. *Crystallography and crystal defects*. Longmans.
KLASSEN-NEKLYUDOVA, M. V. 1964. *Mechanical twinning of crystals*. Translated by Consultants Bureau, New York.
PHILLIPS, F. C. 1963. *An introduction to crystallography*. Longmans.

ANSWERS TO EXERCISES

Chapter 2.6
4. pmg; m, $\pm(\frac{1}{4}, y)$: 2, $(0, \frac{1}{2})$; $(\frac{1}{2}, \frac{1}{2})$: 2, $(0, 0)$; $(\frac{1}{2}, 0)$.

Chapter 3.5
2. (iv) Angle between great circles $\sim 89\frac{1}{2}°$; (vi) Angle between great circles $\sim 58°$.
3. (a) 4-fold axis, m-planes and a centre; (b) 4-fold axis, two 2-fold axes; (c) 4-fold axis, two m-planes.
   New group contains 4-fold axis, four 2-fold axes, five m-planes and a centre.
4. (i) 2770 miles range; (ii) 190 days; 73° W. of S.; (iii) (a) yes, (b) no, (c) yes, (d) yes.

Chapter 4.5
3. mmm, $6/mmm$, $m3m$; they are all centric.
4. 3 (the triad), 32 (the diads); 3 (the triad).

Chapter 5.5
1. (ii) 126°; (iii) 13·6 mm; 8·2 mm; (iv) 25°; (v) $a' = 6\cdot95$ cm, $b' = 3\cdot00$ cm, $\gamma' = 126°$;
   (vi) $(\bar{1}2)$; $(\bar{8}1)$.
2. (i) a is $[50\bar{3}]$; b is $[\bar{3}11]$; (ii) $(\bar{3}35)$ in (a), the rest in (b); (iii) (345).
3. $(112)_n$; $[011]_n$.
4. (i) $(100)\widehat{(001)} = 77°$; $(010)\widehat{(001)} = 86°$; $(100)\widehat{(010)} = 74°$; (ii) $(100)\widehat{(110)} = 34°$;
   $(100)\widehat{(1\bar{1}0)} = 51°$; (iii) $(110)\widehat{[110]} = 11°$.
5. $0\cdot66 : 1 : 0\cdot55$; $\beta = 116°$.
6. (315); $25\frac{1}{2}°$.
7. $(4\bar{1}1)$; (310), (532); $c/a = 2\cdot228$.
8. 26° 31′; $h = 3, l = 1$.
9. 66° 48′; (i) $[\bar{1}\bar{2}.1]$, $[1\bar{1}.\bar{1}]$, $[2\bar{1}.\bar{1}]$; (ii) $[01\bar{1}1]$, $[1\bar{1}0\bar{1}]$, $[\bar{1}01\bar{1}]$.

Chapter 6.6
1. (i) 3; $\{h0l\}$, $\{010\}$ special, $\{hkl\}$ general; (ii) 7; $\{100\}$, $\{010\}$, $\{001\}$, $\{hk0\}$, $\{0kl\}$, $\{h0l\}$ special, $\{hkl\}$ general; (iii) 10; $\{0001\}$, $\{000\bar{1}\}$, $\{10\bar{1}0\}$, $\{01\bar{1}0\}$, $\{11\bar{2}0\}$, $\{hki0\}$, $\{h0\bar{h}l\}$, $\{0k\bar{k}l\}$, $\{hh\bar{2}\bar{h}l\}$ special, $\{hkil\}$ general; (iv) 8; $\{001\}$, $\{100\}$, $\{110\}$, $\{hk0\}$, $\{h0l\}$, $\{hhl\}$, $\{h\bar{h}l\}$ special, $\{hkl\}$ general; (v) 3; $\{0001\}$, $\{hki0\}$ special, $\{hkil\}$ general; (vi) 8; $\{100\}$, $\{110\}$, $\{hk0\}$, $\{kh0\}$, $\{111\}$, $\{hll\}$, $\{hhl\}$ special; $\{hkl\}$ general.
2. (i) $\bar{3}$, 32 and $\bar{3}m$; (ii) $4/m$, 422, $\bar{4}2m$ and $4/mmm$; (iii) $\bar{6}$; (iv) 432, $\frac{2}{m}$ 3, $m3m$; $\frac{2}{m}$ 3.
3. (ii) [010] zone: $(001)\widehat{(\bar{1}01)} = 50°\ 06′$; $(101)\widehat{(00\bar{1})} = 129°\ 54′$; $(00\bar{1})\widehat{(\bar{1}0\bar{1})} = 50°\ 06′$;
   $(10\bar{1})\widehat{(001)} = 129°\ 54′$; [001] zone: $(010)\widehat{(\bar{1}10)} = 59°\ 22′$; $(\bar{1}10)\widehat{(\bar{1}\bar{1}0)} = 61°\ 16′$;
   $(\bar{1}\bar{1}0)\widehat{(0\bar{1}0)} = 59°\ 22′$; $(0\bar{1}0)\widehat{(1\bar{1}0)} = 59°\ 22′$; $(1\bar{1}0)\widehat{(110)} = 61°\ 16′$; $(110)\widehat{(0\bar{1}0)} = 59°\ 22′$; (iv) 24° 05′.
4. (i) $4/mmm$; (ii) $c/a = 0\cdot537$ with forms $\{100\}$, $\{110\}$, $\{111\}$ and $\{311\}$.
5. $0\cdot734$; 143° 42′.
6. $6/m$, 62 or $6/mmm$; $\{21\bar{3}1\}$, another hexagonal bipyramid, class $6/m$.
8. $\{110\}$, $\{014\}$; $a_X = b_M$, $b_X = 2c_M$, $c_X = a_M$; $\{101\}$, $\{180\}$.

Chapter 7.5

1. Cubic F-lattice for Au, tetragonal P-lattices for SnO_2 and $CO(NH_2)_2$; (a) S_{110}, S_{111}, S_{112} are 2·89, 7·06, 5·00 (Å); d_{110}, d_{111}, d_{222} are 2·89, 2·36, 1·18 (Å) respectively; (b) S_{110}, S_{111}, S_{112} are 6·71, 7·43, 9·26 (Å); d_{110}, d_{111}, d_{222} are 3·35, 2·31, 1·15(5) (Å) respectively.
2. Cd: (0, 0, 0) Cl: $\pm (u, u, u)$.
3. Screw diads non-intersecting in $P2_12_12_1$; $I222$.
4. $C2/m$; (0, 0, 0; $\frac{1}{2}$, $\frac{1}{2}$, 0) + : $\pm(x, y, z)$; $\pm(x, \bar{y}, z)$.
5. $Ccca$; for G.E.P's and S.E.P's see Vol. 1, *International tables for X-ray crystallography*, p. 157; $Ccca$ (*abc*); $Abaa$ (*cab*); $Bbcb$ (*bca*); $Bbab$ (*acb*); $Cccb$ (*bac̄*); $Acaa$ (*c̄ba*).
6. $R\bar{3}c$: 12 G.E.P's (on rhombohedral axes); 2 $\bar{3}$ 0, 0, 0; $\frac{1}{2}$, $\frac{1}{2}$, $\frac{1}{2}$;
   $\qquad\qquad\qquad\qquad\qquad$ 2 32 $\frac{1}{4}$, $\frac{1}{4}$, $\frac{1}{4}$; $\frac{3}{4}$, $\frac{3}{4}$, $\frac{3}{4}$.

   $P6/mcc$: 24 G.E.P's; 2 $6/m$ 0, 0, 0;   0, 0, $\frac{1}{2}$;
   $\qquad\qquad\qquad\qquad$ 2 62 0, 0, $\frac{1}{4}$;   0, 0, $\frac{3}{4}$.
7. 16 G.E.P's; 2 $4mm$ $\frac{1}{4}$, $\frac{1}{4}$, z; $\frac{3}{4}$, $\frac{3}{4}$, $\bar{z}$;
   $\qquad\qquad$ 2 $\bar{4}2m$ $\frac{3}{4}$, $\frac{1}{4}$, $\frac{1}{2}$; $\frac{1}{4}$, $\frac{3}{4}$, $\frac{1}{2}$;
   $\qquad\qquad$ 2 $\bar{4}2m$ $\frac{3}{4}$, $\frac{1}{4}$, 0; $\frac{1}{4}$, $\frac{3}{4}$, 0.
8. $Pmcn$; Ca: 4 S.E.P's on m; C: 4 S.E.P's on m; O: 4 S.E.P's on m, 8 G.E.P's; $Pnma$.

Chapter 8.6

1. (i) 25 KV; $\frac{1}{2}$ Å; (ii) 0·0019 cm; 43%; (iii) To avoid fluorescence, wavelengths greater than 1·6 Å must be used; (iv) 0·10 cm.
2. (iii) 60°.
3. (101) planes, $\theta_{101} = 15\frac{1}{2}°$; (202) planes, $\theta_{202} = 32\cdot1°$; (303) planes, $\theta_{303} = 52\cdot9°$.
4. (a) $A = 2\cos 2\pi (hx + ky) \cos 2\pi lz$; $B = 2 \sin 2\pi (hx + ky) \cos 2\pi lz$. No conditions.

   (b) $A = 8 \cos 2\pi\left(hx - \dfrac{k-l}{4}\right) \cos 2\pi\left(ky - \dfrac{l-h}{4}\right) \cos 2\pi\left(lz - \dfrac{h-k}{4}\right)$; $B = 0$

   $0kl$ reflections observed only if $k + l = 2n$
   $h0l$,,   ,,   ,, ,, $h + l = 2n$
   $hk0$,,   ,,   ,, ,, $h + k = 2n$
   $h00$,,   ,,   ,, ,, $h = 2n$
   $0k0$,,   ,,   ,, ,, $k = 2n$
   $00l$,,   ,,   ,, ,, $l = 2n$

   (c) $A = 16 \cos^2 2\pi \dfrac{h+k}{4} \cos 2\pi\left(hx - \dfrac{k+l}{4}\right) \cos 2\pi\left(ky + \dfrac{l}{4}\right) \cos 2\pi\left(lz + \dfrac{k}{4}\right)$; $B = 0$

   hkl reflections observed only if $h + k = 2n$
   $0kl$,,   ,,   ,, ,, $l = 2n, k = 2n$
   $h0l$,,   ,,   ,, ,, $l = 2n, h = 2n$
   $hk0$,,   ,,   ,, ,, $h = 2n, k = 2n$
   $h00$,,   ,,   ,, ,, $h = 2n$
   $0k0$,,   ,,   ,, ,, $k = 2n$
   $00l$,,   ,,   ,, ,, $l = 2n$
5. (i) 100, 200, 300, 110, 330 are absent.
   (ii) 100, 300, 110, 330 are absent.
   (iii) For diamond, hkl reflections are absent unless $h + k, k + l, l + h = 2n$ and $h + k + l = 2n + 1$ or $4n$
   $\qquad$ $0kl$ reflections are absent unless $k + l = 4n$ ($k, l, = 2n$), etc.
   $\qquad$ $h00$ reflections are absent unless $h = 4n$, etc.
   $\quad$ For NaCl, hkl reflections are absent unless $h + k, k + l, l + h = 2n$.
6. (i) Diffraction symbol $2/mP2_1/-$; space groups $P2_1$ or $P2_1/m$.
   (ii) Diffraction symbol $mmm\ Fdd-$; space group $Fdd2$.
   (iii) Diffraction symbol $\bar{3}mR-c$; space groups $R3c$ or $R\bar{3}c$.
7. Diffraction symbol $mmm\ Pnna$; space group $Pnna$. Intense reflections with $h = 2n$, $k + l = 2n$ due to heavy scatterers (Sn) in S.E.P's at $\bar{1}$.

8. Space group $Fd3m$; Zn atoms in S.E.P's at $\bar{4}3m$.
   Al atoms in S.E.P's at $\bar{3}m$.
   O atoms in S.E.P's at $3m$.

Chapter 9.5

1. (a) [100], $mm2$; [011], m; [111], $\bar{1}$.
   (b) In Laue group $4/m$, [100], m; [011], $\bar{1}$; [111], $\bar{1}$.
   In Laue group $4/mmm$, [100], $4mm$; [011], m; [111], $\bar{1}$.
   (c) In Laue group $m3$, [100], $mm2$; [011], m; [111], 3.
   In Laue group $m3m$, [100], $4mm$; [011], $mm2$; [111], $3m$.
2. $a = b = 4{\cdot}50$, $c = 7{\cdot}17$ (in Å); lattice type, A.
3. 40 KV.
4. $a = 5{\cdot}15$, $b = 7{\cdot}71$, $c = 12{\cdot}35$ (in Å), $\beta = 107° 54'$; lattice type, B.

   Matrix for change of axes to smaller conventional cell, $a = 6{\cdot}20$, $b = 7{\cdot}71$, $c = 12{\cdot}35$ (in Å), $\beta = 156° 52'$, lattice type P:

$$\begin{vmatrix} \bar{1} & 0 & \bar{1} \\ 2 & & 2 \\ 0 & \bar{1} & 0 \\ 0 & 0 & 1 \end{vmatrix}$$

   $[100] \rightarrow [20\bar{1}]$; $[010] \rightarrow [0\bar{1}0]$; $[001] \rightarrow [001]$; $[101] \rightarrow [100]$; $[011] \rightarrow [0\bar{1}1]$; $[110] \rightarrow [211]$; $[111] \rightarrow [\bar{2}12]$.
5. $4/mmm$, $c = 9{\cdot}50$, $a = 3{\cdot}75$ (in Å), lattice type I, 4 TiO_2 units per cell.
6. $36° 38'$ or $107° 10'$.
7. Morphological description corresponds to B lattice type. A description in terms of P lattice type is:
   Class $2/m$, $a:b:c = 0{\cdot}811:1:1{\cdot}714$, $\beta = 155° 32'$, showing the forms $\{010\}$, $\{001\}$ and $\{02\bar{1}\}$.

Chapter 10.5

1. $82°$.
2. $h^2 + k^2 + l^2 = 2, 3, 4, 6, 8, 11$; $a = 4{\cdot}27$ Å (for lattice type P); (521).
3. $h^2 + k^2 + l^2 = 3, 4, 8, 11, 12, 16, 19$ (the last two lines are doublets); lattice type, F; $a = 4{\cdot}290$ Å.
4. Compound A: lines 1, 2, 3, 5, 7, 8, 9, 11, 12, 13, 15, 18 in the list, with $h^2 + k^2 + l^2 = 1, 2, 3, 4, 5, 6, 8, 9, 10, 11, 12, 19$ respectively to give lattice type P, $a = 4{\cdot}13$ Å.
   Compound B: lines 4, 6, 10, 14, 16, 17, 19, 20 in the list, with $h^2 + k^2 + l^2 = 3, 4, 8, 11, 12, 16, 20, 24$ respectively to give lattice type F, $a = 4{\cdot}04$ Å.
5. $CaTi_2O_4$.
6. Laue group $4/mmm$, lattice type I with $c = 9{\cdot}90$, $a = 3{\cdot}46$ (in Å).
7. Indices of lines are: 002, 021, 101, 120 and 111, 121 and 002, 130, 022, 131, 112, 200 and 041. 22% Fe_2SiO_4 approx.

Chapter 11.7

2. An ellipse with one principal axis $\sqrt{2}$ times the other; the shorter axis is $\|y$-axis.
3. (ii) $67° 58'$
4. (i) $n_o = 2{\cdot}61$, $n'_e = 2{\cdot}68$; o vibration direction parallel to trace of x-axis.
   (ii) $\{001\}$ cleavage fragment will show inclined $\{210\}$ cleavage traces; $n_\gamma(=1{\cdot}649)$ is associated with vibration direction parallel to bisector of obtuse angle between cleavages, $n_\beta(=1{\cdot}639)$ is associated with other vibration direction parallel to bisector of acute angle.
   $\{210\}$ cleavage fragment will show perpendicular cleavage traces; parallel to one of these is vibration direction with associated $n_\alpha(=1{\cdot}637)$, parallel to the other there is intermediate refractive index $(1{\cdot}643)$.
   (iii) (100) face: vibration directions parallel (and perpendicular) to cleavage trace, with $n_\alpha(= 1{\cdot}586)$ parallel and $n_\beta(= 1{\cdot}590)$ perpendicular.

    (010) face: vibration directions inclined to cleavage trace with $n_\gamma = 1 \cdot 600$ at 25° and $n_\alpha = 1 \cdot 560$ at 65°.

5. (i) Planes of groups all parallel normal to z-axis.

    (ii) Planes of molecules likely to be in several directions.

    (iii) Molecules all parallel to z-axis.

6. (i) (a) Classes 32 or $3m$; (b) Class 32.

    (ii) $P4nc$.

    (iii) 432; Space group $P4_232$ which is uniquely determinable.

INDEX